Transposable Elements

Transposable Elements

A Guide to the Perplexed and the Novice
With Appendices on RNAi, Chromatin Remodeling and Gene Tagging

by

Esra Galun

The Weizmann Institute of Science,
Rehovot, Israel

KLUWER ACADEMIC PUBLISHERS
DORDRECHT / BOSTON / LONDON

A C.I.P. Catalogue record for this book is available from the Library of Congress.

ISBN 1-4020-1458-9

Published by Kluwer Academic Publishers,
P.O. Box 17, 3300 AA Dordrecht, The Netherlands.

Sold and distributed in North, Central and South America
by Kluwer Academic Publishers,
101 Philip Drive, Norwell, MA 02061, U.S.A.

In all other countries, sold and distributed
by Kluwer Academic Publishers,
P.O. Box 322, 3300 AH Dordrecht, The Netherlands.

Printed on acid-free paper

Printed in the Netherlands.

CONTENTS

Preface xi

Chapter 1   Introduction 1

Chapter 2   Historical Background 5

  1   Hybridization and inheritance before Mendel 5
  2   Mendel's revolution 8
  3   Between Mendel and genetics 11
  4   From static genes to mobility 14

Chapter 3   Bacterial Insertion Sequences 25

  1   Simple bacterial Insertion Sequences 26
  2   General features and properties 26
    2.1   The IRs 27
    2.2   The structure of Tpases 27
    2.3   Direct repeats (DRs) 27
    2.4   Effect on neighboring genes 28
    2.5   Tpase expression and transposition 28
    2.6   Host factors 28
    2.7   The reaction mechanism of transposition 28
    2.8   The DDE motif 29
    2.9   Transposition "immunity" 31
    2.10 Target specificity 32
  3   The families of bacterial Insertion Sequences 33
  4   The IS*1* family 33
  5   The IS*3* family 35
  6   IS*911* 36
    6.1   General organization of IS*911* 37
    6.2   The expression of encoded proteins 37
    6.3   The transposition of IS*911* 37
  7   The IS*21* family 40
  8   The IS*21* 41
    8.1   Transposition of IS*21* 41
    8.2   Phylogeny of the IS*21* family 44
  9   Other simple bacterial Insertion Sequence families 44
    9.1   The IS*4* family 45
    9.2   The IS*5* family 45
  10 Some remarks on recombinases 45
  11 The Tn transposons 47

12   The Tn*3* family 48
      12.1   Transposon organization 49
      12.2   Transposition 49
      12.3   Transposon's terminals 50
      12.4   Cointegration 50
      12.5   Cointegrate resolution 50
      12.6   The recombinational process 52
13   The Tn*5* and its transposition 53
      13.1   The overall structure of Tn*5* 53
      13.2   Antibiotic resistance genes 54
      13.3   The structure of the IS*50*s 54
      13.4   An overview of Tn*5* transposition 55
      13.5   The synaptic complex 56
      13.6   Transposon backbone DNA cleavage 59
      13.7   Target DNA capture and strand transfer 59
      13.8   Transposase release 59
      13.9   The regulation of Tn*5* transposition 60
      13.10 Tn*5* as a genetic tool in bacteria and eukaryotes 60
14   The Tn*7* transposon 61
      14.1   The general structure of Tn*7* 61
      14.2   The process of transposition 63
      14.3   TnsABC+E transposition 66
      14.4   Target "immunity" 66
      14.5   The Tn*7* as an experimental tool 66
      14.6   A final general remark 67
15   Transposon Tn*10* 67
      15.1   Origin and overall structure 67
      15.2   The biology of Tn*10* 68
      15.3   A summary of the transposition 70
      15.4   The mechanism of transposition 70
      15.5   The hairpin cleavage 71
      15.6   The Tn*10* transposase 72

Chapter 4   Retrotransposons 75

 1   General characteristics 75
      1.1   Terminology 75
      1.2   Two, rather different, modes of integration 77
 2   The LTR retrotransposons 79
      2.1   The transposable elements of the budding yeast
             (*Saccharomyces cerevisiae*) 81
      2.2   The structure of Ty1 81
      2.3   Transcription and transcriptional regulation in Ty1 83
      2.4   The transcript and the VLP of Ty1 83
      2.5   Reverse transcription in Ty1 84
      2.6   Recombination in Ty1 85
      2.7   The reintegration of Ty1 86
      2.8   Other Ty elements in yeast 88
      2.9   The main features of Ty3 89

2.10   LTR retrotransposons in other fungal organisms 91
        2.10.1 LTR retrotransposable elements of the fission yeast
                (*S. pombe*) 91
        2.10.2 LTR transposons in the pathogenic yeast
                Candida albicans 92
        2.10.3 Additional Ty3/*gypsy*-like retrotransposons in
                hyphal fungi 92
2.11   LTR retrotransposons of *Drosophila* 93
        2.11.1 Hybrid dysgenesis 93
2.12   The *gypsy* LTR retrotransposon 96
        2.12.1 Basic features of *gypsy* 96
        2.12.2 The effect of *gypsy* on gene expression 98
        2.12.3 The similarity of *gypsy* to retroviruses 99
        2.12.4 *Flamenco* and *gypsy* mobility 100
2.13   A brief overview of additional retrotransposons of *Drosophila* .. 102
2.14   A survey of LTR retrotransposons in metazoa 102
        2.14.1 Insects other than *Drosophila* 103
        2.14.2 LTR retrotransposons of worms 104
        2.14.3 LTR retrotransposons in Echinoids 106
        2.14.4 The emergence of the tail and a rich repertoire of T.E.
                in the urochordate *Ciona intestinalis* 107
        2.14.5 LTR retrotransposons in amphibia 108
2.15   Fishing for LTR retrotransposons in fishes 109
        2.15.1 The Sushi family of LTR retrotransposons from
                the *Fugu* fish 109
        2.15.2 Fishing for LTR retrotransposons in herring, salmon
                and *Xiphophorus* 110
2.16   LTR retrotransposons in plants 112
        2.16.1 LTR retrotransposons in *Arabidopsis thaliana* 113
        2.16.2 LTR retrotransposons of maize 116
        2.16.3 LTR retrotransposons in wheat, barley and oats 121
        2.16.4 LTR retrotransposons in rice 123
        2.16.5 LTR retrotransposons in tobacco and other species
                of Solanaceae 125
        2.16.6 LTR retrotransposons in legumes 128
3   The *DIRS1* retrotransposons 129
4   Mobile introns 131
5   Non-LTR retrotransposons 132
    5.1 The early studies on non-LTR retrotransposons of mammals 133
    5.2 *SINE* elements in animals 137
    5.3 *SINE* elements in plants 141
    5.4 *LINE* elements in invertebrates 142
    5.5 A *LINE* element in an amphibian organism 145
    5.6 *LINE* elements in fishes 146
    5.7 *LINE* elements in mammals 147
         5.7.1 The *LINE* elements of primates 147
         5.7.2 The *LINE* elements in rodents 149
         5.7.3 *L1* elements transposition in cultured cells 150

5.7.4   The impact of *LINEs* on the mammalian genome 151
5.7.5   Concluding remarks on mammalian *LINEs* and
        a note on fungal *LINEs* 151
5.8   *LINE*-like elements in plants 154

Chapter 5   Telomeres and Transposable Elements 159

1   Characteristics of telomeres 159
2   The telomeres of *Drosophila* 161

Chapter 6   Class II Transposable Elements in Eukaryotes .. 163

1   General characteristics 163
2   The Class II T.E. first discovered in maize 164
  2.1   The *Ac/Ds* controlling elements of maize 165
      2.1.1   Sexual reproduction in maize 165
      2.1.2   The main findings of McClintock on the
             Ac/Ds system 165
      2.1.3   The mutator function of *Ac* 167
      2.1.4   Mechanism of transposition 167
      2.1.5   *Ac* can undergo reversible inactivation and "mutate"
             into a *Ds* element 169
  2.2   Molecular-genetics of *Ac/Ds* elements: Pioneering studies 169
  2.3   Further molecular studies on *Ac/Ds* elements 171
  2.4   Transposition of *Ac/Ds* elements 174
      2.4.1   Sites of insertion 174
      2.4.2   Transposition and excision requirements 175
      2.4.3   Features of *Ac/Ds* transposition 176
      2.4.4   The transposase of *Ac* 176
      2.4.5   Is there an *Ac* transposome? 177
      2.4.6   Transposition frequency 178
      2.4.7   A special note on the *P* gene of maize 178
  2.5   The *Ac/Ds* elements in transgenic plants 179
      2.5.1   Transgenic tobacco and potato 179
      2.5.2   Transgenic *Arabidopsis thaliana* 181
      2.5.3   Transgenic tomato 181
      2.5.4   Insertion of *Ac/Ds* into rice and other cereals 182
      2.5.5   The introduction of *Ac/Ds* into other transgenic
             plants 183
3   The *Spm/En* elements 183
  3.1   Discovery and nomenclature 184
  3.2   Basic characteristics of the *Spm* family 185
  3.3   Adding molecular methods to *Spm* research 188
      3.3.1   The overall structure of the autonomous *Spm* 188
      3.3.2   The subterminal repetitive regions 189
      3.3.3   Exons, introns and translation 190
      3.3.4   Structure and function of *Spm* and *dSpm* 190
      3.3.5   Transposition 191
      3.3.6   Methylation and the regulation of transposition 194

4   The family of *MuDR/Mu* transposable elements 197
     4.1   Early genetic studies 198
     4.2   The structure of *Mutator* elements 201
     4.3   Transposition 204
            4.3.1   Germinal transposition in the *MuDR* system 207
            4.3.2   The functions of *MURA* and *MURB* 208
            4.3.3   The timing of transposition 208
     4.4   Epigenetic regulation of *Mutator* 208
            4.4.1   Silencing in somatic and germinal tissues 209
            4.4.2   Molecular mechanisms of silencing 209
     4.5   Applications of the *MuDR/Mu* system 209
5   The transposable elements of snapdragon 209
     5.1   The structures of the *Tam* elements 212
     5.2   Transposition 214
            5.2.1   Excision 214
            5.2.2   Integration 215
     5.3   No non-autonomous and mobile elements in the *Tam* system ... 216
     5.4   Gene tagging 217
6   Class II transposable elements in *Arabidopsis thaliana* 218
     6.1   Class II elements that populate *A. thaliana* 218
     6.2   The *Tag1* transposable element 219
     6.3   The "hidden" Class II elements of *A. thaliana* 220
            6.3.1   The resurrection by demethylation 221
            6.3.2   Fishing for T.E. in the DNA sequence data
                     of *A. thaliana* 222
     6.4   Alien transposable elements inserted into *A. thaliana* 223
7   The *Tc1/Mariner* superfamily of transposable elements 224
     7.1   Discovery of transposable elements in *C. elegans* 224
     7.2   The transposition of *Tc1* 226
     7.3   Discovery of the *Mariner* elements in *Drosophila* 226
     7.4   The main features of *Mariner* 226
     7.5   Structure and function of the *Tc1/Mariner* superfamily 229
            7.5.1   Transposition 229
            7.5.2   The transposases of the *Tc1/Mariner* superfamily 230
            7.5.3   Regulation of transposition 230
     7.6   The *Tc1/Mariner* elements as genetic and biotechnological
            tools 231
     7.7   Examples of *Tc1/Mariner* and other Class II T.E. in diverse
            organisms 232
            7.7.1   *MITEs* and *Tc1/Mariner* elements in angiosperm plants . 234
            7.7.2   Class II transposable elements in hyphal fungi 236
            7.7.3   *Tc1/Mariner*-like elements in protozoa 238
            7.7.4   A note on *Minos*, *Hobo*, *Pogo* and other Class II T.E.
                     that reside in insects and in other invertebrate metazoa .. 239
            7.7.5   Class II T.E. in non-mammalian vertebrates: *Tc1*-like
                     transposable elements in amphibia 240
            7.7.6   Fishing for transposable elements in fishes 241
            7.7.7   *Tc1/Mariner*-like transposons in man and
                     other mammals 243

8   The *P* element of *Drosophila* 246
     8.1   Introduction 246
     8.2   The discovery of the *P* element 246
     8.3   The structure of *P* and its truncated versions 247
     8.4   Transcription and translation of *P* 249
     8.5   Transposition of *P* 250
            8.5.1   Additional information on hybrid dysgenesis 250
            8.5.2   Structural and biochemical aspects of *P* transposition ... 251
            8.5.3   Transposition and gap repair 252
     8.6   Regulation of transposition 253
            8.6.1   *P* and *M* cytotypes 253
            8.6.2   Transposition control by antisense and RNAi 254
            8.6.3   Pre-mRNA splicing 254
     8.7   The *hobo* elements 255
     8.8   Use of the *P* element in *Drosophila* genetics 256

Chapter 7   Epilogue 259

Chapter 8   Appendices 261

1   Appendix I: RNA silencing 261
     1.1   Early studies on RNAi 262
     1.2   PTGS and RNA silencing in plants 263
     1.3   RNAi in animals 265
     1.4   Selected reviews on RNAi and PTGS 266
2   Appendix II: Chromatin remodeling 266
     2.1   Chromatin acetylation 268
     2.2   Phosphorylation, methylation and ubiquitination 268
     2.3   Chromatin in plants and animals 271
     2.4   Heterochromatin and centromeres 271
     2.5   Recent reviews: The overall picture of chromatin remodeling 273
3   Appendix III: Gene tagging 275
     3.1   General remarks on gene tagging 275
     3.2   Examples of gene tagging investigations 278

Chapter 9   References 283

Index 323

PREFACE

ΟΥ ΦΡΟΝΕΟΥΣΙ ΤΟΙΑΥΤΑ ΠΟΛΛΟΙ ΟΚΟΣΟΙΣΙ

ΕΓΚΥΡΕΟΥΣΙ ΟΥΔΕ ΜΑΘΟΝΤΕΣ ΓΙΓΝΩΣΚΟΥΣΙ

ΕΩΥΤΟΙΣΙ ΔΕ ΔΟΚΕΟΥΣΙ

*Many fail to grasp what they have seen,
and cannot judge what they have learned,
although they tell themselves they know.*

Heraclitus of Ephesus, 500 BC

כל אשר איננו נקוד וטלוא בעיזים

וחום בכבשים גנוב הוא אתי

*"... everyone that is not speckled and
spotted among the goats and brown
among the sheep, that shall be counted
stolen with me."*

Genesis Chapter 30

From Heraclitus of Ephesus and later philosophers, we can deduce that observation of natural phenomena, even when keen and accurate, will not result in meaningful knowledge unless combined with analysis of the mind; just as analysis of the mind without acquaintance with natural phenomena will not suffice to grasp the perceivable world. Only familiarity with phenomena combined with mental analysis will lead to additional knowledge.

The citation from Genesis, Chapter 30, is part of an unusual story. It tells how Jacob received, as payment for his service to Laban, the bulk of Laban's herds. By agreement, Jacob was to receive "only" the newborn speckled and spotted goats and the newborn brown sheep that differed completely from their parents. Did Jacob know that there was instability (transposable elements?) in the pigmentation of Laban's herd? It is reasonable to assume that Jacob combined his keen observation with analysis of his mind in order to predict the outcome: most of the newborns were indeed speckled, spotted or brown.

There is an ongoing dispute among historians as to what was most decisive in shaping the trend of history. Did human individuals have a decisive effect or was history shaped by myriad conditions and the specific personalities merely filled the role that was created by these conditions? With respect to transposable elements, we have learned an important lesson. We have a clear combination of decisive contributions by specific *individuals* but also, in a later phase, *conditions* accumulated to bring a deeper understanding of transposable elements.

The two key personalities, who opened the fields of genetics and transposable elements were, respectively, Gregor Mendel and Barbara McClintock. Both had an outstanding capability to devise the appropriate experiments as well as to correctly interpret the results of these experiments and then come up with unexpected, but correct, conclusions. Doing so, Mendel (Henig, 2000) and McClintock (Keller, 1983) combined the teaching of Heraclitus and the wisdom of Jacob.

For about 20 years (i.e. 1946 to 1966) McClintock was the sole carrier of the banner of transposable elements (she termed them **"Controlling Elements"**); but then, quite independently, unstable genetic elements ("Insertion Sequences") were revealed in a bacterium (*Escherichia coli*) by investigators in the USA and Germany. Swift advances in the methodologies of molecular genetics then opened the way for vast progress in the understanding of transposable elements and they were revealed in virtually all living organisms.

This book will not deal with transposable elements from the viewpoint of the history of science, though I shall provide an historical background that led to the early investigation of this subject. I shall then divert from the historical aspect and deal with these elements in three parts of this book. I shall first describe the bacterial Insertion Sequences. The retrotransposons will then be handled. These transposable elements (also termed Class I transposable elements) differ from the other (Class II) elements by having an RNA intermediate during their mobility. A subsequent part shall be devoted to the Class II transposable elements of eukaryotes. Three appendices will deal with emerging subjects that are relevant to the understanding of transposable elements and their utilizations: *RNA Silencing*, *Chromatin Remodeling* and *Gene Tagging*. The subject of programmed rearrangement of DNA, which occurs in the immune system of mammals and in other systems, although being related to transposable elements, is beyond the scope of this book.

The "target" audience of this book is rather broad. It should include a wide range of biologists whether they are still students, or well-trained and experienced, but who have no intimate knowledge of transposable elements. Thus, many terms will be explained in this book before being used. Nevertheless, readers who are not acquainted with these fields of endeavor may benefit from consulting the relevant text books.

There is an ancient Jewish maxim that says: "The one who gives credit to the original author brings salvation to the world". I shall follow this advice and rather than providing only general bibliography, the publications on which this book is based, will be listed with full citation of the respective titles.

I acknowledge the help received from several people during the writing of this book. Staff members of my department at the Weizmann Institute of Science provided valuable help. Professor Avraham Levy, Professor Gad Galili, Professor Dan Atsmon and Dr Gideon Grafi, each read parts of the manuscript and made valuable remarks. I received generous assistance from the office personnel of our department, but I am especially indebted to Mrs Renee Grunebaum, who was very devoted during typing and re-typing the manuscript. I am grateful to my wife, Professor Margalith Galun and to my son, Professor Eithan Galun, who read chapters of the manuscript, made corrections and provided valuable advice.

Several authors provided preprints of research articles and reviews, among them authors of review articles in *Mobile DNA II*. I am grateful for their generosity. I am also grateful to the publishers and authors who permitted me to use their figures and tables in this book. The source of each figure and table is indicated in the caption and the legend, respectively and I express my gratitude to each of these publishers and authors. I enjoyed good collaboration with the publisher of this book, Kluwer Academic Publishers, and especially with Dr Ir J.A.C. Flipsen, the Publishing Manager. I am grateful for this collaboration. Special thanks go to Mrs S. Trauffer Ramakrishna, who did excellent and professional work in editing and page-setting of the manuscript of this book. My thanks are due to the Graphics Department of the Weizmann Institute of Science for their professional work with the figures and the cover of this book.

My final gratitude is a very special one and it goes to Dr Nillie Weinstein. The decision to write a book on Transposable Elements, rather than on another subject resulted from discussions with her. I benefitted immensely from her wisdom and knowledge of philosophy during my writing. Since the decision to write this book was made, I have enjoyed constant encouragement from Dr Weinstein.

Esra Galun
Rehovot, 7th of April 2003

CHAPTER 1

INTRODUCTION

One who climbs the Dolomite mountains, or any similar mountain range, is usually not aware of the fact that he (or she) steps on the skeletons of sedimented marine organisms. We are even less aware of the fact that our own genome is composed mostly of "skeletons", meaning silenced **transposable elements** (T.E.). Our capability to appreciate and enjoy, the alpine mountain-view and the beauty of the lily is not disturbed by the fact that both are loaded with skeletons: the lily genome, even more than the human genome, is filled with "skeletons" of T.E. However, resurrection is indeed possible. We learned from the prophet Ezekiel (Chapter 37) that the Lord set him down "in the midst of the valley which was full of bones". But, then the Lord caused breath into the dry bones and the bones resurrected!

Indeed, we now know that the "skeletons" of silenced T.E. can be reactivated by the introduction of another T.E. that has the capability to induce their transposition. Thus, for example the maize genome may include a great number of *Mutator* type T.E. but, most of them have lost the capability to cause their own transposition. Adding the *MuDR* element that is an autonomous *Mutator* element and expresses the enzyme transposase, can cause "silent" *Mutator* elements to "jump". There are additional mechanisms that may cause "jumping". For example, mutation of a gene (*DDM1*) that codes for an enzyme that causes remodeling of the histone/DNA complex (chromatin) can also induce the mobilization of *Mutator* transposons. Several stresses will also cause transposition of "sleeping" T.E. We shall see in our future deliberations several similar mechanisms and shall deal with them in some detail. Here I shall only note that investigators of T.E. have a great talent in coining fancy names for T.E. A synthetic element that could be "resurrected" was named *Sleeping Beauty*. Other T.E. were named *Gypsy*, *Athila*, *Mariner*, *Tourist*, *Castaway*, *Stowaway*, *Wanderer*, *Explorer* and *Mrs*. This is merely a partial list.

Special attention to the research of Barbara McClintock, from the mid-forties to the early fifties of the 20th century cannot be avoided in a book devoted to T.E. because she was the one who "single-handedly" revealed T.E. However, a chronological description of her studies and the consequences she drew from her experimental work may be confusing to those readers who are not experts in the field of T.E. Many or even most of the geneticists of her time got her main message wrong. They believed that her intention was merely to prove that **controlling elements** (her term for T.E.) can change location in the genome. This was not her prime intention. She intended to reveal the spatial and temporal regulation of gene expression that leads to differentiation.

Also, in this book, it is not intended to provide an abbreviated biography of Barbara McClintock. Extended biographies of her have indeed been written. One of these had a personal-humanistic approach (Keller, 1983) and a more recent biography stressed the scientific aspect of Barbara McClintock's achievements (Comfort, 2001).

It is rather obvious that the two, Mendel and McClintock initiated a **revolution** in our understanding of inheritance. This author cannot resist mentioning the verification of another **revolution**. This verification took place about 250 years earlier, during the dawn of the 17th century, and only about 100 miles north-west of Brno, near the city Prague. It concerns the idea of Aristarchus of Samos (about 310–230 BC) on planetary movements, which was substantiated after about 1780 years by Nicholas Copernicus in what was coined **The Copernican Revolution** (i.e. putting the sun, rather than the earth, in the "center" of planetary movements; the heliocentric theory). We owe the verification and further amendment of this **revolution** to two scholars; Johannes Kepler, who made the mathematical calculations and formulated the consequences and Tycho Brahe, who accumulated the data and hosted Kepler (but the former did not accept the heliocentric theory). It required two scholars to reach this achievement.

In the case of the **Genetic Revolution**, that took place in Brno (Brünn in Mendel's time), and was achieved by a single person: Gregor Mendel. He furnished the huge data and used mathematical equations to interpret these data. Moreover Mendel was not a prestigious court-mathematician of the Emperor of the Holy Roman Empire (Rudolf II) as were Brahe and Kepler, but rather a humble monk in a remote Augustinian monastery, who failed twice in his efforts to receive a certificate as highschool teacher. More about the contributions of Mendel that are relevant to this book will be presented in the Historical Background.

The enormous contribution of McClintock to T.E. will become clear in a later part of this book because after a brief historical background, I shall not deal with T.E. studies in chronological order. I shall "jump" a few decades and describe in some detail the Insertional Sequences (IS) in prokaryotes. IS was the name given to T.E. that were revealed in bacteria, without reference to the T.E. studied in maize by McClintock. During these elapsing years there was much progress in molecular genetic methodologies. These methodologies were applied to IS at an early stage. Many talented investigators then joined the study of IS, thus our understanding of these mobile elements progressed quickly and illuminated the earlier studies of McClintock.

The study of T.E. in eukaryotes then entered the molecular level. It was split into two areas. One area was the study of T.E. that are mobilized to an additional location in the genome, through RNA intermediates that result from transcription of DNA sequences. This RNA was found to be transcribed back (by retrotranscription) into a DNA sequence and the latter sequence was reinserted in the eukaryotic genome. These are the retrotansposons; also termed Class I, T.E. The other area of investigation was directed at T.E. that are mobilized without an RNA intermediate. These were termed Class II T.E. I shall deal first with Class I T.E. and then with Class II, T.E.

T.E. should not be regarded as only a kind of disturbance to the eukaryotic genome, interfering with gene-expression, causing chromosomal aberrations and filling the eukaryotic genome with unnecessary DNA sequences. First, we should recall that any genetic aberration (including apparently regular mutations) can be the cause of a genetic variation that serves subsequent selection of the fittest. However, more specifically, it was found that mobile DNA is involved in the vital functions of organisms, such as maintaining the integrity of the distal ends of chromosomes (telomeres) and processes termed programmed-rearrangement of DNA. The latter include the immune system of mammals and the mating types of yeast. However, these latter processes are beyond the scope of the present book.

CHAPTER 2

HISTORICAL BACKGROUND

The purpose of this *Historical Background* is to furnish stepping stones that will lead the reader to the "world" of transposable elements. The metaphore "stepping stones" is appropriate here because they are useful for crossing the stream of events; they do not cover all the events and guide the one who steps on them in a defined direction. During a visit in Kyoto, Japan, I saw stepping stones over a small river that were a work of art, featuring various marine animals. Not in our case, our stepping stones will be plain and to the point. Scholars and their contributions will be mentioned but we shall not even come close to providing the full history of genetics. The choice of the scholars and their respective deeds is according to this author's taste; some readers may feel that grand omissions were made in this *Background*.

1. HYBRIDIZATION AND INHERITANCE BEFORE MENDEL

Mendel was born in 1822 as Johann Mendel. His first name was changed to Gregor when he became a monk. As shall be indicated below, Gregor Mendel caused a revolution in the understanding of inheritance. However, correct concepts of inheritance did already exist much before the time of Mendel. I shall mention only a few of these. It was obvious, since ancient times, that cats will breed cats and not lions and roses will breed roses and not lilies. It was also known that interspecific hybridizations either result in no progeny at all or if they result in progeny the latter is frequently sterile. Thus, for example hybridizing a jackass (male donkey) with a mare (female horse) will result in mules that are sexually sterile. However, there was no scientific definition of species until Carolus Linnaeus (1707–1778) established his taxonomic classification. Thus, what in ancient times was considered a species is not necessarily the same in our present botanical or zoological definitions. This discrepancy can cause confusion in adhering to religious laws. For example the Hebrew Bible prohibits hybridization between "species". This prohibition is further detailed and explained in the Oral Bible (i.e. the Talmud). But the "species" of today are not the same species as understood in the Bible. An interesting indication that 2000 years ago certain trends of inheritance were already well known, emerges from a specific case. The Talmud recommends not to perform circumcision when the maternal cousin had bleeding problems when he was circumcised; such a recommendation was not suggested for the paternal cousin. This indicates that it was known that the trait of hemophilia is passed through the mother to some of her sons without showing symptoms herself. In short, it was already known that hemophilia is inherited and that the expression of this trait can be by-passed by one generation.

5

Cross breeding for the improvement of crops and domesticated animals probably took place soon after domestication. However, written records of such hybridizations are scarce; what did reach us indicates that the ancient breeders probably knew their trade. One example should illuminate this point. It concerns the early breeding of various edible *Citrus* "species" (the definition of species in *Citrus* cultivars is problematic, see Galun, 1988 and Moore, 2001). Very old Chinese records reveal that crosses between *Citrus* species were done unidirectionally. The mandarin cultivars (*C. reticulata*) were always used as the pollen parent and the pommelo (*C. grandis*) or the citron (etrog, *C. medica*) cultivars were used as the female parent. This unidirectional pollination clearly indicates that these ancient breeders knew that mandarin seeds are frequently polyembryogenic and that these embryos, and the plants that develop from them are purely maternal, hence the pollen does not contribute to the inheritance of these mandarin seeds. If you want to obtain real hybrids, you should use the pommelo or the citron that are monoembryonic, as female and the pollen of the mandarin as male parent.

There are numerous examples that indicate that real leaps in understanding biology emerged after the application of concepts and methodologies that came from mathematics, physics, chemistry and/or technology. We shall see that mathematics played a major role in the investigations of Gregor Mendel. However, much earlier there came the possibility to look at the structure of organisms beyond what could be achieved by the naked eye. The microscope was instrumental in opening a vast new world to biologists. Whether invented by Zacharius Janssen in 1590 or by Galileo Galilei in 1610, two investigators soon used the microscope intensively: Robert Hooke (1635–1703) of England and Antony van Leeuwenhoek (1632–1723) of The Netherlands. Both studied plant tissues. Incidentally, van Leeuwenhoek was born in Delft, very near to the place where the Amsterdam-born philosopher Baruch (Benedict) Spinoza (1632–1677) was working. Moreover Spinoza and van Leeuwenhoek were born in the same year. Spinoza, who refused a position of professor in Germany, earned his living by the polishing of lenses for optical instruments. Was the dust of his polishing detrimental to his health? He lived only half the number of years of van Leeuwenhoek, who used polished lenses in his microscopical investigations.

Generations of scholars used the microscope to familiarize themselves with the sex life of plants. Records on "scientific" hybridization in plants, however, only came a century later. One of the investigators was the German botanist Josef Kölreuter (1733–1806), who was trained in medicine. He started interspecific hybridizations in the tobacco genus *Nicotiana*. His efforts were not directed towards producing better crops but to understanding inheritance. Luckily, he found a very favorable environment for his hybridization experiments: the royal gardens of Karl Friedrich, the Margrave of Baden under the patronage of Caroline, the wife of the Margrave. Fortunately for us he recorded his many interspecific hybridizations in a small book. Notably he keenly observed that parental characters may "disappear" in the hybrid generation but reappear in the next sexual progeny. While this observation was also recorded about 100 years later by Mendel, Kölreuter did not draw any conclusions from his observation (see the motto of this book!). The irony is that like the discoveries of Mendel, those of Kölreuter were also neglected for many years and he received his fame mainly after the work of Mendel was rediscovered.

Prize-questions were common in the 19th century. Academies of several countries issued a scientific question and scholars were invited to submit, before a given deadline, answers. Answers that satisfied the respective academy were rewarded with a very great (money) prize. These prize-questions were quite effective in promoting science. The academies did not waste money on research proposals that in many cases do not yield the expected results. Here was an efficient system: money was only delivered *after* results. Needless to say that the system was efficient for the academies, not so for the many scholars who toiled in vain. The Dutch Academy of Sciences also issued a contest. Not surprisingly, it concerned flowers and commercial breeding of new flower types. The question was: what does experience teach us about the production of new ornamental plants of commercial value through sexual hybridization.

Here another Karl Friedrich enters the scene. This time K.F. von Gärtner. Gärtner, a German botanist working in the Black Forest region was not aware of the Dutch Academy prize-question. Nevertheless he studied exactly this question and on a huge scale. Thus, he did not submit his work to the Dutch Academy before the deadline. He did submit it 6 years later, as no one before him ever did submit an answer. The submission of Gärtner was the only one and he received the prize in 1838. The work of Gärtner was probably useful to the Dutch ornamental industry but did not contribute directly to the understanding of inheritance in plants. It was possibly contributing indirectly. When Gregor Mendel was studying at the University of Vienna he heard from Professor Unger on the hybridizations of Gärtner noticing that details of his individual experiments were missing.

The river Rhine connects the Black Forest area, where Gärtner was working, with the location of the Dutch Academy. The same river, upstream, reaches the city of Basel. In this city we meet another *Friedrich* who had a major role in the history of genetics and T.E. This person is Friedrich Miescher (1844–1895). He was born before Mendel started his pea-hybridization experiments; Miescher was still a medical student when Mendel concluded his work with peas. Due to hearing problems Miescher abandoned the medical profession and started his research. He focused on leucocytes and intended to isolate leucocyte nuclei. The laudable pus of used human bandages, that could be obtained freshly from a Basel hospital was the source of his leucocytes. With procedures he had learned mainly during his studies at the universities of Göttingen and Tübingen and his training with Wöhler at the Hoppe-Seyler Institute, Miescher was able to isolate large quantities of nuclei. Following additional extraction procedures he finally isolated a high molecular weight compound that he termed *nuclein*. Nuclein was then also isolated in abundance by Miescher, from the sperm of fish from the river Rhine. This nuclein was later found to contain deoxyribonucleic acid, or DNA. Thus the DNA was first isolated only a few years after Mendel's publication, which is long before the "rediscovery" of Mendel's findings.

We shall note in the following that the prime role of DNA in inheritance was indicated in 1944 and its double-helix structure was published in 1953 but DNA had no role in the early studies of Barbara McClintock, who discovered T.E. It took more than 100 years until Miescher was put into the Garden of Eden. He was immortalized by the chemical and pharmaceutical company Ciba-Geigy (now Novartis) of Basel that established the *Friedrich Miescher Institute*, where important biomedical research is being conducted.

2. MENDEL'S REVOLUTION

Could it be that the confrontation with Professor Eduard Fenzl (who failed Mendel in his second attempt to receive a highschool teaching certificate) sparked the enthusiastic studies of Mendel with pea hybrids? Fenzl advocated the *spermist* theory, claiming that the plant embryo is entirely preformed (although in a submicroscopical form) already in the pollen and it is merely passed to the female part that serves as a foster-mother. The early experiments of Mendel convinced him that the male as well as the female gametes contributed equally to the embryos and the future plants. If this confrontation between Fenzl and Mendel caused Mendel to fail in his second examination, it is reasonable to assume that this very confrontation urged Mendel to prove that he was right, leading to seven years of hard physical and mental labor. Seven years were required to formulate the proof of Mendel's laws of inheritance; should we remember Laban, Rachel and Jacob (Fenzl, laws of inheritance and Mendel, respectively)?

In the following I shall attempt to present the essence of Mendel's discoveries that led to *Mendel's revolution* in understanding inheritance. Details of Mendel's work with many useful remarks were provided in the book of Henig (2000) and technical details of Mendel's experiments are given in textbooks on genetics. I shall use conventional English terms rather than the original German terms. It should be noted that some of these terms were coined many years after Mendel.

What did Mendel know about hybridization and inheritance at the time he abandoned hybridization of mice and started to work with peas (in about 1850)? We have no records but he most probably knew that peas are commonly self-fertilizers: each flower is pollinated with its own pollen and does not require cross-pollination with another plant. Hence if one prevents alien pollen from reaching the flowers of pea plants, one should obtain a "selfed" progeny and this can go on for many generations. After he started to grow pea lines, which were self-pollinated in his garden, Mendel most probably witnessed that some lines bred true, meaning that successive progeny had exactly the same morphological features as their parents. In other lines, the progeny segregated after selfing. He retained only the former lines.

When Mendel came back to his monastery in Brünn (now Brno) from two years of study at the University of Vienna (in 1853) he was "equipped" with a lot of knowledge in biology and in the "exact" sciences. Thus he became familiar with plant reproduction and the development of the plant embryo that includes the plumule, the root initial and the cotyledons. He also learned about the cellular composition of plants. What was very important for Mendel's future analysis of his data was his acquaintance with physics and the mathematics of the mid-19th century. His teachers were Christian Doppler (Mendel became an assistant demonstrator at the Physical Institute) and Andreas von Ettingshausen. The latter had developed the combination theory and his teaching most probably urged Mendel to use the mathematical approach to make sense of his breeding results.

Because a fire destroyed the detailed recordings of Mendel (luckily this happened after Mendel published the summary of his results) we do not know exactly which pea experiments Mendel conducted until his second examination in Vienna (1856). Obviously by then he had several generations of peas. He was also given a new greenhouse and had ample space in the garden. By 1856, he was sure that both parents contributed inheritance factors to the hybrids. He also had in his possession many lines of peas that were pure-breeding for their respective inherited characters.

Out of these Mendel retained 22 pure lines. As a monk with only modest teaching tasks, he had ample time for his explorations. Mendel was thus well equipped with all the "tools" for his research.

He focused on only seven features:

- Angular (wrinkled) *versus* round seeds
- Green *versus* yellow seeds
- White *versus* grey translucent seed coat
- Pinched *versus* inflated pods
- Yellow *versus* green unripe pods
- Terminal *versus* axial flowers
- Dwarf *versus* tall plant height

When crossed, the first generation always expressed only the dominant trait, that is, on the right of the above list. This means, cross pollinating a plant having *angular* seeds (*recessive*) with a plant having round seeds (*dominant*) resulted in F_1 all having *round* seeds. Thus *round* was dominant and so were yellow seed, *grey* seed coat, *inflated* pods, *axial* flowers and *tall* plants. The trait that appeared in the F_1 is the dominant one.

Mendel self pollinated the F_1 to obtain the F_2. Here we should recall the difference between the first two features of the list above, and the other five features. Shape and color of seeds are already characteristics of the *next* generation. This means their inheritance can be scored shortly after the pollination. For the other features to be revealed, the seeds that resulted from the cross-pollination (or the self-pollination) have to be planted and the germinated plants should be grown to maturity and then scored.

Mendel did not stop at crossings between pure-breeding plants that had either yellow or green seed, but continued with crosses between plants that differed in the other five pairs of traits. He consistently recorded that in the first hybrid progeny (F_1) all plants had the dominant trait and upon self-pollinating F_1 to obtain the F_2 generation there was always a segregation of 3:1 (dominant:recessive). When he went further and self pollinated the dominants of the F_2 generation, one third, with the dominant trait, gave only plants with the dominant trait, while two-thirds segregated again (as the F_1 plants) in a 3:1 ratio. Selfing of the plants that had the recessive trait resulted in a uniform progeny of plants with the recessive trait. From the binominal equation noted below, Mendel deduced that for each trait the plants always have two inheritance factors, one obtained from the pollen and one from the egg cell in the pistil. Thus, the gametes have a single inheritance factor. The segregation of the self-pollinated F_1 (Aa in the case of yellow/green seeds) fitted the binomial:

Aa × aA can be written as $(A + a)^2$

and $(A + a)^2 = A^2 + 2Aa + a^2$

or $(A + a)^2 = AA + Aa + aA + aa$

This is actually the same as the original binomial equation of Sir Isaac Newton except that Newton used $(a + b)^2$ rather than $(A + a)^2$. It is rather obvious that Mendel had this equation in mind when he repeatedly obtained the above-mentioned ratios among the segregants resulting from the self-pollination of the F_1 plant and certain plants of the F_2 generation. For verification of his assumption, Mendel went on to back-crosses. In these he pollinated plants of the F_1 progeny and plants of the F_2 progeny with pure-breeding plants that had either the dominant trait or the recessive trait. For example, after pollinating a yellow-seed F_1 plant that resulted from a cross of a yellow-seed plant with a green-seed plant (he assumed that it was Aa) with pollen from a green-seed plant (aa) he expected a segregation of 1:1 of yellow (Aa) and green (aa). Such ratios were indeed obtained.

In other hybridization experiments Mendel used two parents that differed in two traits, such as one having yellow-seeds (AA) and also having round seeds (let us define it as BB), the other parent was with green seeds (aa) and wrinkled seeds (bb). Mendel observed that the yellow *versus* green and the round *versus* wrinkled segregated independently from each other. This was the *dihybrid* cross. Mendel continued to the *trihybrid* cross and obtained the same independent segregations. Now we know that such independent segregations among seven traits should be very rare unless their respective elements of inheritance (now called genes) are located on different chromosomes, or at least on different arms of the same chromosomes. However, chromosomes were not known to Mendel. They were revealed only several years after Mendel reported his results; and then only as a cytological observation, no role was assigned to them. There is an extended literature that attempted to come up with an explanation of how Mendel could select seven traits that segregated independently. Whatever the reason, the reported results of Mendel are correct. We shall note below cases in which two traits did not segregate independently; these cases were handled more than 40 years after Mendel reported his results. Returning to the basics, the *revolution* initiated by Mendel can be summarized as follows:

– The seven features of pea examined by Mendel were inherited as discreet units; crossings and selfings did not blur them.
– The mode of inheritance can be described by mathematical equations.
– For each feature the pollen contributes one inheritance factor and the egg cell contributes one inheritance factor, so that the resulting plant contains two such factors for each feature.
– A feature may be *recessive* or *dominant*; if recessive and dominant meet in the same plant, only the dominant will be expressed, but the inheritance factor (gene) of the recessive gene does not disappear, it can reappear when not "covered" by the dominant trait.
– Traits of different features may segregate independently from each other.

All these "laws" were completely different from what was assumed in Mendel's time, namely that there are gradients rather than discrete units of inheritance.

Mendel presented his results in two lectures in February and March 1865. The "Brünn Society" published these lectures, as was the custom, in its "Proceedings" in 1866. He ordered 40 reprints of which 15 were traced during subsequent years as having arrived at different addresses. Many of these reprints were not opened by the recipients. One of these uncut reprints was later found in the library of Charles Darwin. However, one reprint was opened by the Dutch biologist Martinus Beijerinck (who thought to be the first to discover viruses but was actually scooped by the Russian Dmitry Ivanovsky, who discovered them 6 years earlier). Beijerinck sent the reprint of Mendel to his young Dutch colleague Hugo de Vries, who at that time was engaged in experiments on plant hybridizations.

3. BETWEEN MENDEL AND GENETICS

During several years, before 1900, two talented biologists were actively engaged in plant hybridizations. One was Hugo de Vries in Amsterdam and the other was Karl Correns, who was working in the University of Tübingen, Germany. We noted above how de Vries received the reprint of Mendel. Correns was married to the niece of Professor Nägli to whom Mendel sent his reprint and with whom Mendel had a long correspondence about his hybridization experiment. It happened that de Vries and Correns did similar experiments and even on the same plants (e.g. maize) but de Vries published before Correns. This infuriated Correns who later claimed that while de Vries knew about Mendel's work, he failed to mention him in his publications and was also not aware of their full meaning. At this point I refer the reader to the motto of my previous book (Galun and Galun, 2001), where I cited the Talmudic maxim: "The Envy of Scholars Will Increase Wisdom". The two scientists were previously involved in a rivalry. They both worked on the phenomenon of *xenia*, in which seed components of the mother plant are affected by the pollen, and both discovered "double fertilization" in plants. De Vries scooped Correns also with respect to double fertilization. Ironically, the discovery of double fertilization should probably be attributed to Navashin, who published about it in 1898. A few months after de Vries and Correns rediscovered Mendel's publication, a third person published the rediscovery (in June 1900); this was Erich von Tschermak. Interestingly Tschermak, who was cross-breeding peas in Ghent, Belgium, and in Vienna, was the grandson of Eduard Fenzl, the very person who failed Mendel in his teacher-certificate examination, in 1856.

The drama of the post-Mendel period of inheritance studies was transferred to England. The "transfer" was initiated by de Vries. He, among several other subjects in botanical studies came forward with his intracellular *pangenesis* theory. He suggested that *pangens* are units of inheritance *outside* of the cell nuclei and that the pangens for the same basic character, but in different organisms – are the same. For example he claimed that the lack of hairs on *Lychnis respertina glabra* can be transferred to the hairy *L. diurna* and render the latter species hairless. de Vries also handled mutations that he termed in the beginning "monstrosities" and studied the very frequent monstrosities of *Oenothera lamarckiana* (the monstrosities were later found to occur because of chromosomal aberrations). In July 1899, de Vries went to London to lecture on his "Hybridization Monstrosities" at the First International Conference on Hybridization and Plant Breeding, organized by the Royal Horticultural Society. The audience found some discrepancies in his report, relative to his previous reports, but one of the audience liked the lecture. This person was William Bateson. It fitted with the concept of Bateson that changes in organisms during evolution are not "smooth" and gradual (as Francis Galton claimed) but occur in "bits". Thus what de Vries lectured seemed to Bateson a proof of his "discontinuous" variation. In his lecture at the same conference Bateson demanded that future studies should handle particular traits with a statistical analysis. Actually he demanded the kind of studies that Mendel already had 35 years earlier! The name of Mendel was mentioned only on the next day of the Conference when a lecturer reviewed the history of plant hybridization. It should be noted that de Vries did not mention Mendel in his French *Comptes Rendus* publication. But in the German translation of it (*Berichte der Deutschen Botanischen Gesellschaft*) de Vries put a footnote with the name of Mendel, claiming that he finished his experiments before he was aware of Mendel. Correns did not believe this; he thought that de Vries had committed plagiarism. Correns claimed that he also became aware of Mendel's work *after* he obtained similar results, but refrained from claiming priority. Actually Correns witnessed that while in several cases he verified the conclusion of Mendel, there were many cases in which the results did not fit the "laws" of Mendel. Correns also claimed that *Anlage* (what we call now genes) must reside inside the nucleus while de Vries' *pangens* were claimed by him to be outside of the nucleus. The harsh rivalry between Correns and de Vries reminds one of the ballad of H. Heine, on a medieval dispute between a monk and a Rabbi that was ended with an epilogue:

"Welcher Recht hat, weiss ich nicht
Doch es will mich schier bedunken
Dass der Rabbii und der Mönch,
Dass sie alle beide stinken"

(meaning: I do not know whom of the two was right but I am sure that both stink).

Back to England, between 1900 and 1902 Bateson became familiar with the original work of Mendel and translated it into English. From this period Bateson was a fierce defender of Mendel. Bateson believed that "discontinuous" changes from generation to generation were the basis for evolution (rather than "gradual" changes). He gathered several associates to work on plant and animal hybridization in order to substantiate Mendel's laws and to refute the claims of the "biometricians". A fierce fight developed between Bateson and the "biometrician" W.F. (Raphael) Weldon. It descended into personal rivalry. The relation between Waldon and Bateson underwent a dramatic change. In youth they were close friends but in 1894 Weldon wrote (in *Nature*) a critique on a major book of Bateson and the latter was convinced that Weldon had intended to destroy his academic career. The two found additional subjects for their fights. Bateson actually converted his home, near Cambridge, into an experimental station where many associates were conducting their breeding experiments. One of the upper-floor bedrooms even served as an egg incubator. As one who as a young farmer's apprentice was lodged for several months in a bedroom with such an incubator, this author knows what it means. Actually the Bateson family and the whole estate were recruited to provide evidence against the "biometricians". Unfortunately the fight between Weldon and Bateson ended in the death of the former. Weldon, who saw the manuscript of an associate of Bateson, a colonel Hurst, on the inheritance of coat color in horses sensed his occasion. He was sure Hurst had made mistakes; another dispute started and Weldon devoted days and nights to showing that Hurst and Bateson were wrong. He died, probably of heart failure, before the dispute was ended.

In 1906, at the Third International Conference on Hybridization and Plant Breeding, about a year after the death of Weldon, Bateson came forward with the term *Genetics*, consequently this conference was renamed: *Third International Conference on Genetics*. Possibly only a name but by this Bateson opened a new branch of science: *Genetics*.

After 1906, it became evident that a good name can have a huge impact. Research departments, research institutes and journals adopted the name *genetics* or a version of it. As for Bateson, he was finally offered a professorship at Cambridge University (in 1908) to become the first Professor of Genetics. He stayed there for only two years and then accepted to direct the newly opened John Innes Horticultural Institute. This institute was located southwest of London. During the next eighty years "John Innes" went through transitions: it was transferred to the north, still near London. It then moved to Norwich and its name was changed twice. First it became the John Innes Institute, losing its horticultural character and finally it was named the John Innes Centre and became part of a plant biology complex. (Can one suggest that *John Innes* became a *transposable element*?)

Back to terminology. In 1909 the term *gene* was coined by the Danish plant physiologist, Wilhelm Johannsen. When it was invented the term *gene* had an abstract meaning. The real meaning was understood only many years later when the roles of DNA and RNA (ribonucleic acid) became evident and the controlled translation of RNA into polypeptides (proteins) was understood. Johannsen came up with additional terms: *phenotype* as the organisms appearance and *genotype* as the genetic composition of the organism. Several additional terms were adopted. The result of the fertilization of the egg cell with the male gamete was termed *zygote.* Consequently the *homozygote* is one that received the same gene from both parents while *heterozygote* is the one that received different genes. Another term was *allelomorph*, which was subsequently shortened to *allele*, and describes the status of a gene.

Although the term *genetics* was invented in England, much of the genetic research was then conducted in Germany (e.g. Correns and Erwin Bauer) and in the USA (as will be narrated below).

In 1907 Bateson toured the USA and Canada, giving lectures and meeting fellow *geneticists.* Among the latter was Thomas Hunt Morgan (1866–1946) who was already an established researcher at Columbia University (New York city). The meeting did not bring Bateson close to Morgan. Morgan, who was working with different animals (e.g. mice) found cases where crossings of two heterozygotes did not result in the Mendelian 1:3 ratio, but rather in a 1:2 ratio. In such cases no true-breeding dominants were revealed in the progeny. Although Morgan accepted units of inheritance, he was contemplating the explanations of results that differed from the classical laws of Mendel. On the other hand, Morgan and Bateson shared the same attitude towards the emerging *Chromosome Theory*; they did not accept it. This *theory* deserves clarification. Chromosomes were observed already by Karl von Nägeli in 1840 but the name was given only in 1888 by W. von Waldeyer. They were then recorded frequently, especially by German investigators, in dividing cells, namely at the stages of *mitosis* and *meiosis.* They disappear during the other phases of the cell cycle. Walter Sutton (also of Columbia University) who followed the chromosomes closely, found that gametes bring half the chromosome number to the zygote so that the zygote and subsequent embryonal cells again have the full number of chromosomes. These and other considerations caused Sutton to suggest that chromosomes constitute the physical basis of the Mendelian laws of inheritance. On the other side of the Atlantic Ocean, Theodore Boveri at the University of Würzburg, also studied chromosomes. He found that each species has a fixed number of chromosomes. Whenever the zygote loses a chromosome the further development is impaired. Boveri also assigned the chromosomes a role as carriers of inheritance factors. Thus, Sutton and Boveri were linked to the *Chromosome Theory of Inheritance.* Many biologists of the time had a hard time accepting that all the genetic code of an organism is packed into tiny microscopic structures that appear and disappear in dividing cells. While Bateson retained his disbelief in the *chromosome theory* for almost the rest of his life, Morgan was gradually convinced of its validity. It was then found that the lack of dominant/homozygous progeny does not require an amendment of Mendel's laws but rather such a combination is lethal. Indeed dead embryos were detected in heterozygous/yellow mice mothers that were fertilized with heterozygous/yellow male mice. These accounted for the missing segregants. Morgan tended to accept that diversions from Mendel's laws can be explained by specific interference and did not require an amendment of the basic laws.

Morgan then concentrated on one experimental organism, the fruit fly *Drosophila melanogaster*. His relatively small laboratory was converted into a "fly room". Using milk bottles and a simple feeding mix many thousands of flies could be maintained and crossed in a small space. But genetic studies also required mutations. Morgan tried several treatments to obtain mutations. In the beginning, in vain, but then one day he saw a male fly with white (rather than the wild-type red) eyes. This triggered the most important discoveries. As happens in such cases, crosses of normal females with the mutated, white-eyed males had unexpected results. The flies of the first (F_1) generation were all red-eyed as expected if *white* was recessive. But in the next generation all the females were red-eyed. Only among the males half the progeny was white-eyed and the other half was red-eyed. At that time it was already known that fruit flies have four pairs of chromosomes and that females had two X chromosomes while males had one X and one Y. It was quickly verified that the mutation resided in an X chromosome and a male with such a chromosome should develop white eyes. In females the mutated X is complemented by the other, wild-type X chromosome and develops the wild-type (red) eyes. Other mutations were then revealed. Some of them were traced to the X chromosome. Many crosses could then establish which of the traits (mutations) had a tendency to segregate together. They were considered to be linked.

Intensive activity in genetics then spread all over the USA. It included plants and especially crop plants of economic importance to the USA, such as maize. Departments of Genetics were established in the Land-Grant universities and in agricultural experimental stations. Soon linkage maps were established for the chromosomes of several eukaryotic organisms and in the early thirties the consensus was that genes were stably located on chromosomes like beads on a string. Each "bead" was believed to have its eternal position on its "string", relative to the other "beads". Thus, by 1915 sufficient evidence had accumulated that enabled T.H. Morgan, A.H. Sturtevant, H.J. Muller and C.B. Bridges to publish the text: "The Mechanism of Mendelian Heredity", which opened a new era.

True, variegations were amply detected but these were explained by activation and inactivation of genes, not by their mobility. Genes were reported to change places but only with the chromosomal fragment upon which they are located. This means that exchanges of chromosomal fragments by translocations were reported and even losses of chromosomal arms were revealed but the "jumping" of individual genes from one location to another was unthinkable until a new revolution was started by Barbara McClintock.

4. FROM STATIC GENES TO MOBILITY

This section will be rather extensive because the main achievements of Barbara McClintock, which resulted in the novel concept of *Controlling Elements*, will be presented as part of the *Historical Background*. As we shall see below, McClintock made good use of the genetic and cytogenetic tools that were available at the time of her studies. Moreover, important cytogenetical methodologies were also developed by McClintock herself. However, there was a vast increase in molecular biological understanding as well as in efficient methodologies that were subsequently developed. This understanding and these methodologies were instrumental when the theme of mobile DNA moved into bacterial systems.

A vivid account of the early days of maize genetics and cytogenetics at Cornell University was provided by Rhoades (1984). Rhoades is a reliable and impartial source of information on maize research activity at Cornell and elsewhere. However, much of what he narrated is based on his memory of events that took place more than forty years earlier! Though he arrived at Cornell several years after the establishment of the Department of Plant Breeding, he studied maize genetics and cytogenetics during most of his life and maintained close connections with all the maize investigators. He was amply informed on the research activities before his arrival at Cornell and followed those activities very closely after he left. Rhoades and Barbara McClintock kept a life-long close friendship that spanned about sixty years. In 1914 Rollins Adam Emerson, who had a faculty position in Nebraska, was appointed Head of the Department of Plant Breeding at Cornell University. Emerson took with him two graduate students from Nebraska to Cornell (E.G. Anderson and E.W. Lindstrom). Two additional Nebraskans (G.F. Sprague and G.W. Beadle) also came to Emerson's department. Emerson established a very devoted and prolific group of maize investigators. The talent of this group became evident not only by rendering Cornell a main center of maize genetics and cytogenetics but also by the fact that two of the group later received Nobel Prizes and many others occupied leading positions in genetic research all over the USA. The Emerson department operated successfully for over 20 years as a "clearing house" for maize genetics. Although it should be noted that maize genetics was studied in several other locations in the USA that today are not at all considered agricultural research centers, e.g. at Harvard University (E.M. East) and Columbia University in Manhattan (R.A. Harper).

While already rather early in his investigations Emerson reported on variegation in maize, it was more of a descriptive study than the solving of the genetic mechanism of this phenomenon. A great deal of the work in the department was devoted to studying linkage between maize genes. Here, some explanations are due. Emerson, as others before him, encountered cases in which the segregants of certain crosses did not conform with the classical Mendelian laws; there was no random segregation of the genes, following test-crossing the F_1. Suppose one performs a cross of *AABB* with aa bb and obtain the F_1. The latter should be *AaBb* and this F_1 is expected to produce equal gametes of *AB, Ab, aB* and *ab*. This means, if there are 1000 gametes, there should be about 250 of each kind, namely that *AB* gametes and *bb* gametes should total 500 and the sum of *Ab* and *aB* gametes, which are the *recombinant* compositions (not found in either parent) should also total 500. In other words: 50% recombinants. In practice it was found that the recombinant gametes were much less than 50%. The gamete composition was easily revealed; the F_1 plants were test-crossed (or "back-crossed") to the double recessive (in this case *aabb*). Thus, parental composition and recombinant composition could be calculated. The percentage of recombination is an important number. It was the suggestion of the geneticist A.H. Sturtevant to use the percentage of recombination as a quantitative index of the linear distances between genes: the lower this index the closer the genes are to each other. If the percentage is 1% the genes are considered to be closely-linked and if it is approaching 50% there is no linkage at all. The above description is obviously over-simplified. There are also other means to obtain linkage values. For linkage evaluation in plants that are easily self-pollinated the F_1 is self-pollinated rather than back-crossed and the percentage of recombination is calculated from the F_2 progeny. In addition to obtaining an $AaBb$ F_1 one can also obtain F_1 with three different genes as $AaBbCc$ and $BbCcDd$. Such F_1 plants will provide, respectively, information on the linkages between A and B, B and C, and C and D. In short, there is a simple, though tedious, procedure to reveal which genes are linked and then to determine their *sequence* in each *linkage-group*.

As noted above, after the first decade of the 20th century the chromosome theory was generally accepted. The work at Cornell and elsewhere (compiled by Emerson et al., 1935, though much of this information was already known at Cornell ten years earlier) indicated that there are 10 linkage-groups in maize. This fitted well with the results of cytological investigations that revealed 10 pairs of chromosomes in this species. It should be noted that not all crosses yielded simple and expected results. In many cases there were interferences such as genes, in some maize lines, that affected other genes under study. Lethalities were also revealed and some genes were not at all located on chromosomes but rather in the "cytoplasm" (meaning in the genomes of the chloroplasts or the mitochondria). All these exceptions slowed down the mapping of genes but also rendered maize genetics much more interesting. One additional note is due. While we are dealing here with maize genetics and cytogenetics, a very important part of the work in all the centers of maize research, that were active in many parts of the USA (a great number of them in the Mid-West), was devoted to maize improvement; this means primarily, "corn breeding". Indeed these breeding activities had enormously impressive results: the yields increased drastically. Much of this increase was achieved by the development of hybrid corn cultivars. On the other hand, maize research was conducted not only in *Breeding* departments but also in *Botany* departments. This was also the case at Cornell. The maize genetic work was also paralleled with genetic studies in other plants and in animals.

We shall now turn to Barbara McClintock, the pioneer of T.E. The following will deal with the research achievements of Barbara McClintock with a focus on the research that led to the new and very broad field of T.E. A detailed account on the personality of McClintock was provided by Evelyn Fox Keller (Keller, 1983); personal recollections on the work of McClintock were summarized by Rhoades and a kind of summary that includes recollections, reviews by investigators in other T.E. fields as well as transcripts of a choice of McClintock's publications and a full list of her publications was published by Fedoroff and Botstein (1992) as a kind of "Festschrift" for the 90th birthday. Additional information can be found in Peterson and Bianchi (1999). A recent book on the work of Barbara McClintock was written by Comfort (2001). In this Historical Background, I shall adhere to the "facts" of her research achievement; although I put "facts" in citation marks, remembering the claim of Nietzsche (in his Will to Power) that there are no facts but merely interpretations of facts.

McClintock enrolled in Cornell in 1919, when she was 17 years old and graduated in 1923. She then became a graduate student of the cytology professor L.

Sharp (Botany Department). During her graduate study she was assigned to assist L.F. Randolph, who started as maize cytologist of the USDA at Cornell. Randolph was successful in producing maize polyploids and in the rescue of embryos from difficult-to-obtain maize hybrids. McClintock swiftly became very proficient in maize cytology and one collaborative paper (on maize triploids) was published (Randolph and McClintock, 1926). There was then a misunderstanding and McClintock departed from Randolph to work independently. At that time the methodologies of maize cytology were gradually improving. Much of this improvement was due to the work of McClintock, who could then identify clearly each of the 10 maize chromosomes. The final identification was based on the pachytene stage of meiosis. Thus, 10 linkage groups of genes emerged and 10 identifiable chromosomes were revealed. However, which linkage group was in which chromosome had still to be established. One way to assign linkage groups to chromosomes was to obtain "odd" chromosomes. If instead of a pair of chromosomes you have three chromosomes, the genes are now in triplets and can be recognized. As triploids of maize were available the task was then to obtain trisomics. This is a normal set of 2n chromosomes plus one extra chromosome. This could theoretically take place in each of the 10 pairs of chromosomes. McClintock assigned numbers to each chromosome. In the final designation, the longest was no. 1 and the shortest no. 10. She became an expert in characterizing each maize chromosome at pachytene, according to its unique features as length of the arms, knobs and other markers. Consequently, McClintock obtained trisomics for all the chromosomes with the exception of nos. 1, 8 and 4. So that for seven (out of 10) chromosomes McClintock could determine which linkage group they include. Other investigators assigned the other three chromosomes to respective linkage groups. This work was performed swiftly: in 1928 there were no assignments and by 1931 all linkage groups of maize were assigned to specific chromosomes (Figure 1). McClintock received her Ph.D. in 1927 so this work was done mostly after she received her title. Indeed, the years 1928 to 1935 were considered the "golden age" of maize genetics and cytogenetics at Cornell. By 1935 most of the brilliant, ambitious and productive young people of the maize research group had left Cornell to pursue their respective careers. McClintock was at that time considered an expert in maize cytogenetics, more familiar with this field than anyone else, but she was left behind at Cornell. The status of maize cytogenetics up to this period was reviewed by Rhoades and McClintock (1935). As Rhoades recollected many years later (Rhoades, 1984), there were 17 major achievements and discoveries in maize cytogenetics by 1935. In ten of these McClintock was either the sole investigator or an active participant in the research. Not all of these were published by 1935 but McClintock already had the data in her possession. An example of the latter is the phenomenon of the "breakage-fusion-bridge cycle" (BFB cycle) of dicentric chromosomes that shall be reported below. While spending most of her time at Cornell, McClintock also spent a short time at Stanford University, where she swiftly resolved the cytology of the fungus *Neurospora crassa* that served in the genetic studies of Beadle. These studies, continued when Beadle moved to the California Institute of Technology, culminated with a Nobel Prize to Beadle. She also visited Germany in the winter of 1933/4. There were extended visits to the University of Missouri and to the California Institute of Technology.

One of the advantages of maize for genetic and cytogenetic studies is the abundant information one can obtain from a small experimental plot. To appreciate

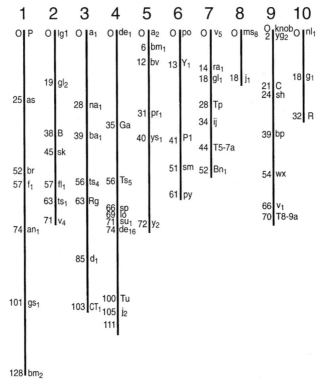

Figure 1. Linkage map of the ten chromosomes of maize as determined by 1935. (From Emerson et al., 1935).

that advantage we have to recall two features. First, what was already alluded to above: the double fertilization. One pollen nucleus fuses with the egg cell to produce the embryo and the other pollen nucleus fuses with the two nuclei of the central cell, to produce the grain's endosperm. The latter is the main component of the grain. It is a triploid tissue that starts as a syncytium but latterly is composed of cells. The outer layer of the endosperm is the aleurone. Moreover, several characters of the endosperm can be seen while the grain is still on the ear. Both the endosperm and the embryo are already the next generation. Thus, for example, when an F_1 plant is self-pollinated, the F_2 generation is already developing on the ear. If there are two ears per plant and 250 grains on each ear, then for endosperm characters a "population" of about 500 segregants can be visualized a few weeks after pollination. The ears can be stored easily and specific grains can later be replanted as required. Hence, for endosperm genes one gets quick results while for other plant characters (leaf shape and color etc.) one has to wait for the next season. There is an additional advantage. An endosperm with a peculiar pattern provides a "warning" to the cytogeneticist. Such peculiarities indicate that a genetic change may also have occurred in the germinal cells. Thus, interesting seeds can be marked for test-planting. The other advantage of maize is the size of the field required for experiments. A hectare of a maize field can contain 20,000 plants and each plant may produce 500 grains. This amounts to 10,000,000 grains per hectare. Consequently, if one has one hectare of a maize field the limitation is the ability of the investigator to collect the huge amount of data and to "digest" it mentally, rather than space limitations. Ironically, the "winter-thinking" could be too short!

During her stay at the University of Missouri, Columbia, McClintock participated in teaching some courses but had no regular teaching obligations and was not overloaded with the submission of research proposals. She devoted ample time to a special type of maize chromosome breakage. Columbia was an appropriate place to study the effect of X-irradiation on maize chromosomes. McClintock found that when breakage occurred in a line having a given chromosomal constitution, a cycle of BFB took place. If a specific zygote contains homologous chromosomes with breaks, there may be an intra-chromosomal fusion of the two sister chromatids. When, in anaphase, the two centromeres are moving away from each other, the fused sector of the chromosome stretches until it is ruptured. Now, again, two ruptured ends are available. Intrachromosomal healing may be repeated, and so on. This can go on, in the endosperm cells, until cell division ceases or the ends are healed. Because the rupture could occur at each of several points between the two centromeres, and the respective chromosomes were well marked with genes, McClintock could have a detailed picture of where the breaks took place and which genes were affected. In fact, these studies on the BFB cycle and previous studies on ring structures in maize chromosomes were effective preludes for the future studies in Cold Spring Harbor. McClintock became so familiar with maize cytogenetics that she later claimed that she could look at a maize plant and tell what microscopical observations will show at the chromosome structure.

The study of Activator (*Ac*) and Dissociation (*Ds*), the first pair of T.E. investigated by McClintock has its roots many years before she revealed these elements. In studies that date back to the thirties of the 20th century, McClintock devised cytogenetic methods that would result in mutations. She then focused on one chromosomal arm in maize: the short arm of chromosome 9 (Figure 2). There are several advantages in focusing on this arm. She could identify it cytologically and the arm contains several genes that can be scored on the grains while still attached to the husk; these genes affect the endosperm characters. Thus, a lot of information was derived from this arm.

The following is a list and description of the genes:

C   Anthocyanin production in aleurone (endosperm)
C/I   Inhibition of anthocyanin, dominant over *C*
c   No anthocyanin in aleurone, recessive to *C*
Sh   Normal starch synthesis in endosperm with an active sucrose synthase, causing full-round grains
sh   Reduced starch synthesis, recessive to *Sh*
Bz   Anthocyanin in aleurone *and* in the vegetative part of the plant, encodes UFGT
bz   Aleurone is brownish or bronze due to lack UFGT, recessive to *Bz*
Wx   Endosperm contains amylose and stains blue-black with a I$_2$-Kl solution
wx   Endosperm waxy, with only amylopectin, stains reddish-brown with I$_2$-Kl, recessive to *Wx*.

Thus by using the appropriate cytogenetic material, McClintock could investigate, in detail, the results of the BFB cycle in the short arm of chromosome

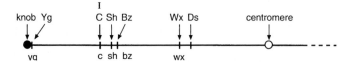

Figure 2. Approximate locations of the genetic markers in the short arm area of chromosome 9 that were amply used in the studies of B. McClintock. (From McClintock, 1951).

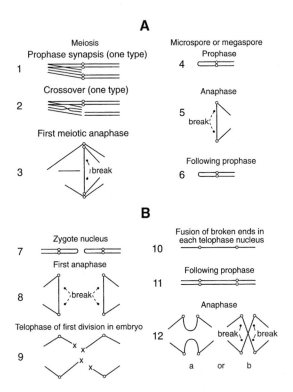

Figure 3. The origin of a newly broken end of a chromosome at the meiotic anaphase and its subsequent behavior. A – the chromatid type of breakage-fusion-bridge (BFB); B – the chromosome type of BFB. (From McClintock, 1951).

9 (Figure 3). Starting from the centromere, towards the arm's terminal knobs, these genes are in the following order: *Wx, Bz, Sh* and *C*. Plants that had one regular short arm of chromosome 9 with the recessive genes and the other chromosome 9 with the respective dominant genes, but undergoing the BFB, are very suitable for the determination of locations on the short arm of chromosome 9, where the breaks take place. At least theoretically, if the BFB goes on in the endosperm of a given grain that initially had all the four dominant genes, these may be gradually lost, meaning that they will lose the genes that are on the distal side of the arm, now without a centromere. Hence, first the *C* may be lost (and the cells affected will show *c*), then

the *Sh* will be lost and the cells will be *sh* (shrunken), thereafter *Bz* to *bz* and finally *Wx* to *wx*. The studies on the BFB were actually performed during a considerable number of years and detailed reports were presented by McClintock about these studies (e.g. McClintock, 1938, 1941, 1942a, 1942b). All these studies did not yet handle the *controlling elements*, but McClintock was able to cause BFB by her cytogenetic plant material. Moreover, this was the first time a rigid association was found between a physical location and genetic data. There were several important conclusions from the BFB studies. Let us itemize them in the following.

1. By the use of certain cytogenetic stocks it was possible to obtain functional gametes carrying a short arm of a chromosome that terminates with a broken end.
2. The broken end results from a dicentric chromatid whose centromeres move away from each other at the meiotic anaphase.
3. The broken ends of sister chromatids fuse so that in the next cell division of the gametophyte the dicentric chromosome is ruptured again in the next late anaphase.
4. Chromosomes with BFB that are introduced (through fertilization) into the endosperm, may continue with BFB. However, in the embryo the chromosomal broken ends "heal" and the BFB stops. Once healed the BFB is not renewed.
5. The breakage is not confined to a specific "hot spot" of the short arm of chromosome 9.
6. Once broken ends fuse to each other, the healed location behaves just as any other location of this arm of chromosome 9. Hence the fusion is "seamless".
7. For fusion of broken ends the two chromosomes do not have to be in contact *while* they break. Thus, two broken-end chromosomes may enter the zygote (or the central cell, for the endosperm) and fuse in the egg cell or in the central cells.

The extensive studies of McClintock culminated at the time of her transition from the University of Missouri, Columbia, to Cold Spring Harbor. Consequently, the two extensive publications on BFB were submitted from Columbia (McClintock, 1941) and from Cold Spring Harbor (McClintock, 1942a), respectively. By this time, the transitions of McClintock ended and of the transitions of her *controlling elements* were initiated.

In Cold Spring Harbor McClintock revealed that a break in the short arm of chromosome 9 could occur without an inverted-repeat at the distal end of this arm. But there was a decisive difference. The "new" break was at one specific location: proximal of the *Wx* gene. When such a break takes place the chromosome "loses" the genes that are distal to the break. Due to the use of appropriate cytogenetic material McClintock could relate the loss of genes (e.g. *Wx*, *Bz*, *C*) to the loss of a fragment of the chromosomal arm.

McClintock concluded that the breakage was related to a genetic element; she called it *Dissociation* (*Ds*), meaning that this element causes the dissociation between the two parts of the chromosomal arm. However, *Ds* will not cause breakage unless another element is present in the genome. This additional element was termed by her *Activator* (*Ac*). She discovered that *Ac* was *autonomous* with respect to transposition. It could *move* from one location to another. This transposition of *Ac* was a completely new concept! For over 30 years geneticists developed the concept that genes are fixed at specific locations on the chromosomes. Now, McClintock provided clear evidence that a "genetic element" does move!

Moreover, she provided convincing evidence that in the presence of *Ac*, *Ds* can also move (transpose) from one location to another. However, once *Ac* was removed from the genome the *Ds* became stationary. McClintock revealed additional phenomena related to *Ac/Ds*. One of these seemed paradoxical. Using her cytogenetical material she could combine *Ds* with different doses of *Ac*. Three doses could be entered into the triploid endosperm tissue. Strangely when three doses of *Ac* were in a cell the transposition of *Ds* occurred later and less frequently than in the presence of only one *Ac* dose. True, about 60 kilometers to the west of Cold Spring Harbor, in Manhattan (Rockefeller Institute) the proof was provided that genes (and therefore also "controlling elements") are made of DNA; but at that time McClintock had no chemical definition for her *Ac* and *Ds*. She could therefore not come up with an explanation for this paradox. It was also found by McClintock, that when *Ds* is mobilized (in the presence of *Ac*) from its "standard" position, proximal to *Wx*, to another position such as to *C*, the latter gene changes its expression (from dominant to recessive). Furthermore, upon a further transposition, away from the affected gene, the latter can regain its dominant expression. These and additional results indicated clearly that *Ac* and *Ds* are not merely mobile elements but have the capability to control the expression of known genes. Nina V. Fedoroff is an excellent source for the contributions of McClintock to maize transposable elements. Fedoroff (1989a) reported as follows: "The results of McClintock's early genetic analyses suggest that there are additional heritable factors that affect the developmental timing of *Ac* transposition. She identified elements which underwent either a transposition or a functional change during either the first or second nuclear division in endosperm development in over 90% of the kernels examined. She also noted heritable differences among *Ac* elements in the frequency and developmental timing of *Ac*-activated excision of *Ds* elements or *Ds*-mediated chromosome breakage. She designated these *states* of the element signifying that they were heritable but relatively labile". These findings were also later substantiated by other maize cytogeneticists.

McClintock also revealed another family of transposable elements, the *Suppressor-mutator* (*Spm*) elements (independently identified by Peterson, 1961, and termed by him *En* system). This family of T.E. will be handled, together with other System II T.E. of eukaryotes, in a later part of this book. Several of McClintock's publications are noted in the References (e.g. McClintock, 1943, 1945a, 1945b, 1946, 1947, 1948, 1951, 1952, 1953a, 1953b, 1956, 1961a, 1961b, 1968, 1978, 1984).

There are conflicting opinions on the importance of McClintock's discoveries. Some scholars maintain that the relevance of McClintock's *Controlling Elements* was not merely in showing that these elements were mobile (intra-chromosomal and inter-chromosomal). These people believe that McClintock paved the way for understanding how differentiation (especially in plants) can be controlled. Her basic concept was that structural genes (e.g. genes affecting form, biochemical composition and pigmentation) are controlled by controlling genes (such as the *controlling elements*) and that these controlling genes are themselves affected by internal signals (e.g. of specific tissues during different stages of development from zygote to mature plant), as well as by environmental effectors (e.g. stresses). Others scholars claim that while McClintock initiated the concept of mobile genes and discovered the genetic and cytogenetic mechanisms of this mobility, she didn't provide the *proof* for the role of *controlling elements* in the regulated differentiation and development in plants. In short, these scholars claimed that McClintock hoped to be recognized for her contributions in gene regulation but was awarded for her contribution to gene transposition. Even her close colleague and admirer, Rhoades, praised McClintock for her "transposition" work but was almost mute about the *regulation* issue. Some claim that authors who write on history should maintain objectivity, but this is rarely so. Thus, also the author of this book, albeit an "outsider", is impartial and tends to agree that *gene transposition* was the outstanding breakthrough for which McClintock should receive her fame. The possible hopes and disappointments of McClintock are not relevant to science; it is obvious that she caused the second revolution of our understanding of inheritance. In this connection the recent review of Lönning and Saedler (2002) is rather enlightening. This review provided an historical angle to the study of variegation in plants and discussed the impact of McClintock's experimental results and visions with the recent findings on transposable elements.

Understanding of the phenomenon of *controlling elements* in maize and in additional eukaryotes, on the molecular level, had to wait till the era of molecular genetics. At the beginning of this era bacteria were a favorite object of research. Thus the first elements of this kind that were investigated at the molecular level were the *Insertion Sequences* of bacteria. These will be handled in the next part.

CHAPTER 3

BACTERIAL INSERTION SEQUENCES

Studies with bacteria and viruses, in the late fifties and early sixties of the last century, substantiated the dogma that the genetic information is coded by DNA and that messenger-RNA transcribed from DNA is translated into proteins. It also became apparent that DNA can enter a bacterium and may then integrate into the bacterial *genophore* (I prefer this term over bacterial chromosome, since the structure of eukaryotic chromosomes differs substantially from the bacterial genophore). Therefore, when the first reports appeared in the literature that fragments of bacterial DNA were apparently mobile: they could be excised and then reinserted into a new locus of the genophore, the investigators in the field of bacterial genetics accepted these reports without much excitement. It was not considered a "revolution".

Evidence for insertions of DNA into the bacterial genophores was derived from polar mutations in the *gal* operon or in the *lac* operon of *Escherichia coli*. It was found that these mutations reverted upon the removal of the inserts. The studies were conducted by Starlinger, Saedler and their associates (e.g. Saedler and Starlinger, 1967; Jordan et al., 1968) as well as by Malamy (1966) and Shapiro (1969). The evidence for insertion and removal of a DNA fragment from a given location of the bacterial genome required laborious experimentation. Molecular methods such as Southern blot hybridization and efficient sequencing of DNA fragments became available only from the mid-seventies. The methodologies then available to Starlinger, Saedler and others are now considered laborious. One of the early ways to compare two fractions of DNA was to perform density gradients in an analytical ultracentrifuge. The heavier fraction was recognized by a shift in the gradient (e.g. Jordan et al., 1968; Shapiro, 1969).

Another early method was to perform heteroduplex hybridization. For that, two single strands of DNA are isolated for comparison. They are then hybridized to provide a double strand. If one of the strands has an extra sequence this sequence will not hybridize into a double strand but rather form a loop. The hybridized DNA strands are spread for electron microscope observation and such loops can be detected (e.g. Fiandt et al., 1972; Malamy et al., 1972; Hirsch et al., 1972).

A third method can be termed the "Limited Head Capacity". This method is based on bacterial phages. The genome of the phage is packed into its head. The head has a limited capacity. Thus, if too much DNA is packed into it, the head is fragile and will explode when put in solution with a detergent. The DNA fragment to be tested can be inserted into a "space" of a phage DNA. If the introduced DNA is longer than the normal "space" – this will lead to a fragile head of the phage.

The last method is especially noteworthy for this author. It shows that the human head differs substantially from the head of a bacterial phage. According to a "Law" of Amiel Ben David Halevi (a pseudonym of the author of this book, used for non-scientific writings), the capacity of the human head for memory is *increased* by adding more information into it. Not so with the head of the phage: it explodes when too much information is compacted into it. Salinger, Saedler and collaborators used all three methods to show that indeed a fragment of DNA was inserted near the operon that showed a polar mutation and that no insertion could be detected in the revertants.

It is noteworthy that no reference to McClintock's *controlling elements* was made by the investigators who reported on the discovery of Insertion Sequences (IS) in the *gal* and the *lac* oprons of bacteria. However, by 1972 both the German investigators (e.g. Starlinger and Saedler, 1972) and the USA investigators (e.g. Malamy et al., 1972) indicated that there could be an analogy between the IS of bacteria and the controlling elements of maize.

From the evidence for only a very few IS that were detected in bacterial genomes in the late sixties, the number of reported IS increased substantially. The last count (Chandler and Mahillon, 2002) of IS is about 800 and this number is continuously growing. The IS are being updated in http://www-is.biotoul.fr.

The bacterial IS shall be handled below under two headings. We shall first handle the "regular" or relatively simple IS. These are compact elements that have the minimal features that are required for their mobility. They range in length from about 700 base pairs (bp) to about 2,700 bp (most are in the range of 1,200 to 1,500 bp). An additional type of insertion sequences are the *Tn Transposons*. These are more elaborate elements that encode enzymes for resistance to antibiotics. Their nomenclature starts with Tn (e.g. Tn*3*, Tn*5*, Tn*7* and Tn*10*). There is a third group of insertion sequences, the *Conjugative Transposons*. These typically move from a donor bacterium to a different recipient bacterium. The latter will not be discussed in this book.

Before starting to describe the bacterial Insertion Sequences, there is a warning. Although I have tried to simplify this subject and have referred readers to the relevant literature for more details, the subject is indeed complicated. For some readers it may even appear to be boring. However, it is a fascinating subject and one tends to become enchanted by the intricate systems of nature.

1. SIMPLE BACTERIAL INSERTION SEQUENCES

The IS were thoroughly reviewed by Starlinger and Saedler (1972, 1976), by Ohtsubo and Sekine (1996) and by Chandler and associates (Galas and Chandler, 1981; Mahillon and Chandler, 1998; Chandler and Mahillon, 2002). The following description of the simple IS will mostly be based on these reviews.

2. GENERAL FEATURES AND PROPERTIES

The organization of a typical simple IS is provided in Figure 4. This IS has all the components required for its mobility – but nothing more. There are recombinationally competent DNA sequences that define the two ends of the element. These are inverted repeats (IR) of nearly the same DNA sequence. The "outside" base pairs of the IRs are required for the transposase (Tpase) cleavage. The Tpase is a key enzyme that binds to the flanks of the IS and activates the transposition. The rest of the IRs are necessary for the sequence-specific recognition

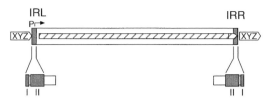

Figure 4. Organization of a typical insertion sequence. The IS is represented as an open box in which the terminal IRs are shown as gray boxes labeled IRL (left inverted repeat) and IRR (right inverted repeat). A single open-reading frame encoding the Tpase is indicated as a hatched box stretching over the entire length of the IS and extending within the IRR sequence. XYZ enclosed in a pointed box flanking the IS represents short directly repeated sequences generated in the target DNA as a consequence of insertion. The Tpase promoter, p, partially localized in IRL, is shown by a horizontal arrow. A typical domain structure (gray boxes) of the IRs is indicated beneath. Domain I represents the terminal base pairs at the very tip of the element whose recognition is required for Tpase-mediated cleavage. Domain II represents the base pairs necessary for sequence-specific recognition and binding by the Tpase. (From Chandler and Mahillon, 2002).

and binding of the respective Tpase of the IS. The upstream and downstream IRs are termed IRL and IRR, respectively. The promoter for the Tpase is located in the IRL. The long central sequence is the code for the Tpase. This extends into the IRR.

2.1. The IRs

With a few exceptions the length of the IRs are in the range of 10 to 40 bp. As indicated above, the IRs have two *Domains*. The outer, Domain I, of two or three terminal bp, is involved in the cleavage and elements transfer reactions (the transposition). The inner, Domain II, is involved in Tpase binding. There are exceptions to this IR organization, as shall be noted when we deal with IS*21*. The DNA sequences of the two IRs (i.e. IRL and IRR) are not exactly in the same (inverted) bp order. This difference probably provides the functional distinction between the two ends of the IS element.

2.2. The Structure of Tpases

The Tpase proteins are involved in both the recognition of the elements boarders as well as in the catalysis of the transposition. The domain for recognition is frequently located in the N-terminal region of the Tpase, while the catalytic domain is located in the C-terminal region. In several cases each of these two domains is encoded in a different open-reading-frame (*orf*) but by translational frame shifting they are assembled into one protein.

2.3. Direct Repeats (DRs)

During the cleavage of the target DNA for the insertion of the IS, directly repeated (DR) short sequences are generated. This is caused by the staggered method of cleavage. The length of the DR may be 2 to 14 bp, but the exact length is characteristic for each given IS. Hence, a defined IS will be flanked with the same DR bp. There are exceptions to such DR sequences. In a few ISs no DRs were revealed, while in some ISs the DR is longer than the range given above.

2.4. Effect on Neighboring Genes

It was observed that the insertion of an IS element can activate downstream genes. Such an effect was reviewed by Galas and Chandler (1989) for IS*1*, IS*2*, IS*3*, IS*4*, IS*5* and IS*10*. It was suggested that this effect was due to the presence of an outwardly directed promoter within the IS or to the formation of a new promoter caused by the insertion.

2.5. Tpase Expression and Transposition

If an IS has all the components required for transposition, including the complete code for an active Tpase, transposition should be rather frequent. In fact, transposition activity is generally maintained at a low level. Chandler and Mahillon (2002) discussed this issue and suggested several mechanisms that reduced the expression of the Tpase gene. Some of these mechanisms involve regulation of transcription but there are also indications that the stability of the Tpase protein controls the transposition activity. It was also observed that a Tpase is much more effective when it is encoded by a gene that is located within or nearby the sequence that is undergoing transposition. When this gene is far away from the transposed sequence (i.e. in *trans* position), the transposition is substantially reduced.

2.6. Host Factors

The detailed chemistry of the transposition of IS is now emerging. While the details are not identical for all ISs there is no doubt that several components of this process are provided by the "host". In this case the "host" means the genetic information available in the bacterial genome outside of the IS itself. These "host" factors are of different kinds, such as DNA chaperons and protein chaperons as well as other factors that are regular components of DNA replication and DNA supercoiling of bacterial DNA. A summary of "host" factors affecting IS transposition was provided recently by Chandler and Mahillon (2002).

2.7. The Reaction Mechanism of Transposition

This reaction can be divided into several steps. The first step is the binding of the Tpase to the IS ends. The second step is the elaboration of a synaptic complex. This complex involves the recombinase, possibly accessory proteins and both transposon's ends. The recruitment of the target DNA is in some elements concomitant with this step but in other elements it is a subsequent process. The target DNA is cleaved and there is a strand transfer of the transposon ends into the target DNA. The processing of the strand transfer complex is the final step.

The basic chemistry of cleavage and strand transfer is very similar in most analyzed ISs. Therefore we shall describe here some of the processes but shall provide more details when we shall deal with specific insertion sequences such as IS*1* and Tn*5*. Also, some important remarks on transposases are provided in a special section, below.

The Tpase first catalyses the cleavage of the IS at its 3' ends by an attacking nucleophile residue (generally H_2O) in the DNA to expose a free 3'-HO group. This hydroxyl, in turn, acts as a nucleophile in the attack of a 5'-phosphate group in the target DNA in a single-step transesterification reaction. A concerted transfer of both transposon ends to the target site, while maintaining the correct strand polarity, results in the joining of each transposon strand to the opposite target strands and leaves a 3'-OH group on the cleaved target strand. The reaction does not require an external energy source and probably does not involve a covalently linked enzyme-substrate intermediate, as do certain site-specific recombination reactions in bacteria. It should also be understood that since it is the donor-strand itself that performs the cleavage-ligation step in the target DNA, no cleaved target molecule is detected in the absence of the strand transfer.

As indicated above there are several detailed differences in the chemistry of the above schematized process. Some of these will be described below, where certain IS elements will be detailed. The overall scheme is presented in Figure 5.

2.8. The DDE Motif

In recent years it has emerged that many of the Tpases involved in the IS transposition actually belong to a large family of phosphoryltransferases. The latter are active in a variety of reactions like RNaseH and the RuvC "Holiday junction resolvase". The latter *junction* is a special polynucleotide configuration occurring during recombination. These enzymes contain an acidic amino acid triad that is intimately involved in catalysis and its role is presumably in coordinating divalent metal cations (especially Mg^{2+}), implicated in assisting the various nucleophile attacking groups, during the course of the reaction. For many IS (as well as for retroviral integrases) this amino acid triad is known as the DD(35)E motif. The consensus of this motif in several IS as well as in other elements (and the retroviral integrase of HIV) is presented in Figure 6, and a comparison of the DDE domains of the IS3 family and a retroviral integrase is shown in Figure 7. A proposed role for the DDE motif is schematically presented in Figure 8.

The release of the IS from the donor DNA can proceed in two ways. In one way the 3' end is released first while the 5' end is still attached to the donor DNA; or, the release (i.e. cleavage) at the 5' end occurs concomitantly with the cleavage of the 3' end. In the latter way the IS is physically separated from the donor DNA. The transfer of the IS strand can then proceed directly (this is a *direct insertion* of the element). If the 5' end cleavage occurs only after the 3' end is released, the donor and the target DNAs become covalently linked. Only upon 5' end cleavage will the IS be separated completely from the donor DNA and thereafter direct insertion will take place. On the other hand, while 3'-strand transfer joins the transposon and the target, it leaves a free 3'-OH in the target DNA, at the junction. This can act as a primer for replication of the element and thus generates *cointegrates* where donor and target molecules are separated by a single transposon copy at each junction. Without going into further detail here, it should be noted that several variants of this process are used by different families of IS and related mobile DNA elements (in prokaryotes

A)

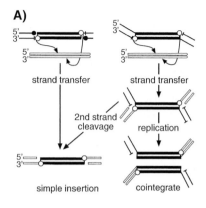

B)

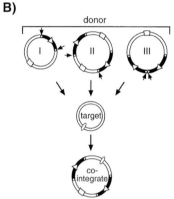

Figure 5. Simple insertions and cointegrate formation. (A) Strand transfer and replication leading to simple insertions and cointegrates. The IS DNA is shown as a shaded cylinder. Liberated transposon 3'-OH groups are shown as small circles. The 5' phosphates are indicated by bars. Strand polarity is indicated. Target DNA is shown as open boxes. The left panel shows an example of an IS which undergoes double-strand cleavage prior to strand transfer. The right panel shows an element which undergoes single-strand cleavage at its ends. After strand transfer, this can evolve into a cointegrate molecule by replication or a simple insertion by second-strand cleavage.
(B) Replicative and non-replicative transposition as mechanisms leading to cointegrates. Three "cointegrate" pathways are illustrated: (I) by replicative transposition, (II) by simple insertion from a dimeric form of the donor molecule, and (III) by simple insertion from a donor carrying tandem copies of the transposable element. Transposon DNA is indicated by a heavy line, and the terminal repeats are indicated by small open circles. The relative orientation is indicated by an open arrowhead. Square and oval symbols represent compatible origins of replication and are included to visually distinguish the different replicons. Arrows show which transposon ends are involved in each reaction.
(From Mahillon and Chandler, 1998).

and eukaryotes). We shall describe such variants when we present more detailed descriptions of some specific mobile DNA elements.

	94			116			152	
HIV-1 (IN)	wqi **D** cth	(51)	vht **D** ngsnf	(35)	ynpqsQgvi **E** smkKel**K**
	299			336			392	
Mu (MuA)	ing **D** gyl	(66)	iti **D** ntrga	(55)	kgwgqaKpv **E** rafgvg
	273			361			394	
Tn7 (TnsB)	yei **D** ati	(87)	lla **D** rgelm	(34)	rrfdaKgiv **E** stfRri
	196			246			276	
Tn552	wqa **D** htl	(73)	fyt **D** hgsdf	(35)	gvprqRqki **E** rffQtv
	589			766			396	
Tn3	asa **D** gmr	(75)	imt **D** tagas	(129)	riltqlNrq **E** srKava**R**
	207			287			323	
IS911 (IS3)	wcq **D** vty	(59)	fhs **D** qgshy	(35)	gncwdNspm **E** rffRsl**K**
	97			161			292	
IS10 (IS4)	viv **D** wsd	(63)	ivs **D** agfkv	(130)	niyskRmqi **E** etfRdl**K**
	119			184			326	
IS50 (IS4)	siq **D** ksr	(67)	avc **D** readi	(136)	diythRwri **E** efkKaw**K**
	121			193			299	
IS903 (IS5)	lvi **D** atg	(71)	asa **D** gaydt	(65)	tdynrRsia **E** tamyrv**K**
	78			134			173	
IS26 (IS6)	whm **D** ety	(59)	int **D** kapay	(36)	qikyrNavi **E** cdNgkl**K**
	122			184			230	
IS21 (IstA)	lqh **D** wga	(61)	vlv **D** nqkaa	(46)	rrartKgKV **E** rmvKyl**K**
	237			293			327	
IS30	wwq **D** lvs	(55)	ltw **D** rgmel	(33)	qspwqRgtn **E** ntNqii**R**
	157			233			341	
IS256	lmt **D** viy	(65)	via **D** ahkql	(107)	nrlkstNii **E** rlNgev**R**
	138			217			363	
IS481	lhi **D** ikk	(78)	llt **D** ngsaf	(36)	yrpqtNgka **E** rfiQsal**R**
	131			261			297	
IS630	fye **D** evd	(80)	liv **D** nyilh	(35)	vyspwvNhv **E** rlwQalN
	112			182			237	
IS982	aii **D** sfp	(79)	vlg **D** mgylg	(45)	nfakrRKvi **E** rvfsfl
	166			236			342	
IS1380	ivl **D** vda	(74)	vrg **D** sgfar	(112)	rryckRgtm **E** nriKeq
	168			240			384	
ISL3	laf **D** efr	(69)	lvi **D** rfhmv	(129)	kmsysNgcl **E** gvNRki**K**
	84			177			296	
Tc1	iws **D** esk	(90)	fqq **D** ndpkb	(108)	sqspdlNpi **E** hmqcele**R**
	N2			**N3**			**C1**	

Figure 6. DDE consensus of different families. Individual representative members of each family are shown. Amino acids forming part of the conserved motif are indicated by large bold letters. Uppercase letters indicate conservation within a family and lowercase letters indicate that the particular amino acid is predominant. The numbers between parentheses show the distance in amino acids between the amino acids of the conserved motif. Conservations indicated were derived from previously published alignments or from alignments generated. (From Chandler and Mahillon, 2002).

2.9. Transposition "Immunity"

This is an ambiguous term. In this case "immunity" is used to describe a specific phenomenon: if a target DNA already carries an inserted element, it exhibits a very much reduced affinity for an additional insert. This reduction of affinity is increased with the proximity between the two inserts. So that a site that is far removed from the initial insert will have very little or no "immunity". The immunity is rare or

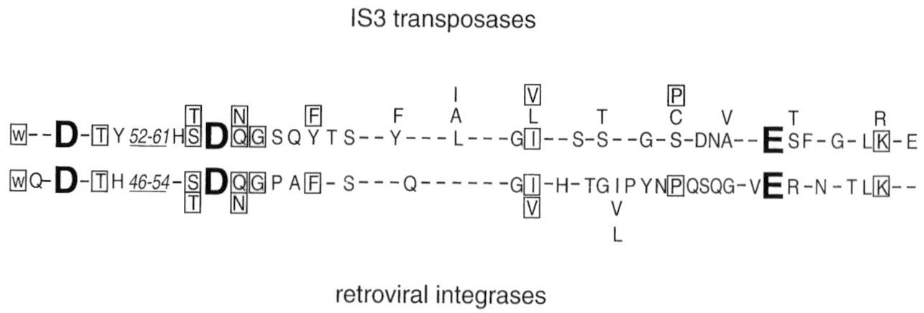

IS3 transposases

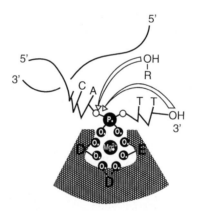

Figure 7. Consensus sequence of IS3 family transposase (upper) and retroviral integrase (lower) DDE domains. The DDE motif is shown in large bold letters.
A dash (–) indicates no real amino acid preference. Amino acids shown in normal script are present in between 50 and 75% of cases and those indicated in bold are present in more than 75%. Boxed amino acids show additional similarities between the IS3 family transposases and integrases. The sequence represented can be divided into two parts: an initial conserved region which contains the first D residue separated by a spacer (of 52–61 and 46–54 residues for the IS and retroviral proteins, respectively) from the conserved region which carries the second D and E residues. The consensus is based on 30 members of the IS3 family and 20 retroviruses. (From Polard and Chandler, 1995b).

Figure 8. Proposed role of the DDE motif. The DDE triad is schematically shown as part of a catalytic pocket (enclosed white area) in which the acidic groups co-ordinate a divalent magnesium ion. This co-ordination complex is capable of sharing the divalent metal ion with a backbone phosphate group of the retroviral DNA end and renders the associated phosphodiester bond labile. Nucleophilic attack by a component in the reaction buffer or by the 3'-terminal OH group of the DNA molecule itself is indicated by open arrows. (From Polard and Chandler, 1995b).

non-existent for simple IS but was observed in compound transposons as Tn7 and the bacteriophage Mu, and also in the IS21 element.

2.10. Target Specificity

The various bacterial IS differ substantially with respect to target specificity. In some transposons (e.g. Tn7) the insertion is restricted to unique genophore sites. For Tn7 it is the *attTn7* site and for IS*91* it is a GAAC/CAAG target sequence. In other cases there may be preferred locations but no exact requirements for the sites of insertion. For example, it was reported that some IS prefer GC- or AT-rich DNA segments. A true picture of preferences will emerge with the further total sequencing of bacterial genomes and the location of IS in these genomes (see below). In the update presented by Mahillon and Chandler (1998) the sequences of 52 bacteria species were analyzed for the inclusion of IS but no statistics for target preferences for specific IS has yet been calculated.

3. THE FAMILIES OF BACTERIAL INSERTION SEQUENCES

When the IS were first discovered, in the late sixties and early seventies, the investigators could not guess how many IS would be found in bacteria. Thus, the first IS was given a number IS*1*. Then came IS*2* and IS*3* and IS*4* ... but the numbers kept increasing so that the last count (Chandler and Mahillon, 2002) was about 800! Obviously this is not the final number. The investigators then started to put "order" into the vast collection of IS. The IS were grouped into "families" and guidelines for nomenclature were adopted. The grouping into families was based mainly on four criteria: (1) similarity in genetic organization (such as the arrangement into open reading frames); (2) similarity in the sequences of their Tpases; (3) similar features of the ends of the IS (i.e. the terminal IR); (4) fate of nucleotide sequence at the insertion site (generation of DR and length of the DR). The most detailed information that became available was the amino acid sequence of the Tpases. These sequences thus provided further information on individual IS in each family. Table 1 provides a summary of 19 families of bacterial IS. The search for ISs in bacteria benefited from the complete sequences of bacterial species that have become available in recent years.

4. THE IS*1* FAMILY

The IS*1* was the first IS to be detected by the observation of "polar mutation", as noted above. It was also the first to be isolated and sequenced (Fiandt et al., 1972; Hirsch et al., 1972). The nucleotide sequence of many variants of IS*1* from *Escherichia* and *Shigella* species has been determined. Some members of this family, such as IS*1*N and IS*1*H differ significantly in their nucleotide sequence from other members of the IS*1* family. The IS*1* was also found in the antibiotic-resistant plasmid R100. It is also a component of compound transposons such as Tn*9* and Tn*1681*.

The integration of IS*1* is accompanied by a duplication (DR) that is commonly 9 bp, at the target DNA site of the insertion. However, DRs of 8, 10 and 14 bp were also reported for this family. Typically there are two, approximately 23 bp, imperfect inverted repeats (IRs), termed IRL and IRR, located at the ends of the element. There are also typically two, partly overlapping, open reading frames (*ins*A and *ins*B') located in the O and −1 relative translational phases, respectively, as shown in Figure 9. The IS*1* shows preference of insertion into AT-rich target regions. It should be noted that IS*1* was also found in a prokaryote that is far removed from *Escherichia*, namely in the cyanobacterial genus *Synechocystis*.

Table 1. Major features of prokaryote IS families.

Family	Length range[a]	DR[b]
IS*1*	770	9 (8–11)
IS3	1,200–1,550	3–5
IS4	1,300–1,950	9–12
IS5	800–1,350	2–9
IS6	750–900	8
IS21	1,950–2,500	4 (5,8)
IS30	1,000–1,250	2–3
IS66	2,500–2,700	8
IS91	1,500–1,850	0
IS*110*	1,200–1,550	0
IS200	700	0
IS605	700–2,000	0
IS256	1,300–1,500	8–9
IS481	950–1,100	5–6
IS630	1,100–1,200	2
IS982	1,000	ND
IS*1380*	1,650	4
IS*AS1*	1,200–1,350	8
IS*L3*	1,300–1,550	8

[a]*Length range in base pairs (bp) represents the typical range of each family.*
[b]*Direct target repeats in base pairs; ND, not determined. Less frequently observed lengths are included in parentheses. (Summarized from Chandler and Mahillon, 2002).*

The transcript of IS*1* is initiated from a promoter, pIRL, which is partially located in the IRL and is translated to give two products: InsA and InsAB'. The first, InsA, is more abundant and is encoded by the *ins*A frame (Figure 9). The small, basic InsA protein, binds specifically to the IRs and also represses transcription from the pIRL. It appears also to inhibit transposition by competing with the Tpase binding to the ends of the element. The second IS*1* protein, InsAB', is the *Tpase* of the element. Its production results from a programmed translational frame shifting between the *ins*A and *ins*B' frames. The transposition of IS*1* has a low frequency: about 10^{-7}. The insertion of an additional A residue (see Figure 9) within the AAAAAAC motif, or replacing it with GAAGAAAC, fuses the two reading frames, leading to constitutive production of the Tpase while eliminating the production of InsA. This results in higher levels of transposition: between 0.1 and 1% *in vivo*. No independent product of the downstream frame, *ins*B', alone has been detected. Generally the transposition activity appears to depend on the ratio of InsAB'/InsA. The level of Tpase may also be controlled by transcription termination.

IS*1* generates both simple insertions and replicon fusions (cointegrates) composed of two directly repeated copies of the IS; one at each junction between target and donor replicons. The occurrence of stable cointegrates as transposition end-products led to the suggestion that transposition of IS*1* can proceed in a replicative manner, while simple insertions may occur without replication. Thus IS*1* may be capable of both replicative and conservative transposition. Circular copies of

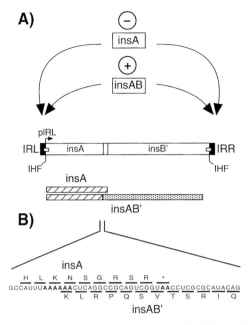

Figure 9. Organization of IS1. (A) Structure of IS1. Left (IRL) and right (IRR) terminal Ins are shown as solid boxes. The relative positions of the insA and insB reading frames, together with their overlap region, are shown within the open box representing IS1. The IS1 promoter pIRL, partially located in IRL, is indicated by a small arrow. IHF binding sites, located partially within each terminal IR, are shown as small open boxes.
The InsA protein is represented as a hatched box beneath. The InsA and InsB components of the InsAB' frame shift product are shown as hatched and stippled boxes, respectively. Arrows indicate the probable region of action of InsA and InsAB' proteins. The effect of InsA and InsAB' on transposition is shown above. (B) RNA and protein sequence in the crossover region between the two open reading frames. Codons shown above the RNA sequence show the product of direct translational readout. Those below show the product of a –1 translational frame shift. The heptanucleotide A₆C frame shift sequence involved in production of InsAB' from the wild-type IS1 coding sequence is indicated in boldface type, as is the UAA termination codon for InsA.
(From Mahillon and Chandler, 1998).

the IS1-derived transposon were also detected and these may also play a role in "simple" insertions. In addition to the IS1 family members mentioned above, which are autonomous IS, a non-autonomous element was also revealed. This element spans only 116 bp; it has IRs of IS1 and some internal sequences of IS1. This latter element was found to be widespread in Enterobacteria.

5. THE IS3 FAMILY

This is a very widespread family of IS. The IS3 was found in 19 of about 50 bacterial genomes that were sequenced. In addition to its presence in Enterobacteria (e.g. the genus *Escherichia*) it was found in bacteria that belong to diverse genera

IS3

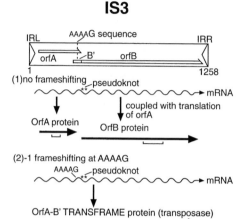

Figure 10. Structure and gene expression of IS3. (From Ohtsubo and Sekine, 1996).

such as *Bacillus, Mesorhizobium, Mycobacterium, Pseudomonas, Mycoplasma, Staphylococcus, Synechocystis* and *Thermoplasma*. The existence of IS3 in *Mycoplasma pulmonis* is noteworthy because this is a bacterium with a very small genome (less than 10^6 bp), very rich in AT and a non-universal genetic code.

The IS3 family members range in length between 1,200 and 1,550 bp. The terminal IRs are in the range of 20 to 40 bp. Most elements terminate with 5'-TG and 3'-CA and have an internal block of G/C residues.

This family includes several IS groups such as IS2, IS51, IS150 and IS911. As IS911 was the subject of detailed study it will be handled in more detail below. While the expression of their Tpase differs in details, the overall scheme of this expression is as presented in Figure 10. There are two consecutive and partially overlapping open reading frames, *orf*A and *orf*B in relative translational reading phases 0 and –1, respectively. In some cases, by the addition of a product, OrfA, there is also an OrfAB fusion product. Moreover, in contrast to the case in IS1 the OrfB is also produced. It appears that there are also exceptions in some members of this family in which there is no frame shift; these have only a single, long, open reading frame. The respective protein is thus equivalent to the OrfAB (the Tpase).

There are differences in the DR among members of this family. Thus, in the IS2 sub-group there are 5-bp of DR while in other members of this family there are 4- or 3-bp DRs. Unlike IS1 the IS3 does not mediate cointegration and it is thus supposed to transpose in a non-replicative manner. The amino acid sequence of the Tpases of this family shows that this enzyme belongs to those that have the conserved DDE motif (see Figure 6).

6. IS911

As indicated above, IS911 is a member of the IS3 family. It was initially isolated from *Shigella dysenteriae* as an insertion into the cI gene of bacteriophage λ.

Sequencing showed that the cI gene was interrupted by a 1,250 bp DNA segment that had the characteristics of an insertion sequence and that it had similarities to the IS3. Once defined it could serve to search for this element in additional bacteria. Indeed IS911 (or its truncated versions) was not only found in multiple copies in the genome of the original host but also in the genomes of *Escherichia*, *Klebsiella* and *Pantoea* (i.e. enterobacteria). The IS911 was amply studied by Chandler and collaborators (e.g. Polard et al., 1992, 1996; Ton-Hoang et al., 1997, 1998, 1999; Haren et al., 1998); and was recently reviewed by Rousseau et al. (2002).

6.1. General Organization of IS911

The IS911 is bordered by imperfect, 36-bp of terminal inverted repeats, designated as in other IS, as IRL and IRR, respectively. This element is flanked by a 3-bp target duplication (DR), but in some cases there are 4-bp of DR. Most IS911 elements terminate in a 5' CA3' and may show a short GC-rich stretch of about 8 bp into the IR. As with most other members of the IS3 family, the IS911 has two consecutive, partially overlapping, open reading frames (*orf*A and *orf*B). These are under the control of a weak promoter, pIRL, which is partially located in the IRL.

6.2. The Expression of Encoded Proteins

The 5' end of *orf*B overlaps the 3' end of *orf*A. The details are presented in Figure 11. There are three proteins that can be produced: OrfA, OrfB and (the Tpase) OrfAB. The 101 codon *orf*A starts with ATG and terminates with a TAA. The OrfA protein has a calculated molecular mass of 11.5 kDa. There is a helix-turn-helix (HTH) motif that is probably involved in the sequence-specific binding of this protein to the IRs of this element. There is also a C-terminal leucine zipper (LZ) motif. With this respect the IS911 is similar to other members of the IS3 family.

The OrfB protein contains 299 amino acids and has a predicted molecular mass of 34.6 kDa. Its TAA terminal codon lies within the IRR. The initiation codon for this protein is unusual: AUU. Such an initiation codon was formed for the translation of the initiation factor IR3 of *Escherichia coli* (the *inf*C gene). The OrfB amino acid sequence has significant similarities with retroviral integrases. This means the OrfB has a conserved sequence of the amino acid triad DDE, common to all IS3 family members and to many phosphoryltransferases. On the other hand, OrfB contains neither the HTH nor the LZ motif.

The OrfAB is the Tpase of IS911. It is assembled from *orf*A and *orf*B transcripts by a programmed −1 translational frame shift near the 3' end of *orf*A. The combined OrfAB protein thus contains the HTH motif, the LZ motif as well as the DDE domain. The total length of this OrfAB is 382 amino acids. The "slippering" that causes the translation of OrfAB has a frequency of about 15%.

By causing specific changes in the bp sequence it is possible to produce the OrfAB constitutively. This can be done by adding an A residue to the AAAAAAG.

6.3. The Transposition of IS911

The detailed chemistry of IS3 and IS911 transpositions was actively pursued by Sekine and Ohtsubo and their associates (e.g. Sekine et al., 1999) as well as by

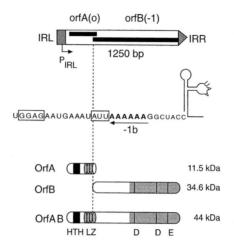

Figure 11. Organization of IS911. A cartoon of the IS is presented at the top of the figure. This includes the overlapping reading frames orfA and orfB (black rods) together with their relative reading phases (0 and −1). The point of translational frame shifting is shown as a cloned vertical line. The 36 bp left and right terminal inverted repeats are shown as a heavy gray box and triangle, respectively. The endogenous promoter PIRL is also indicated. The RNA sequence encompassing the frame shift window is shown below. The potential orfB ribosome binding site is boxed, as is its AUU initiation codon. The ribosome engages the A6G slippery codons and slips one nucleotide to the left, as shown by the arrow. The structured downstream region is shown to the right and the orfA termination codon is indicated. The bottom of the figure shows the organization of IS911 proteins together with their molecular mass. The α helix-turn-α-helix motif (HTH) is shown as a black square. Individual heptads of the leucine zipper (LZ) are represented by four elipses. The unfilled elipse shown in OrfA represents the heptad, which differs between OrfA and OrfAB. The catalytic domain is shown as a gray box including the DDE signature; black upright lines. (From Rousseau et al., 2002).

Chandler, Ton-Hoang, Turlan, Polard, Haren and their associates (see Rousseau et al., 2002). Thus only an overview of this process will be provided below, based on Rousseau et al. (2002). The cell-free system that can accomplish the transposition in IS*911* (Polard et al., 1996) was very useful in analyzing the details of this process. As with other reactions in which the DDE class of Tpases is active, the chemical steps of IS*911* transposition involve single-strand cleavage of the transposon end strand transfer to the target DNA. Two additional species were observed when the plasmid of *Escherichia coli* carried a Tpase and a transposon donor. One species is a covalently closed transposon circle. The other species is a molecule in which only one of the two transposon strands is circularized to form a figure-eight structure. The process of transposition is schematically shown in Figure 12. The process can be described by the following steps.

– *Synaptic complex A.* This initial step involves recognition of the terminal IRs by the Tpase (OrfAB), and their assembly into a synaptic complex where the IRs are correctly positioned for subsequent chemical steps.

– *Cleavage.* Once assembled into the complex, one transposon (donor) strand, at

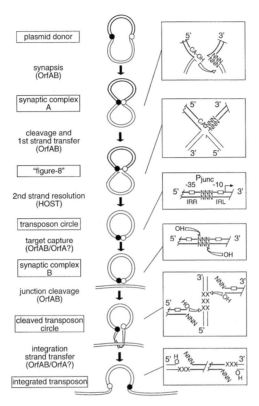

Figure 12. Steps in the overall IS911 transposition pathway. The center shows a global view of the transposition process. The transposon is shown in bold, donor backbone sequences as fine lines, and target DNA as dotted lines. The small filled and unfilled circles represent IRL or IRR. The panels on the right of the figure show a finer view of the various steps. They illustrate (descending): cleavage at the terminal 5' CA 3' to generate a 3'-OH and attack 3 bases (NNN) 5' of the opposite (target) end. The junction of the figure eight with a free 3'-OH in the vector strand are indicated (half arrow), together with the IRR-IRL circle junction and the associated promoter junc with −35 and −10 regions (unfilled boxes), the position of nucleophilic attack on both strands of the junction (arrows), target attack by the two liberated 3'-OH groups either side of the prospective 3-bp flanking target repeat (arrows), and the final insertion product ready to undergo repair. (From Rousseau et al., 2002).

one or the other end, is cleaved; probably using H_2O as a nucleophile, to generate a free 3'-OH.

– *Strand transfer*. This transfer leads to a figure-eight formation. The liberated 3'-OH then acts as a nucleophile, which directs the strand transfer to the same strand, 3 bases 5' to the other end of the element. This generates a molecule in which a single transposon strand is circularized, giving rise to a figure-eight structure when the terminal IRs of the transposon are joined by a single-stranded bridge and are separated by three bases that are derived from flanking DNA. The reaction can be viewed as a one-ended site-specific transposition event.

The above noted three steps can be accomplished by the OrfAB (Tpase) alone.
- *Figure-eight conversion to a circular form*. Kinetic data suggest that the figure-eight then gives rise to a circular transposon form, which carries terminal abutted IRs that are separated by three base pairs of DNA, flanking the original insertion. This step may involve host factors.
- *Promoter assembly*. A remarkable consequence of the formation of the circular transposon is the assemblage of a new promoter, *pjunc*, in which a −35 hexamer is contributed by IRR and a −10 hexamer by IRL. This assembled promoter is significantly stronger than the indigeneous promoter (pIRL) and is correctly placed to drive a high level of transposase synthesis. This plays an active role *in vivo* in controlling IS*911* transposition. The integration then results in the disassembly of the strong novel promoter.
- *Synaptic complex B and integration*. The formation of a circle junction brings both transposons' ends together in an inverted orientation. This reactive junction must then participate in a second type of synaptic complex with the target DNA. Two single-strand cleavages, one at each abutted IR, would linearize the transposon circle, permitting the two liberated 3'-OH groups to direct coordinated strand transfer. The final step also requires OrfAB but it is greatly stimulated by OrfA. It is not yet known whether strand capture occurs before or after cleavage of the circle junction.

It should be noted that the description of the above listed steps is only a summary of the work presented by Rousseau et al. (2002). Much more detail has been revealed; but many more details await clarification. Moreover, the transposition pathway outlined above – a simple insertion of the IS*911* element into a target DNA – is happening in the majority of the transpositions. A small (about 5%) number of cases end in cointegration. In these cases fused donor and recipient molecules are separated by directly repeated copies of the IS. It is also noteworthy that the scheme of IS*911* transposition differs from the "cut and paste" scheme of transposition that involves the processing of both strands at each end of the element.

7. THE IS*21* FAMILY

The members of this family are the longest among the simple bacterial Insertion Sequences. Their lengths span from about 2,000 to 2,800 bp. Their characteristics were summarized by Berger and Haas (2001). The members of the IS*21* family carry related terminal IRs whose lengths vary between 11 and 50 bp, and generally terminate in 5'-CA-3'. Many members carry multiple repeated sequences at their ends, including part of the terminal IRs. These repeats may represent Tpase binding sites. The "sets" of these repeats may have confused investigators. Thus, in a 1998 report (Mahillon and Chandler, 1998) the IS*21* was presented as lacking such terminal repeats, but in a more recent report (Chandler and Mahillon, 2002) they are presented as a "degenerate set". The insertion in this family results in 4 or (more rarely) 5 bp of DR.

The members of the IS*21* family have two consecutive open reading frames (Orfs): a long, upstream frame, *ist*A and a shorter, downstream frame, *ist*B (Figure 13). The derived proteins carry several blocks of conserved residues. The *ist*B frame maybe in phase −1, slightly separated or overlapping. The "border" between the two frames is commonly preceded by a potential ribosome binding site. This

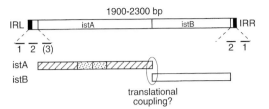

Figure 13. The IS21 family. General organization. Terminal inverted repeats IRL and IRR are shown as filled boxes. The position of the istA and istB reading frames is also shown. The horizontal lines below show the relative positions of the multiply repeated elements. IstA (hatched box) together with the potential: "DDE:" motif (stippled box) and IstB (open box) are indicated below. The possibility of translational coupling between the two reading frames is indicated. (From Chandler and Mahillon, 2002).

arrangement of the two reading frames suggests that translational coupling could occur as indicated in Figure 13.

The members of the IS*21* are widespread and were detected in diverse bacterial genera (and their plasmids), such as: *Escherichia, Pseudomonas, Salmonella, Yersinia, Shigella, Bacillus, Mycobacterium, Agrobacterium, Alcaligenes* and *Lactococcus*. Among the "hosts" of IS*21* family members are bacteria of relevance to human pathogenicity.

8. THE IS*21*

This IS, that carries the "family" name, was discovered over 20 years ago during "chromosome mobilization" experiments with a plasmid of *Pseudomonas aeruginosa*. When a conjugative plasmid undergoes replicon fusion with a bacterial genophore (i.e. "chromosome"), the genophore can be mobilized from the origin of transfer, provided by the plasmid, and conjugative genophore transfer to the recipient bacterium can occur. It was revealed that the ability of many conjugative plasmids to mobilize bacterial genophores is related to the activity of transposable elements (i.e. to IS). *P. aeruginosa* may host the broad-host-range plasmid R68.45. This plasmid promotes genophore mobilization in *P. aeruginosa* and in about 30 additional bacterial species. It was in this R68.45 plasmid where a tandem repeat of IS*21* was discovered. The R68 plasmid contains a single copy of IS*21*. It was assumed that a spontaneous tandem duplication of IS*21* in R68 resulted in the R68.45 plasmid, although the latter could also be obtained under laboratory conditions. For experimentation the R68.45, with its (IS*21*)$_2$, is rather useful: it may integrate into bacterial genophores at many different sites and thus initiate genophore transfer almost at random (see Berger and Haas, 2001, for a review).

8.1. Transposition of IS21

This transposition is manifested in different ways. A single copy of IS*21* can generate a simple insertion by a "cut and paste" (non-replicative) mechanism. This mechanism may involve a stage of circulization. A single copy of IS*21* can also yield cointegrates (though at low frequency). By contrast, cointegrate formation

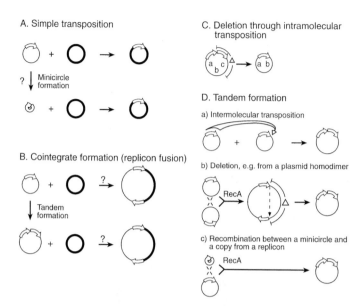

Figure 14. Overview of transposition pathways of IS21. Bold line, target replicon; thin line, donor replicon; arrows represent site-specific recombination events; a cross stands for homologous recombination. (From Berger and Haas, 2001).

between a replicon that carries the (IS21)₂ and a target replicon is very efficient and also follows a "cut and paste" mechanism. These "choices" are presented schematically in Figure 14. A more detailed description of the structure of IS21 is presented in Figure 15.

The istA gene is essential for both simple insertion and replicon fusion. Two ATG start-codons associated with ribosome binding sites (RBS) enable the istA gene to generate two proteins from the same reading frame (as shown schematically in Figure 15). Both these proteins (P46 and P45) have a helix-turn-helix motif at their N-termini, and both have a typical DDE motif of Tpases. When only P45 was overexpressed in E. coli, it promoted replicon fusion (cointegrate formation) when the substrate was a reactive junction of two abutted IS21 ends on a replicon and the target was a multicopy plasmid. The P45 has been named cointegrase.

The P46 displays an intermediate activity (of about 10⁻³) in both simple transposition and the replicon fusion pathways and was named transposase (Tpase). The eight extra amino acids of P46 (relative to P45), MLSERDFY, at the terminus of the transposase may help to bring the two IS21 ends close to each other, so that they can form an IR-IR junction that is subsequently inserted into the target DNA.

The role of istB is probably that of a "helper protein". It is probably required for accurate strand transfer and capture of the target DNA. The IS21 probably has "target immunity" that prevents or reduces multiple IS21 insertions in close-by sites. It is still not clear whether or not IS21 has target specificity. A target preference for regions that display an 11-bp periodicity of the dinucleotide AA, was inferred.

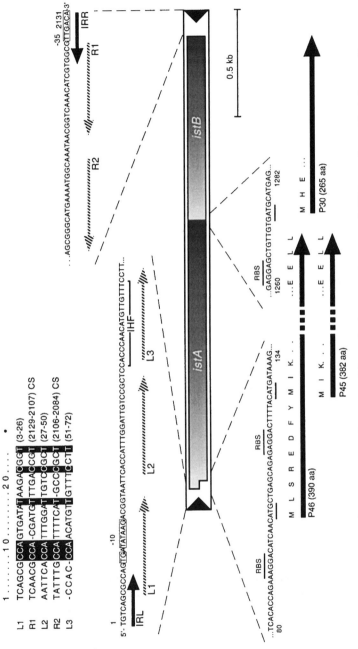

Figure 15. Insertion sequence IS21. A. Alignment of the multiple terminal repeats (MTRs) together with their coordinates; CS, complementary strand; nucleotides outlined in black, identity; nucleotides outlined in gray, pyrimidine or purine; L3 is slightly less conserve and therefore not considered in the determination of the motif. B. Upper part: termini of IS21; IRL and IRR, inverted repeats left and right L1, L2, L3, R1 and R2, multiple terminal repeats; IHF, sequence resembling the consensus sequence for integration host factor binding −10 and −35, sequences forming the promoter of the IR-IR junction. At the center: istA and istB, the ORFs of IS21; black triangles represent the IRs. Lower part: products of the istAB genes; P46, the transposase coded by istA; P45, the cointegrase coded by istA; P30, the helper protein coded by istB; RBS, ribosome binding site; the start codons ATG are underlined.   (From Berger and Haas, 2001).

8.2. Phylogeny of the IS21 Family

As was briefly indicated above, the members of the IS*21* family share several common features. They have ORFs: *ist*A and *ist*B. They have IRs of up to 50 bp that are not conserved within the family but usually end with CA. They have MTRs of 15–20 bp and target duplications (DRs) of 4 to 9 nucleotides, ORF1 is always longer (340 to 585 amino acid residues) than ORF2 (245 to 270 amino acid residues). How much are these two ORFs (i.e. *ist*A and *ist*B) related?

Berger and Haas (2001) summarized their detailed study by two dendrograms, each based on 24 members of the IS*21* family (their Figure 3). Generally, the dendrogram for ORF1 is similar to the dendrogram for ORF2. There are also significant differences: two family members, IS*1532* (of *Mycobacterium tuberculosis*) and IS*1631* (of *Bradyrhizobium japonicum*) are very close in their ORF1 sequences but diverted in their ORF2 sequences. How can this happen? As Berger and Haas furnished no explanation, the present author will also not make a suggestion. But he has a "story" that follows.

In 1987 Wilson and associates published (Cann et al., 1987) in *Nature* that, based on mitochondrial base-sequence dendrograms of about 150 individual humans of diverse ethnic origin, all existing humans in the world had one female progenitor. This "Mother Eve" was calculated to live in Africa about 150,000 to 200,000 YBP. Notably one year earlier George and Ryder (1986) had published in *Mol. Biol. Evolution*, the evolution of the genus of the horse (*Equus*) which was also based on a mitochondrial dendrogram. The calculation indicated that the immediate progenitor of the domesticated horse (*E. przewalskii*) that is still grazing in Mongolia, originated in Africa at about the same time inferred for "Mother Eve".

This promoted my suggestion (Galun, 1988) that about 150,000 YBP the daughters of "Mother Eve" mounted Przewalski horses and were riding from Africa to ~Mongolia... Well, there is still an enigma put forward by Wainscoat (1987) in his editorial remark to the article of Cann et al. (1987), in the same *Nature* issue. Wainscoat pointed out that a dendrogram could be constructed on the sequences of the Y chromosome of man. This chromosome is transferred paternally, has variability among ethnic groups and thus could serve to trace, spatially and temporarily, the original "Father Adam". He then asked what will we say if the respective calculations based on these two dendrograms, will place "Father Adam" in a different continent and in a different period to "Mother Eve".

9. OTHER SIMPLE BACTERIAL INSERTION SEQUENCE FAMILIES

From a recent survey of bacterial Insertion Sequences (Chandler and Mahillon, 2002) it appears that in addition to the three IS families handled above there are "about" 16 additional IS families, as summarized in Table 1. The quotation marks that flanked the word "about" indicate that the grouping into defined families of IS is not precise and with additional studies these, and possibly other IS families, may be re-grouped.

Several IS family members are components of *Compound Bacterial Transposons* and as such we shall handle them below. The following IS families are surveyed in detail by the review of Chandler and Mahillon (2002): IS*4*, IS*5*, IS*6*, IS*30*, IS*66*, IS*91*, IS*110*, IS*256*, IS*481*, IS*630*, IS*982*, IS*1380*, ISAs*1* and ISL3. In addition, this review mentioned a number of "odd" IS such as the IS*200* and the

IS*605* complex as well as "orphan" families.

In the following, we shall not detail these IS families but merely provide remarks on two of them.

9.1. The IS4 Family

As indicated, this is a heterogenous family. It includes the IS*231* "cluster", IS*10* and IS*50*. The IS*231* cluster contains several individual IS. One of these was isolated near the coding sequence for the endotoxin crystal-protein of *Bacillus thuringiensis*. This endotoxin and its coding region became useful for the protection of crops against pathogenic insects. Commercial companies produced transgenic plants of several major crops such as maize, cotton and tomato that express this endotoxin, and these plants were claimed to be resistant to phagocitic insects.

The IS*10* forms a part of the tetracycline-resistance transposon Tn*10* and IS*50* forms a part of transposon Tn*5*. These transposons will be handled below with other compound transposons.

9.2. The IS5 Family

This is also a heterogeneous family. As with the IS*4* family, several family members are associated with compound transposons. Thus IS*903* forms a part of the kanamycin-resistant Tn*903*. This IS*903* is probably one of the best known of the IS*5* family.

10. SOME REMARKS ON RECOMBINASES

Mizuuchi and Baker (2002) divided the recombinases, which are active in DNA cutting and joining reactions and required in mobilizing DNA segments, into two major classes. In one class, which includes the conservative site-specific recombinases, no high energy cofactors are consumed; the total number of phosphodiester bonds remains unchanged by the reaction and no DNA degradation or resynthesis is necessary. The enzyme active in the integration of the bacteriophage λ DNA into the bacterial genophore (λ integrase) is a prototypical example of this class of recombinases. This class includes several additional proteins such as members of the tyrosine recombinase, P1 Cre, and yeast Flp as well as enzymes involved in the mobilization of certain conjugative transposons, and invertases and resolvases. Some of these mobilizations will be handled in later chapters.

The second class of recombinases includes the transposons/retroviral integrase family and its short term is *transposases*. This class is active in the transposition of bacterial Insertion Sequences that were handled above and also in integration of DNA copies of retroviruses into their host genomes and integration of the DNA copies of long terminal repeat (LTR) retrotransposons.

The conservative site-specific recombination will not be handled in this book. Briefly, in this recombination both DNA strands of each recombination partner are cleaved at specific sites, exchanged and religated to new partners. Thus, four recombinase subunits, each associated with one of the four cleaved ends, are probably engaged in this recombination.

The transposases that are amply referred to in this book, catalyze two types of

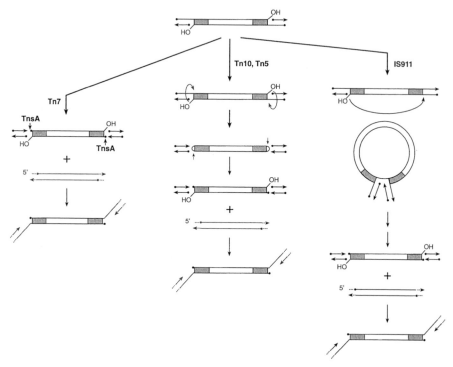

Figure 16. Multiple mechanisms used to cleave the "non-transferred" strand during transposon excision. (From Mizuuchi and Baker, 2002).

chemical steps to mobilize a segment of DNA. In the first step a pair of site-specific endonucleolytic cleavages separate the 3'-OH of the transposon (insertion sequence) DNA ends from the 5-phosphoryl ends of the adjoining host DNA. This donor DNA cleavage occurs at each of the ends of the element's DNA, resulting in a doubly-nicked DNA molecule with the transposon terminating with free 3'-OH ends. The second reaction, catalyzed by the transposase, is the direct "attack" of this 3'-OH DNA end on a new phosphodiester bond – the bond of the new insertion site. The transposon end is thus joined to the new DNA site and a nick is introduced into the target DNA. This transesterification reaction, termed "DNA strand transfer", occurs at the two ends of the element. The "attack" of the two transposon ends on the two strands of the host DNA is separated by 2 to 9 nucleotids, with 5' stagger. The exact nucleotide distance of separation is typical for each element (i.e. insertion sequence). The result is a typical short target-sequence duplication that flanks the newly transposed copy of the element.

Additional details of this process are discussed in Mizuuchi and Baker (2002) but three mechanisms of cleavage are schematically presented in Figure 16. The mechanism of cleavage in IS*911* was handled above and Tn*10*, Tn*5* as well as Tn*7* shall be handled below. It should be understood that the transposases as well as the DNA donor and DNA target segments are all involved in a complex termed "reaction-center". There are various chemical changes going on during the stages of transposition but the components are never completely dissociated during these steps. Thus, while this multiple-reaction in one center is "economical" energy-wise, it requires a lot of flexibility, and is rather complex. The process can still be defined as "catalysis" and the transposase can be defined as an enzyme, but it is a rather different process than the reactions activated by conventional enzymes. The transposases do not release their products repeatedly and frequently. The transposition has a very low frequency, less than once every cell cycle. The catalytic strategies used by transposases (and other recombinases) thus differ fundamentally from those for high-turnover enzymes that are true catalysts. For the transposition to occur all required components should be the appropriate ones and they should be available at the reaction center prior to any recombination process. Assembly "checkpoints" exist to ensure that undesirable transposition reactions are avoided. We are thus faced with biological control mechanisms that evolved with the transposition process.

We should also keep in mind that we consider transposition of insertion sequences (including compound IS) a less complex process than transposition in eukaryotic organisms (e.g. retrotransposons and Class II eukaryotic T.E.). This is because in the latter we face the elaborate complex of the DNA with histons, into the chromatin structures.

11. THE TN TRANSPOSONS

These elements have a similar structure to that of IS elements but they may be termed Composite Insertion Elements (including more than a single IS). Many of the Tn elements carry a DNA sequence that encodes an enzyme that degrades an antibiotic compound. Antibiotic compounds, such as those revealed by the respective pioneers (e.g. A. Fleming, H.W. Flory and E.B. Chain, for penicillin and S.A. Waksman and A. Schatz for streptomycin), were microbial components for eons. It is plausible that both the antibiotic compounds and the enzymes that can degrade them evolved in the microbial world in a kind of a "war between Titans" (though tiny Titans). One microorganism equipped itself with a compound that would suppress (or kill) a "foe" microorganism and the latter microorganism developed an enzyme that would decompose the antibiotic compound. This "war" was going on for eons, without being detected by man, until it was revealed by investigators who attempted to cure humans from bacterial infections. However, only 35 years or more after the detection of antibiotics, the awareness emerged that microbes already had weapons available against antibiotics. Pathogenic microbes even found the means to equip themselves with these weapons from outside sources.

In the Tn elements we meet the genes that encode antibiotic-degrading enzymes. The Tn elements are commonly not an integral component of the bacterial genophore; they are in bacterial plasmids or in bacteriophages and as such can move from one bacterium to another. However, they may occasionally integrate into the bacterial genophore. Ironically, the "war" is still going on, but with different opponents: bacteria against man. Man introduces antibiotic compounds into his bloodstream and pathogenic bacteria acquire plasmids containing genes that produce enzymes that degrade the antibiotic compound. The Tn elements not only contain such enzymes – they have the capability to be mobilized among plasmids.

We shall handle below four types of Tn transposons: Tn3, Tn5, Tn7 and Tn10. We shall provide their essential features and their modes of transposition, but shall not detail all that is known on these transposons. For those readers who are interested in the detailed knowledge, we shall cite reviews and specific publications on each of these elements (or family of elements).

12. THE TN3 FAMILY

The research on the Tn3 family started by the detection of bacterial plasmids that harbored a DNA sequence encoding a gene for β-lactamase. It had the capability to degrade ampicillin. Transfer of ampicillin-resistance by plasmids was reported by Hedges and Jacob (1974) and the element with the code for the β-lactamase was first termed TnA (Heffron et al., 1975a,b). In later years very similar (or identical) elements, Tn1, Tn2 and Tn3, were revealed. Detailed studies that included sequencing were performed with Tn3 (e.g. Heffron et al., 1979a,b; Ohtsubo et al., 1979). The Tn3 emerged as an element that had a sequence of 4957 bp. Three open reading frames were detected. A 1015 bp sequence (the tnpA gene) that encodes the transposase; a shorter sequence (tnpR gene) that encodes the "repressor"; and a medium-length sequence that encodes the 286 amino acids of a β-lactamase. In addition there were three specific regions. There were the two inverted repeats (IRs) of 38 bp each, that flanked the element, and between tnpA and tnpR there was a sequence that was then coined IRS, for internal resolution site (but later named res). During the following years intensive research by several investigators (e.g. F. Heffron, D. Sherratt, N.D.F. Grindley), revealed ample information on Tn3. First it was found that there is a family of Tn3 elements. Some family members had great similarity to Tn3 (e.g. Tn1, Tn2, Tn2601) with respect to the general structure and the code for resistance to ampicillin. In others (Tn501) there was a gene for resistance to mercury or to other antibacterial agents. Research focused on Tn3 as well as on the element Tn1000 (also termed gd). Another sub-family of Tn3 differed in the overall structure: in these elements the transcripts of tnpA and tnpR are not divergent as in the Tn3 sub-family, but are rather in the same direction. Also in the Tn501 sub-family the res site is downstream of tnpR rather than between tnpA and tnpR, in the Tn3 sub-family.

The information on Tn3 and the Tn3 family of transposons was provided by two comprehensive reviews: Sherratt (1989) and Grindley (2002). The following description of Tn3 is mainly based on these reviews.

The general features of the Tn3 family can be summarized as follows.
1. They encode a related and unusually large transposase enzyme of about 1000 amino acid residues.
2. They have similar terminal inverted repeats (IRs) of about 40 bp.
3. They transpose by a *replicative* pathway, forming an intermediate, called *cointegrate*, in which the entire *replicon* is inserted into the target DNA (commonly another plasmid) with the insert flanked by direct repeats of the transposon (Figure 17).
4. The transposition of this family results in a 5 bp target-repeat sequence.
5. There is transposition "immunity": once a target DNA accepted an element of the Tn3 family this DNA shows "immunity" to a further insertion of this element.
6. They commonly encode a *site-specific recombination system*. This system separates the cointegrate intermediate into two components: the target with a simple transposon insertion and the regenerated donor replicon. This process is performed (in Tn3) by a *serine-recombinase* of the *resolvase*/DNA invertase

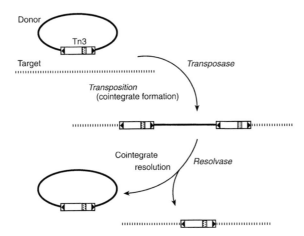

Figure 17. Two stages of Tn3 transposition: the formation of the donor-target cointegrate and its subsequent resolution. Transposase, responsible for the first step, acts at the transposon ends. The site-specific recombinase, resolvase, acts at a site, res, shown as a stippled patch within the transposon. (From Grindley, 2002).

family enzyme complex or by a *tyrosine-recombinase* of the 1 integrase enzyme family (in some other members of the Tn3 family).

12.1. Transposon Organization

Grindley (2002) specified 28 members of the Tn3 family. Most of these members are of the serine resolvase type (as Tn3 itself). But other members such as Tn4430 (of *Bacillus thuringiensis*) and Tn4652 (of *Pseudomonas putida*) have the tyrosine type.

The site of the resolvase action (*res*) is a DNA segment of about 120 bp. In the Tn3 and its related elements, in which the transcription of *tnp*A and *tnp*R is divergent, the promoters for both transcripts are located at this site. The gene for antibiotic-resistance (that is commonly a component of the Tn3 family members) resides between the *tnp*R and the right-side IR. A few members of this family are similar to other members with respect to most features (e.g. *tnp*A, *tnp*R, *res*, IRs) but they lack the gene for resistance to antibiotics or to toxic metals.

The Tn3 family includes also elements that have a rather different general organization than the Tn3. Among these are "composite transposons" such as Tn5271 that is structured by two complete sequences of IS1071, in the same orientation, having between them a chlorobenzoate catabolic gene. In brief, the Tn3 family includes an assortment of dissimilar transposons.

12.2. Transposition

The transposases of the Tn3 family are rather unique among transposons. The former are considerably larger proteins, spanning 950 to 1020 amino acid residues. It was suggested that their larger-than-usual size is related to their sensing of "immunity".

In some other transposons, the immunity-sensing is attributed to a separate protein. In the Tn3 family the transposase has multiple tasks: recognition, strand transfer, target capture and target "immunity".

By DNA sequencing analyses of the transposons of various Tn3 family members, the transposases were grouped into several "clusters". Remarkably, the division into clusters correlates poorly with the overall transposon organization and the association with a specific type of site-specific recombinase (for cointegrate resolution). In general, this enzyme (TnpA) is suggested to bind the IR domain; the central portion (residues 240–630) is suggested to be involved in the non-specific DNA binding. The C terminal half of the TnpA is probably involved in the catalytical activity of this protein. This half of the TnpA is also the most conserved sequence among the Tn3 family members. It is where the catalytic triad, DDE (mentioned above in the description of simple Insertion Sequences in bacteria) resides.

12.3. Transposon's Terminals

The transposons of the Tn3 family have typically terminal IRs of about 38 bp. The outer bp are commonly 5'-GGGG. The inner sequence (of about 34 bp) varies considerable among the family members. This probably ensures that each transposase recognizes its "own" IR. There are two functional domains. The inner domain is involved with specific recognition and binding to the transposase, while the outer bp are required only for subsequent steps of the transposition. The contact of the transposase is probably located at positions 11, 21 and 31, which means once for each turn of the double helix of the DNA.

12.4. Cointegration

The final product of intermolecular transposition of Tn3 and its related transposons, is a simple integration into a target DNA sequence. However, there is a vast difference between the transposition of Tn3 and other elements such as the bacterial IS handled above. In the latter we found a "simple" process of excision and insertion ("cut and paste"). In the Tn3 there is no excision. Rather the first step of the transposition in Tn3 is the formation of a cointegrate that results from the cointegration between the Tn3 of the donor and the recipient DNA (commonly a plasmid). This cointegration is driven by the transposase and by the host-cell replication machinery. The general model was actually suggested for the bacteriophage Mu by Shapiro (1979). A scheme for this process is provided in Figure 18.

12.5. Cointegrate Resolution

As indicated briefly above, the Tn3 family can be split according to the resolution of the cointegrate. Some members of the Tn3 family have an integrase-related site-specific resolution system. We shall not detail this system but focus on the resolvase-related site-specific resolution which operates in Tn3 and most other Tn3 family members. The latter system will be described briefly. A very detailed account of this system was presented in the recent review of Grindley (2002).

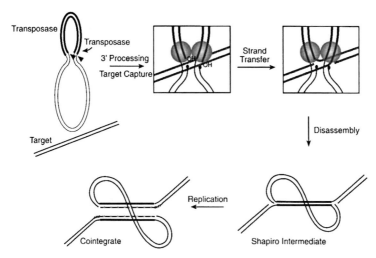

Figure 18. The Shapiro model for replicative transposition. Transposase binds to the transposon ends, pairs them then makes single stranded breaks at both transposon 3' ends (3' processing). The complex captures a target DNA, the 3'-OH Tn ends directly attack the target phosphate backbone, linking them to the target and leaving a free 3'-OH end in each target strand (strand transfer). Following disassembly of the complex, replication complexes are loaded on each 3-way junction, and the entire transposon is duplicated, forming the cointegrate.
(From Grindley, 2002).

The resolution requires the formation of a double-strand break, at the crossover point of each "parental" duplex, the exchange of the DNA partners and finally the joining of the ends in the recombinant configuration. The process was investigated also *in vitro* and Tn*3* as well as Tn*1000* (*gd*) served in these investigations.

The *in vitro* investigations indicated that the processes of breakage and rejoining require no outside energy. During the first process the 5' phosphates of the DNA was covalently joined to a serine residue of the resolvase protein. The following scenario emerged. In the appropriate complex, containing a pair of recombination sites, DNA cleavage at each crossover point results from the direct nucleophilic attack of a pair of activated serine side-chains on the two strands of the DNA duplex. The transesterficiations, four per recombination complex, form a serine-DNA phosphodiester linkage to each 5' end, releasing free 3'-OH groups. Rearrangement of the DNA ends, from parental to recombinant configuration, enables the 3'-OH to attack the serine-DNA phosphodiesters in a second set of transesterification reactions, releasing the DNA backbone to produce covalently-closed recombinant sites, and releasing the serine-DNA linkage.

The resolvase-recombination requires a pair of identical sites, the *res*, of the two cointegrates. The *res* sites have a length of between 90 and 140 (usually about 120) bp. The *res* sequences should be joined in a head-to-tail orientation in the same superhelical DNA molecule. A complex is thus established that includes the *tnp*R, and the *res* sequences in a kind of *synapsis*. The typical *res* has three *tnp*R binding sites. One of these (site I) is the breakage and strand-exchange site. Two additional

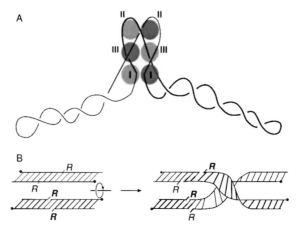

Figure 19. Synapsis and strand exchange in Tn3. A. Cartoon of the resolvase synaptic complex. The two substrate domains, separated by the 2 res sites, are shown as thick and thin lines. The only interdomainal DNA crossings are the 3 (–) nodes entrapped in the complex. One res (thick line) is bound by the stippled resolvase dimmers, the other (thin line) by the striped dimmers. B. Double strand break-rotation model for strand exchange. Site I DNA segments (represented by planar ribbons) are aligned with either both major or both minor grooves facing one another. Double strand breaks are made by resolvase (R) at both crossover points, then one pair of duplexes is rotated by 180° in a right-handed direction about the two fold axis between the aligned sites. This rotation creates one (+) interdomainal node and adds 1/2 a turn for each duplex (relaxing one additional substrate supercoil). For further details, see text.
(From Grindley, 2002).

downstream sites, on *res*, are between site I and the coding sequence for *tnp*R. The divergent transcripts of *tnp*A and *tnp*R are regulated by the respective promoters P_A and P_R that also reside in *res*. When the resolvase binds to *res* the promoters are suppressed. While the exact space between site I and site II is not crucial, the length between site II and site III is crucial; a change introduced in this length will abolish the resolvase activity. These characteristics are true for Tn*3* and Tn*1000*, but some other Tn*3* family members differ in the *res* binding cites.

The resolvase of most Tn*3* family members contains 180 to 210 amino acid residues and can be divided into two domains. The N-terminal domain (about 120 amino acids) contains the catalytic active site of *tnp*R. The C-terminal domain (about 65 amino acids) is responsible for DNA binding. The active site of the resolvase of Tn*3* and of other family members was studied intensively (Grindley, 2002) and the details of its structure were found to differ among the Tn*3* family members.

12.6. The Recombinational Process

The interaction of the resolvase with *res* is complicated. The binding of resolvase causes bending of the *res* DNA; moreover, binding to site I affects the binding (and bending) of the other two sites of *res*. The outcome of this complicated binding and bending is a resolvase-*res* complex termed *resolvasome* that emerges as a closed structure: a loop of DNA closed into a ring by protein-protein interactions. Two resolvase-bound *res* sites must then interact and assemble into an organized *synaptic complex* that was termed *synaptosome*. A model for the synaptosome and for strand exchange thus emerged (Figure 19). The study of this process is still being actively pursued. Probably "outsiders" should show patience and wait until the *res*-resolvase complex is resolved and a consensus solution for the mechanism of strand exchange and resolution emerges.

13. THE TN*5* AND ITS TRANSPOSITION

The following description of Tn*5* is mainly based on the recent review of Reznikoff (2002). 1975 was a good year for Tn transposons. We noted above the pioneering work of Heffron et al. (1975a,b) on the first member of the Tn*3* family. In the same year Berg et al. (1975) published information that led to the identification of Tn*5*. Moreover, the articles of Berg et al. and of Heffron et al. were published head-to-tail in the very same issue of the *Proc. Natl. Acad. Sci. (USA)*. However, these two articles arrived in Washington D.C. from two opposite corners of the world: from Geneva, Switzerland (Berg et al.) and from Seattle, Washington State (Heffron et al.).

As shall be detailed below, Tn*5* differs vastly from the Tn*3* family members. Albeit, they have the term Tn in common and they also share the capability to hop off and on DNA "wheels" (circles). These "wheels" can be the genophores of bacteria or the circular plasmids of bacteria. With a smile we may claim that the "wheel" was not invented in ancient Babylon, nor in ancient Egypt, as a component of a vehicle of war. Bacteria invented the DNA "wheel" much before 2700 BC. Although, the "wheels" (spheres) were already manifested with the creation of galaxies. There is a difference: as we shall see below, there is evidence that for hopping, at least some transposons prefer supercoiled DNA circles ("wheels"), rather than simple circles.

Back from celestial entities to Tn*5*. The Tn*5* is a sequence of DNA that has a length of 5.7 kilobase pairs (kbp). It can be inserted into many sites of bacterial genophores, bacterial plasmids and temperate bacteriophages. It encodes resistance to the aminoglycoside antibiotics kanamycin and neomycin in bacteria and to G418 in eukaryotic cells. The Tn*5* was detected in the bacteriophage λ grown in a strain of *E. coli* that carried an R factor from a clinical isolate (Berg et al., 1975). There is a paradox: although it seems to be able to be inserted into many DNA sites, it is very rare in nature.

In spite of its rarity in nature, Tn*5* became very popular in the laboratory. Because of its relatively simple structure, its random insertions and other features that will be detailed below, Tn*5* became a useful tool for analyzing and manipulating DNA molecules. Due to the establishment of *in vitro* methods Tn*5* also emerged as very favorable for the study of "cut and paste" transposition.

13.1. The Overall Structure of Tn5

This Tn is a composite transposon. It is composed of two inverted IS*50* of 1.5 kbp each (Figure 20). These IS*50* are similar but not identical and are termed IS*50*L and IS*50*R, for the left and the right IS, respectively. Each of these IS is flanked by

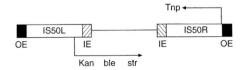

OE: CTGACTCTT**ATAC**ACAAGT
IE: CTGTCTCTT**GAT**CAGATCT
MOSAIC: CTGTCTCTT**ATAC**ACATCT

Figure 20. The structure of the Tn5 transposon. Tn5 is composed of two insertion sequences, IS50L and IS50R. IS50R codes for Tn5 transposase. Each insertion sequence is flanked by 19-bp end sequences termed outside end and inside end. The black box represents the OE and the striped box represents the IE. The OE and IE differ by 7 bases (shown in boldface). A hyperactive, hybrid end sequence, the mosaic end (MOSAIC) was used. The boldface letters in the mosaic end sequence are outside end sequences, whereas the lightface letters are inside end sequences. (From Bhasin et al., 1999).

inverse-repeat (IR) sequences that are termed OE and IE for the outside and inside IR, respectively. The DNA sequence between IS50L and IS50R harbors the codes for kanamycin-bleomycin-streptomycin resistances. The DNA sequence between the two ISs has no apparent role in transposition. Because it is composed of two IS50 elements the Tn5 may be considered as a member of the IS4 family.

13.2. Antibiotic Resistance Genes

The expression of the three genes for resistance to antibiotic compounds is driven by a constitutive promoter. This promoter is located within the right side of IS50L sequence. The streptomycin-resistance gene is cryptic in *E. coli*, but active in some other bacteria. It is noteworthy that the resistance to kanamycin of Tn5 made biotechnological history. This very gene (and a related gene from Tn903) was used in order to confer eukaryotic cells resistant to kanamycin as a selective marker. It became very popular in the *Agrobacterium*-mediated genetic transformation of plants (angiosperms) as detailed in our previous books (Galun and Breiman, 1997; Galun and Galun, 2001).

13.3. The Structure of the IS50s

These IS contain the information required for Tn5 transposition. The IS50R encodes two proteins. One is the transposase (Tnp) enzyme that catalyses the transposition. This enzyme is composed of 476 amino acids. The second protein (Inh) is a transactive inhibitor of transposition. Tnp and Inh are read in the same direction and in the same reading frame but Inh lacks the first 55 amino acids of Tnp. The two proteins are programmed by two different (overlapping and possibly competitive) promoters. The promoters are modulated by a host Dam methylase. Each of the ends of the IS50s (the OE and the IE) has a sequence of 19 bp. There are differences in sequence between the OE and the IE, at 7 of the 19 positions. While the OE and the IE are recognition sequences for the element's Tnp, host regulators as DnaA and Fis recognize OE, and Dam methylase recognizes IE.

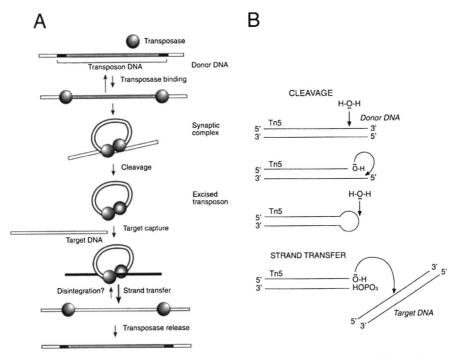

Figure 21. Transposition in Tn5. A. Cut and paste transposition model. B. Hairpin intermediate model for double strand cleavage and strand transfer during transposition. (From Reznikoff et al., 1999).

There is also a difference in the coding region between IS*50*L and IS*50*R. The IS*50*L contains an *ocher* codon that causes premature termination of the two proteins (Tnp and Inh), shortening the proteins and rendering them non-functional. However, in *ocher*-suppressing bacterial strains the IS*50*L also codes for a functional Tnp.

13.4. An Overview of Tn5 Transposition

The process of Tn*5* transposition can be described by several discrete steps as in the diagram in Figure 21.

These steps are the following:

1. Tnp binding to specific sites in the 19-bp end sequences.
2. The dimerization of the DNA-bound transposase that forms the catalytically active *synaptic* complex. An activated water molecule performs a nucleophilic attack, hydrolyzing one strand of the DNA that exposes a 3'-OH group at the end of the transposon. This 3'-OH is then activated to perform a nucleophilic attack on the opposite strand of DNA, forming a "hairpin" structure and excising the transposon from the donor DNA. The result is the release of the transposon (as a synaptic complex) from the donor DNA.
3. Hydrolysis of the hairpin produces a blunt-ended DNA at the transposon end.

4. At a further step the *synaptic* complex binds to the target DNA. This is the "target capture".
5. The process of the final step leads the activated 3'-OH groups, at the ends of the transposon, to perform nucleophilic attacks on the target DNA, causing strand transfer.

Much progress was achieved in recent years in elucidating the detailed chemistry of Tn5 transposition. Details were provided by Davies et al. (1999, 2000), and it was reviewed by Reznikoff (2002). While the fine details of the now known chemistry are beyond the scope of this book, some additional information will be provided below. Much of the recent progress is due to the artificial facilitation of the *in vitro* transposition process. Wild-type Tn5 is down-regulated so that the wild-type Tn5 transposase does not produce detectable levels of the synaptic complex *in vitro*. Progress was possible when mutated, hyperactive, Tnp became available that led to an efficient *in vitro* transposition and enabled chrystalographic analyses.

13.5. The Synaptic Complex

This complex, as noted above, consists of two molecules of Tnp and the two end DNA sequences. The formation of the synaptic complex is probably a disfavored process and thus of low frequency. Also, the catalysis is performed in *trans*, meaning that the subunit that performs catalysis at one end is the subunit that first binds on the other end. Thus, there is first synapsis formation and then the catalysis. Davies et al. (2000) provided a ribbon-presentation of the Tn5 transposase/DNA complex and showed a stereoview of one monomer of transposase (Figure 22). These authors also gave a three-dimensional image of the *cis* and *trans* protein/DNA interaction (Figure 23). This figure clarifies the meaning of *cis* as the Tpn/DNA binding and the *trans* as the catalysis by one domain of the Tnp at a particular transposon end. This latter figure also locates the Mn^{2+} ion in the complex. In short, a relatively simple *in vitro* system enabled one to analyse the interaction of Tnp subunits with short (20 bp) DNA sequences that represent the relevant sequences of the IRs. A central issue in the catalytic core of this interaction is that the active site contains three conserved amino acid residues (the DDE motif), that bind Mn^{2+}. These conserved amino acids are located as amino acids numbers 97 (aspartate-D), 188 (aspartate-D) and 326 (glutamate-E). This basic core is shared by other catalytic processes as in retroviral integrases (e.g. in human immunodeficiency virus-1).

Figure 22. The structure of the Tn5 transposase/DNA complex. (A) Ribbon representation of the transposase/DNA dimer viewed along a crystallographic two-fold axis of symmetry. One protein subunit is colored yellow, the other is blue, and the two 20-bp DNA molecules are purple. The three catalytic residues are represented as green ball-and-stick structures, and the associated Mn^{2+} ion is black. (B) Stereoview of one monomer of transposase. The NH2-terminal domain is yellow, the catalytic domain is blue, and the COOH-terminal domain is red. The active site residues Asp^{97}. Asp^{188}, and Glu^{326} and the associated Mn^{2+} ion are shown as green ball-and-stick structures. The backbone of a double-stranded DNA is represented by transparent ribbons. (From Davies et al., 2000).

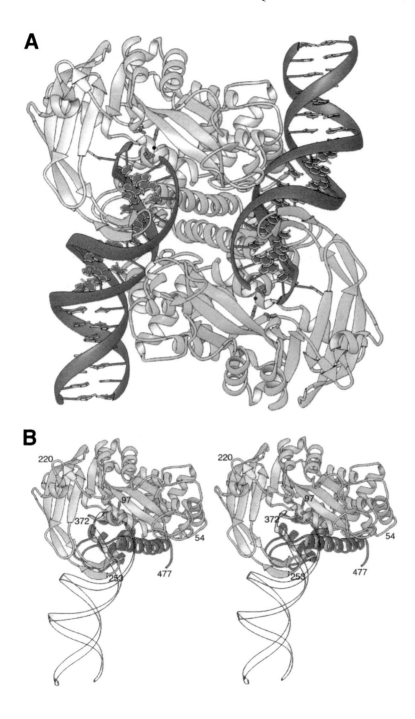

Figure 22. See legend on previous page.

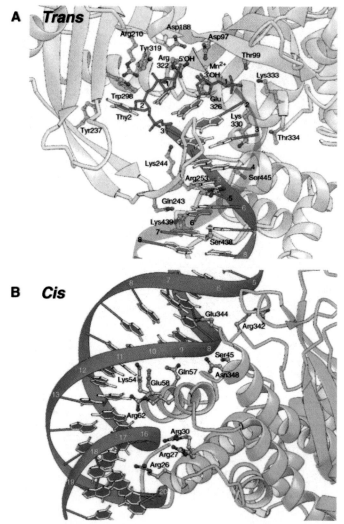

Figure 23. Three-dimensional representation of cis and trans protein/DNA interactions. Cis interactions are defined as those between DNA and the protein subunit that is binding DNA molecule via the NH_2-terminal domain. Trans interactions are defined as those contributed by the protein subunit responsible for catalysis at a particular transposon and for each panel. (A) Trans protein/DNA interactions. Amino acid residues that interact with DNA are shown as predominantly yellow ball-and-stick structures, and the Mn^{2+} ion bound in the active site is pink. The backbone ribbon of the transferred strand of DNA (the strand that contains the 3'-OH responsible for nucleophilic attack on the target DNA) is semi-transparent to show the positions of Lys^{439} and Ser^{445}. (B) Cis protein/DNA interactions. Amino acid residues that interact with DNA are depicted with ball-and-stick representations. In both panels, nucleotide residues are identified according to the convention for transposable elements where base pairs of both strands are numbered starting at the cleavage site. (From Davies et al., 2000).

The structure of Tnp in the synaptic complex is generally similar to the structure of Inh, but with three important exceptions. First, the Inh is missing the N-terminus with its DNA binding domain. Second, the sequence between residues 243 and 255 is disordered in the Inh. Third, the Tnp (but not the Inh) in the synaptic complex has a disordered region from proline 373 to glutamine 391 that is associated with a relocation of the C terminus so that it no longer occupies an overlapping space with the N-terminus.

It is noteworthy that the protein-DNA contacts are very complexed and that these contacts rather than protein-protein contacts provide most interactions that stabilize the complex. Howevert, once the synaptic complex is established it is very stable. Thus *in vitro* formed complexes can be stored (in the cold) for extended periods.

Although the formation of the synaptic complex and its exact role in Tn5 transposition have not yet been solved completely, it is already clear that we are faced with a complex system. For example, the two domains of Tnp (the N-terminus, that is involved in primary DNA-binding and the C-terminus, that is involved in dimerization) inhibit each others' activities. This C-terminus-N-terminus mutual inhibition may be an important mechanism for the down-regulation of the frequency of transposition.

13.6. Transposon Backbone DNA Cleavage

The transposase catalyses the transposon donor backbone (DBB) cleavage through a three-step "hairpin" mechanism which is Mg^{2+}-dependent. Such a mechanism was also revealed in Tn10 (handled below). This is a *trans* mechanism, meaning that the Tnp monomer that initially binds to one end of the transposon performs the catalysis on the other transposon-transferred strand-donor. The DNA-junction is nicked in the first step; the released 3'-OH of the transferred strand then attacks the opposite strand to form a hairpin intermediate; the hairpin is subsequently nicked.

13.7. Target DNA Capture and Strand Transfer

The released transposase-transposon complex should bind the target DNA to initiate the strand transfer. In this transfer the 3' ends are led into the target so that the two ends go into the target DNA and these two penetrations are spaced 9 bp apart from each other. This then results in the Tn5-typical 9-bp duplication of the target sequence.

The entrance into the target is apparently random but there may be preferences which are still not well defined. Although there is evidence that negatively supercoiled DNA is a favored target.

13.8. Transposase Release

The final stage of the transposition is the gap repair of the 9-base single-stranded target DNA. This occurs at both ends. For this to happen, the Tnp has to be released from the transposition products. It was suggested that for that some "outside" factors are required because *in vitro* the Tnp remains bound.

13.9. The Regulation of Tn5 Transposition

Transposition occurs usually once per 10^5 cells, per generation. This means it is a relatively rare event. We should not be surprised to find that regulatory mechanisms evolved to keep the transposition at a low frequency because frequent transpositions, in nature, could have deleterious effects.

The review of Reznikoff (2002) revealed means by which the transposition of Tn5 may be down-regulated. I shall mention these only briefly. First the wild-type Tnp probably has a suboptimal structure. This became apparent when Tnp with certain changes showed a vastly elevated activity. We already mentioned that hyperactive Tnp mutants were useful for establishing *in vitro* transposition. Then, the Tnp has a strong bias to act in *cis*. The transposition takes place when the *tnp* gene is inside the transposed transposon or very close to it. Probably the strongest down-regulation is caused by the Inh protein that was noted above. One of the outside regulatory mechanisms is the Dam methylation. This methylation suppresses the transcription of the messages *tnp* and *inh* in opposite ways. Methylation of DNA fluctuates with the cell cycle. Thus, this may be an additional means to regulate transposition. Methylation of the IR also affects the binding of Tnp to these sequences, thus providing an additional regulatory mechanism for transposition. Several host factors, such as the integration of the host factor (IHF), may also affect transposition frequency in ways that have not been fully characterized.

13.10. Tn5 as a Genetic Tool in Bacteria and Eukaryotes

The relatively simple structure of Tn5, the modest requirements for its transposition and the availability of *in vitro* synaptic complexes of Tn5 render this transposon a versatile and efficient tool for the geneticist. All that is really required are the 19-bp end sequences and a functional Tnp. The fragment of DNA with these two ends can vary vastly in length and still function as a transposon. The space between the two ends can be as short as 264 bp (possibly even shorter, but shorter spaces were not tested) or as long as 11 kbp (possibly even longer, but longer spaces were not tested). These "spaces" can be filled with any DNA sequence of choice. The code for Tnp can be included, but, Tnp can also be furnished in *trans* (with the price of lower efficiency).

The natural Tn5 harbors three genes for resistance to antibiotics. If another selectable marker is required, another gene (flanked with due controlling sequences) can replace these natural genes. Likewise, a sequence serving as reporter gene can be inserted between the 19-bp ends. Reporters such as green-fluorescent-protein (GFP) come to mind.

One can also include into a "synthetic" Tn5 a specific primer for DNA sequencing. Methods were developed to use "nested deletions". In "nested deletions" a defined number of bp are eliminated from a given DNA sequence. One of the specific advantages of Tn5 as a genetic tool is the result of the *in vitro* system that was developed for the production of synaptic complexes. For example, a synaptic complex can be produced with a desired sequence of DNA that is flanked with the 19-bp ends. Such complexes can be introduced into cells by electroporation. The complex will then integrate into the host DNA (e.g. the bacterial genophore).

These and a multitude of additional applications were listed by Reznikoff (2002) who suggested that Tn*5*, its synthetic versions and its *in vitro* produced synthetic complexes may also be useful in eukaryotes such as yeast (*Saccharomyces cerevisiae*) and insects (*Drosophila melanogaster* and *Aedes aegypti*).

A final note is due with respect to the use of Tn*5* as a genetic tool. The apparently random insertion of Tn*5* that is of advantage for many genetic studies has a drawback: this element cannot be aimed into a specific site (sequence). This latter drawback is not stressed by the proponents of the usefulness of Tn*5*.

14. THE TN*7* TRANSPOSON

The "War of (tiny) Titans" was mentioned above to emphasize that in the evolution of microorganisms, some bacteria developed the capability to produce specific antibiotics that suppress (or kill) other microorganisms while some of the latter developed the capability to resist these antibiotics. This "war" went on for eons before man emerged. Several decades ago it surfaced and caused great concern to those engaged in therapeutics against microbial infection in humans and animals. The Tn*7* transposon provides an enlightening example of this "war" and its possible consequences. The Tn*7* transposon was revealed in bacteria that infected a calf that was treated by trimethoprim (against bacterial infection). The isolated bacteria were found to have resistance to trimethoprim and to streptomycin. This led to the finding of Tn*7* that encoded these resistances (Barth and Datta, 1977; Barth et al., 1976). The field of research on bacterial Insertion Sequences, at the time of the discovery of Tn*7*, was very young; two other Tns were discovered only one or two years earlier (and the first IS was reported in 1967).

In subsequent intensive studies the Tn*7* was characterized. It was found to share a basic feature with other Tn elements: it encodes antibiotic-resistance. It has the capability to be mobilized from a plasmid into the bacterial genophore as well as between plasmids. It has inverted repeat sequences and upon excision from an integration site leaves "typical footprints". The Tn*7* differs from IS and even from other Tns by several features that will be detailed below. Tn*7* is not a composite transposon (as Tn*5* mentioned before); it commonly does not insert "at random" but rather has a limited choice of integration sites and its transposition is relatively frequent.

Since the finding that *E. coli* genophores have a unique site for Tn*7* integration (Lichtenstein and Brenner, 1981, 1982) and the determination of the nucleic acid sequence of some Tn*7* genes (e.g. Fling and Richards, 1983), Tn*7* became the subject of very intensive studies by N.L. Craig and her associates as well as by others. These studies were amply reviewed by Craig (Craig, 1989, 1996, 2002). The following description of Tn*7* is mainly based on these reviews.

14.1. The General Structure of Tn7

The overall structure of Tn*7* is schematized in Figure 24. Figure 24a shows that Tn*7* has a total length of about 14 kbp. It has two ends, Tn*7*L (left) and Tn*7*R (right). Between these ends there are genes for resistances to trimethoprim (*dhfr*), to streptothricin (*sat*) and to streptomycin/spectinomycin (*aadA*). On the right side of Tn*7* there are coding sequences for five proteins that are involved in the transposition/integration (*tnsA*, *tnsB*, *tnsC*, *tnsD*, *tnsE*). More information on the

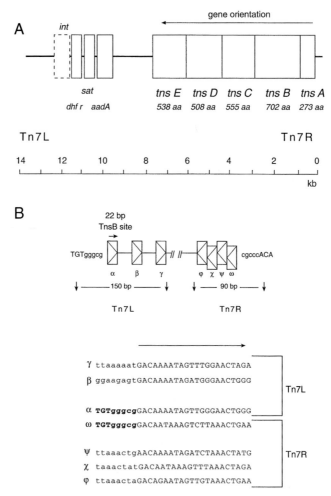

Figure 24. The map and the ends of Tn7. A. The map. The transposition genes – tnsA, tnsB, tnsC, tnsD and tnsE – are all encoded in the right end of Tn7, with their 5' termini closest to the right terminus of Tn7. The left end of Tn7 encodes several antibiotic resistant genes: dhfr which provides resistance to trimethoprim, sat which provides resistance to streptothricin, and aadA which provides resistance to streptomycin and spectinomycin. These antibiotic resistance cassettes are part of an integron element. B. The ends. The Tn7L and Tn7R end segments contain multiple copies of the 22 bp TnsB binding sequence as defined by TnsB foot-printing studies. The 30 bp nearly perfect inverted repeats contain a perfect 8 bp terminal inverted repeat (bold) containing a critical 5'-TGT...ACA-3', a 5 bp spacer region and a TnsB binding site. (From Craig, 2002).

left and right ends of Tn7 are shown in Figure 24b. First, there are eight inverse repeat (IR) bp at the extreme ends. Inwards of these ends there are 150 bp of Tn7L and 90 bp of Tn7R. These sequences are essential for recognition and binding of the protein complex that executes the transposition, as shall be detailed below. These sequences contain several protein-binding sites and those of Tn7L differ substantially from those of Tn7R. This difference provides orientation of the transposition. It was found that a DNA sequence that is flanked by Tn7L and Tn7R (mini Tn7) is competent in *in vitro* transposition, i.e. when the appropriate proteins and cofactors are added and a proper target sequence is available.

Of the 8 bp ends (5' TGTGGGCG...CGCCCACA 3') of the IR the 3 bp of the extreme termini (5' TGT...ACA 3') have a special role; they signal the site of the cleavage. The other, more inward 22 bp of the terminal repeats have probably an essential task of binding the proteins that activate the transposition. The most interior bp of these binding sites have the highest binding affinities to TnsB.

Cells that contain a Tn7 are resistant to antifolates such as trimethoprim. The gene *dhfr* encodes a dihydrofolate reductase that is not sensitive to inhibition by trimethoprim. Such cells are also resistant to streptothricin because the *sat* gene encodes a transacetylase that inactivates this antibiotic compound. The resistance to streptomycin and spectinomycin results from an adenyltransferase, encoded by *aadA* that inactivates these antibiotics. It is said that very careful people use both suspenders and a belt to keep their trousers in place. It seems that Tn7 is even more "careful" – it added "buttons"; it equipped itself with *three* resistances. How this triple-resistance evolved is enigmatic.

14.2. The Process of Transposition

There are two pathways of Tn7 transposition. In both cases the Tn7 is mobilized by the cut and paste reaction. In one, the probably more prevalent pathway, this transposon is inserted into a very specific site in the bacterial genophore. This site is termed *attTn7* and it exists in the genophores of many bacteria (including *E. coli*). The insertion into this site has no deleterious effects on the bacteria. The other pathway can be defined as insertion that is not into the *attTn7*. The latter insertion is into conjugating plasmids and into the lagging-strand of replicating DNA. The conjugating plasmids can move between bacteria by mating. It is actually the state of conjugation that attracts the insertion of Tn7 in this latter pathway.

The choice of these two pathways is related to the transposition proteins encoded by Tn7. Thus an appreciation of these pathways will follow a short description of these proteins.

As shown schematically in Figure 24, there are five Tn7 proteins: TnsA, TnsB, TnsC, TnsD and TnsE. Their coding sequences are clustered at the right half of Tn7. The *regulation* of expression of these genes is not very clear but TnsB is a repressor of the transcription of *tnsA* and *tnsB*. It is not known whether or not the other three genes are part of the same operon.

The two proteins TnsA and TnsB, together form the transposase of Tn7. Hence, they bind specifically at the ends of Tn7 and execute the chemical process of the transposition: double strand cuts that release the Tn7 from the rest of the donor DNA and the subsequent joining of the ends of the Tn7 to the target strands. Of these two proteins TnsA resembles a restriction enzyme (an enzyme that cleaves DNA at a specific short sequence of deoxyribonucleotides). The TsnB is a member of the retroviral integrase superfamily of recombinases (having an DDE motif mentioned above). The TnsC interacts with ATP and controls the activity of the transposase. The TnsD directs the insertion into the *attTn7* site and binds to this location. These four proteins are involved in the insertion of Tn7 into the *attTn7* site. The TnsE is a DNA-binding protein and is involved in the insertion of Tn7 into locations that are not *attTn7* sites (e.g. conjugating plasmids).

Much of the investigations of Craig and her collaborators were performed with a mini Tn7. This engineered element has 165 bp of the left end of Tn7 and 90 bp of the right end of Tn7. As shown in Figure 24b these ends have several binding sites for the TnsB protein. The "correct" orientation of the transposition results from the differences in the binding sites at the left and the right ends of Tn7. By correct orientation it is understood that the right end is in the direction of the bacterial gene *glmS* (glutamine synthetase). In the *in vitro* transposition of mini Tn7 by TnsABC+D, with cofactors (ATP any Mg^{2+}) as devised by Craig and collaborators, the donor (mini Tn7) can be relaxed DNA (not supercoiled, a requirement known for the transposition of other insertion sequences).

From numerous experiments the following scheme emerges for the *attTn7*-located transposition of Tn7. The first steps involve the cutting of the Tn7 from the rest of the donor DNA. This is performed by the TnsA and B (TnsAB). Each of the two proteins has its own role but they must act in coordination. It is assumed that TnsB binds first to the ends of Tn7 and then TnsA can join the DNA binding. TnsB cleaves the 3' ends of Tn7 and at a later stage links these ends to the target DNA. TnsA handles the 5' ends. The gap that is caused in the donor DNA, after the excision of the Tn7 from a genophore, can be repaired via double-strand break-induced homologous recombination.

There is a notable phenomenon: all the Tn7 transposition components should be "in place" before the initiation of transposition. This probably means that prior to the double-strand cleavages there must be the nucleoprotein complex that includes the DNAs of the donor and the target as well as the TnsABC+D. Clearly the target DNA should be ready and in the complex *before* the ends of the Tn7 are actually cleaved. The target DNA which has the bound TnsC (recruited by TnsD) is required for joining the donor-bound TnsAB. In this respect Tn7 differs from other IS (e.g. IS*10*) where there is first transposon excision and only thereafter the recognition of the target DNA. Albeit when specific TnsA and TnsB mutants, that had a "gain of function", were utilized, the recombination could proceed even in the absence of the other proteins. The binding of the Tns proteins to the Tn7 DNA and to the target DNA are schematically shown in Figures 25 and 26. The upper part of Figure 25 shows the bindings when *attTn7* is the integration site, while the lower part of this figure shows binding to a site of the triplex DNA, where the TnsD is not necessarily involved.

Because of the similarity of TnsB to other transposases (that have the DDE motif) it was assumed that its transposition reaction is also similar. Thus a hydrolysis step is assumed that exposes the 3'-OH of the Tn7 and that during the strand transfer this 3'-OH "attacks" the target DNA via one transesterification reaction. The 5' end cleavage by the Type II restriction-enzyme, resembling TnsA, requires the presence of Mg^{2+}.

Considering now the target (*attTn7* containing) DNA we shall first mention the TnsD protein. This protein binds at the *attTn7* site and thereby brings the whole complex into the appropriate location on the target DNA. The binding of TnsD is at a very specific location: within the *glmS* coding sequence, where the C-terminus of the Glms protein is encoded. The recognition site is adjacent to the exact insertion site but not on the insertion site itself. The TnsD interacts mainly with the major

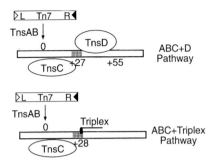

Figure 25. Scheme of attachments of Tns components in Tn7. (From Craig et al., 2002).

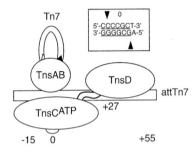

Figure 26. Attachment of the Tn7 ends, TnsAB, TnsC and TnsD, to the attTn7 target.
(From Craig et al., 2002).

groove of the target DNA and causes there a modest bend (see Tn5 for similar bends).

The TnsD/attTn7 binding recruits the binding of TnsC to the target DNA (Figure 25). But TnsC probably interacts with the minor groove of the target DNA. The Tn7/TnsAB complex can thus occupy the major groove of the target DNA at the insertion site (Figure 26). There is evidence that the binding of TnsD to the target DNA causes an adjacent conformational change that is a signal for the binding of TnsC. While the TnsC is "guided" by TnsD to the insertion site, the TnsC also has its own preference: in the absence of TnsD and attTn7 it will bind near pyrimidine triplex DNA (Figure 25, lower part). The role attributed to TnsC is the activation of the transposase. For that to happen ATP is required; under the standard in vitro conditions of TnsABC+D and target attTn7 there will be no transposition if ADP rather than ATP is provided. However, the situation is different when certain mutants of TnsC are used. It is assumed that TnsC is recruited to the TnsD-attTn7 and after ATP is furnished this complex can interact with the incoming Tn7 – bound to the transposase (TnsAB). The awareness of the requirement of ATP for normal functioning of TnsC came from a "gain of function" mutant of TnsC[A225V]. In vitro the latter mutant has a slow rate of ATP hydrolysis but it binds strongly to DNA. It is capable of activating TnsAB transposition in the absence of TnsD and TnsE. It is possible that due to the longer time with ATP, the TnsC[A225V] is able to achieve a tight binding to DNA even in the absence of TnsD and TnsE.

Because TnsAB+C^{A225V} transposes at high frequency and with a low level of target discrimination, this transposition system is very useful for the molecular genetics of bacteria.

Although *in vitro* transposition is possible with minimal components, it seems that *in vivo* there are host factors that stimulate this transposition. One of these factors is the L29 (one of the large-subunit ribosomal proteins); the other is the acyl carrier protein-ACP.

14.3. TnsABC+E Transposition

This is the transposition of Tn7 that is not involved in insertion into the *attTn7* site. Typically such a transposition will mobilize Tn7 from the bacterial genophore to a conjugating plasmid in the same bacterium. This insertion is facilitated by TnsE. Obviously by transfer of Tn7 into a conjugating plasmid, which can move into other bacterial cells, the capability to resist antibiotics will spread. This horizontal transfer of Tn7-containing plasmids has obvious advantages to the host bacteria. Unfortunately for patients, this horizontal spreading is very problematic. While transfer from the genophore into a conjugative plasmid is the prevalent direction of mobility, the other direction, movement from plasmid into the bacterial genophore, also occurs but at low frequency. In the presence of wild-type TnsE the insertion occurs preferentially into the terminus of *E. coli* DNA replication. There are TnsE mutants that promote a high level of transposition with a broad distribution of insertions. However, the orientation of insertion is the same orientation with respect to the lagging strand of DNA replication. This points towards the assumption that a component of the replication-fork directs the attachment of TnsE. It was also observed that binding of TnsE is preferable to DNA containing a recessed 3'-OH end.

14.4. Target "Immunity"

As was noted above for Tn3, the Tn7 avoids insertion into DNA that already harbors Tn7. The components of Tn7 that signal this avoidance are the binding sites for TnsB on the Tn7L and Tn7R. The "immunity" is *cis* acting. The frequency of insertion of a second Tn7 is directly correlated to the distance from the existing Tn7. In practice, a distance of 8 kbp causes very strong "immunity" that is reduced in longer distances. There is some "immunity" even at a distance of 175 kbp, but if the distance is several Mbp, no "immunity" was detected. The bacterial genophores have a total length of about 4±2 Mbp. Craig (2002) suggested that the TnsB in the existing Tn7 interferes with the interactions of the new site of integration and if the new TnsC is not in proper location, there is no ATP hydrolysis and no transposition. This suggestion is supported by experiments in which a non-hydrolysable analogue of ATP is used. Under such conditions, the "immunity" can be bypassed.

14.5. The Tn7 as an Experimental Tool

The wild-type Tn7 itself is not very useful as a genetic tool, but this transposon can be manipulated to render it suitable for several applications. For example, when the TnsABC^{A225V} is used the frequency of insertion is increased and there is relatively low target-site selectivity. This TnsABC^{A225V} system can be elaborated. It is possible to use the TnsABC^{A225V} for *in vitro* mutagenesis of genomic DNA. The mutated genomic DNA can then be used by homologous recombination to introduce the specific mutation into a host bacterium.

Biery et al. (2000) developed a system that is based on a version of Tn*7* that can cause the introduction of a five amino acid segment into a target protein. This system should be useful in the analyses of protein and nucleic acid structure/function relationship. It is possible to add specific genes into a Tn*7*. The engineered Tn*7* can then be introduced at the *attTn7* site. The impact of such an additional gene can then be studied. The engineering can also add flanking restriction sites so that a gene that is added can also be removed swiftly. Mini Tn*7* can also be used and the *tns* genes can be added-on plasmids that can then be eliminated. The studies of Tn*7* provided a by-product: they showed clearly that there is a genophore site (e.g. in *E. coli*) into which alien DNA sequences can be introduced without causing any deleterious effect on the bacterium.

14.6. A Final General Remark

The Tn*7* and some other bacterial IS show us primordial attempts by a simple species (of DNA) to "conquer" a territory in the environment of a much more sophisticated DNA (the bacterial genophore and bacterial plasmids). These IS are paying their entrance-fee by bringing with them a "gift" (e.g. resistances to antibiotics). The host may react to this "patriotic" (attempt to gain territory) effort of the IS, by several means such as restricting the settlement of the incoming DNA, to very specific locations (e.g. *attTn7*), by cleavage (restriction endonucleases) of the immigrating DNA or by down-regulating the propagation capability of the incoming DNA.

15. TRANSPOSON TN*10*

15.1. Origin and Overall Structure

The Tn*10* is one of the "oldest" Composite Insertion Sequences, but its full deoxyribonucleotide sequence was only recently determined. The "roots" of Tn*10* were revealed in Japan, in the early sixties of the 20th century. According to Sharp et al. (1973), it was discovered on a drug-resistance "factor", from the enteric bacterium *Shigella flexeri*; the "factor" was related to the F episome in *Escherichia coli*; it was first isolated by Nakaya et al. (1960) and termed NR1, and coined R222 by Watanabe and Fukasawa (1961). The recognition of Tn*10* as a transposable element came only several years later as published by Kleckner et al. (1975) and Foster et al. (1975). As noted above, the year 1975 was indeed a lucky year for transposons (or rather for the investigators who studied these transposons).

While many parts of Tn*10*, as the two IS*10*, were sequenced over the years, the full DNA sequence of Tn*10* was published only a few years ago by Chalmers et al. (2000). These authors reported that the total length of Tn*10* is 9147 bp. Notably, in 1989, in her comprehensive review of Tn*10*, Kleckner (1989) still estimated this transposon to be about 9,300 bp in length.

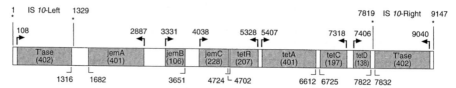

Figure 27. Sequencing strategy and map of Tn10. The sequence of Tn10 was used to construct an ORF map which shows the name of each ORF and the predicted number of amino acids. The direction and location of predicted translational start sites are indicated above the map. The locations of predicted stop codons are indicated below the map. The locations of the ends of the flanking IS elements are indicated by asterisks. (From Chalmers et al., 2000).

As shown schematically in Figure 27, Tn*10* contains a pair of bacterial Insertion Sequences, in opposite orientation, on the left side and the right side, respectively, of Tn*10*. These are termed IS*10* left (IS*10*L) and IS*10* right (IS*10*R), respectively. Each of these IS*10* elements has a length of 1,329 bp. Thus the space between the two IS*10* is 6,489 bp.

The two IS*10* elements encode transposases with promoters on the outer sides, but only the IS*10*R produces this protein, since it contains a long, uninterrupted ORF (open reading frame) of 1,206 bp. The ends of the two IS*10* (the outer end as well as the inner end) contain DNA determinants for binding with transposase. The outer end also contains a binding site for a bacterial protein termed IHF (integration host factor). Between IS*10*L and IS*10*R there are seven ORFs. Four of these are encoding genes for tetracycline-resistance. The other three were termed *jem* genes. Of the latter, *jem*A is a predicted gene for sodium-dependent glutamate permease; *jem*B is a gene with unknown function and *jem*C has some homology to a family of bacterial transcriptional regulators that express the arsenic and mercury resistance operon.

None of these seven genes is involved in the transposition of Tn*10*. The Tn*10* confers high-level tetracycline-resistance to its hosts but only after induction by tetracycline. This is expressed in *E. coli* and *Salmonella typhimurium*. Of the four genes involved in tetracycline-resistance, one (*tet*A), encodes an inner membrane protein that is responsible for the resistance, while another gene (*tet*R) encodes a repressor protein that negatively regulates the transcription of both *tet*A and *tet*R. Tetracycline binds to the repressor and hence induces the resistance. In practice the *tet* determinant is probably not very useful as a selective marker because in the presence of multicopy plasmids the resistance to tetracycline is at a low level.

At the termini of both IS*10* elements there are almost perfect inverse-repeat sequences (23 bp) that serve as the recognition sites and cutting signals for the transposase.

15.2. The Biology of Tn10

Although the "simplest" case of transposition is when the whole wild-type Tn*10* transposes into the recipient site, other transpositions are possible. First, the IS*10*R is "autonomous" and can transpose by itself. IS*10*R or the whole Tn*10* can trigger the transposition of other (defective) transposons such as IS*10*L, either in *cis* (i.e. on

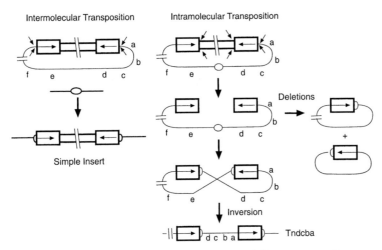

Figure 28. Nonreplicative transposition can generate different types of transposition products. In addition to the formation of simple inserts, Tn10 can generate both deletions and inversions as alternative reaction products; sites of insertion are indicated by a bubble. Inversions of the type shown generate new composite transposons, e.g. Tn-dcba. (From Haniford, 2002).

the same DNA strand) or in *trans* (i.e. in the same cell but on a different DNA strand).

Usually the Tn*10* does not undergo structural changes during (even several) cycles of transpositions. However, various specific exceptions were recorded. Various modifications of Tn*10* can be obtained by genetic manipulation. Thus an element was isolated that had the two IS*10* in the opposite orientations to wild-type Tn*10* (i.e. "inside-out"). The sequence between the two IS*10* can also be replaced, such as providing ampicillin or chloramphenicol resistances rather than tetracycline resistance.

Contrary to some of the other Tn elements, the Tn*10* does not exhibit transposition "immunity". That means a potential recipient DNA (plasmid or bacterial genophore) that already harbors a Tn*10* element can accept additional Tn*10* elements. However, as shall be pointed out below, the Tn*10* integrates preferably into consensus sequences.

In addition to the "simple", non-replicative and *intermolecular* transposition, the Tn*10* can perform *intramolecular* transposition. When this happens, it can lead to deletions and inversions. The terms *intermolecular* and *intramolecular* mean, respectively, that the transposition is into a different DNA or into the same DNA that harbors the Tn*10* (Figure 28).

Moreover because of the lack of "immunity", Tn*10* (and IS*10*R) can even integrate into the transposon itself. Such transpositions can cause DNA rearrangement that changes the host genome. Obviously insertions into or adjacent to bacterial genes can strongly affect gene expression. Transposition events of Tn*10*/IS*10* can also result in integration of exogenous DNA into the bacterial genophore.

15.3. A Summary of the Transposition

The transposition of Tn*10* is by the "cut and paste" mechanism. In this mechanism the Tn*10* DNA is completely separated from the flanking donor-DNA at the first stage of the transposition. This is achieved by the transposase that complexes with the ends of Tn*10*, cleaves these ends and creates "hairpin" structures at these ends. While maintaining this complex with activated hairpins, the complexed Tn*10* is inserted into the host DNA site. This insertion is also performed by the transposases that are in the complex. The insertion is staggered and creates nine nucleotide gaps at the site of the junction. The gap is repaired by DNA replication and ligation, generating a direct repeat of 9 bp of the original target-DNA sequence. This gap repair is not performed by the Tn*10* transposase but rather by host proteins. The "cut and paste" transposition thus has the following chemical steps at each of the Tn*10* ends: nicking of the transferred strand, hairpin formation and hairpin cleavage (or "resolution"). At the end of these three steps the Tn*10* ends have 3'-OH termini. These "attack" the PO_4 groups of the target DNA and the "strand transfer" step is achieved.

15.4. The Mechanism of Transposition

The former description is a short summary of the events of transposition. The mechanism was studied in great detail mainly by Kleckner and her associates (e.g. Kleckner et al., 1996) and in recent years by Haniford and his associates. What was revealed up to the writing of this book was presented in the comprehensive review of Haniford (2002). These concentrated efforts resulted in detailed and interesting information. Albeit, several issues of the details of Tn*10* transposition are still at the level of reasoning rather than firmly established facts. A collection of expressions from the review of Haniford will illustrate the state of these investigations: "... evidence that ... comes from ..."; "... can explain, in part how ..."; "... the available evidence argues that ..."; "... it has been suggested that ..."; "... this observation implies that ..."; "... experiments ... have provided evidence that ..."; "... results are consistent with the idea that ..."; "... it seems reasonable to assume that ...".

However, in spite of such phrases there are many parts of the mechanism of Tn*10* transposition that can be regarded as biological facts. We shall start with some terms and abbreviations that should be helpful in our further description.

The transposition of Tn*10* takes place in a protein-DNA complex termed *transposome*. This complex contains transposases and the two transposons' ends. The transposons' ends are brought together by a series of protein-protein and protein-DNA interactions. In this process there is similarity between Tn*10* and Tn*5*. The mechanism of Tn*5* complexing was noted above and was based on *in vitro* transposition and crystalographic studies. *In vitro* studies were also used in studies with Tn*10*. A very useful *in vitro* system included short DNA (donor) fragments containing the outside-end of IS*10*R, purified Tn*10* transposase and purified IHF. When these purified components were combined, the first stage was the establishment of the PEC (paired end complex) that consisted of two transposase molecules between two DNA ends and attachment of the IHFs inwards of the transposases attachment. In the next step Mg^{2+} is added to the complex and a single-end break complex (SEBC) is formed. In SEBC the flanking donor DNA is separated from one end of the transposon. In a further stage, the double-end break complex (DEBC), is produced. In DEBC the flanking donor DNA is completely removed from both ends. These changes are relatively very slow, when compared to enzymatic conversion of one metabolite to another. The formation of PEC and the conversion from PEC to SEBC takes many minutes.

The next step is the formation of the "target capture complex". This step has probably two stages. In the first there is a sequence-non-specific binding of the complex to the target, and only in the second stage the DNA target is complexed in a sequence-specific manner. Even in the presence of an appropriate divalent ion (Mg^{2+}) only about half of the complexes of the transposomes with the target DNA are suitable for strand transfer. Moreover only the DBEC, but not the SEBC (or the PEC) will associate with the target DNA (even if the latter is in the supercoiled form). This requirement, that the Tn*10* ends should be first completely separated from the donor DNA, is different from what is known in other transposition systems such as Tn7. The additional requirement in Tn*10* is that there should be an opening of the hairpin ends, as detailed below.

We noted above that Tn*10* exhibits a preference for insertion into a site of the recipient DNA that conforms to a consensus sequence. This sequence is: 5'-GCTNAGC-3'. This preference of insertion into a consensus sequence does not mean an exclusive insertion into this sequence. Such an exclusive insertion was noted for Tn7, in the presence of TnsABC+D, when this element inserts specifically into the *attTn7* sequence. It is assumed that the DEBC first makes a "loose" contact with the recipient DNA, and this is the non-specific binding; the DEBC then diffuses along the target DNA, in the vicinity of the initial contact. Then, in the second stage, the DEBC finds the consensus sequence to establish the stable binding. Again, this is a long process. It takes about 1 hour from the formation of the excised transposon until the appearance of the strand transfer product. The two-stage association of the transposome with the target DNA probably causes a conformational change in the target site. First the transposome (with its reactive Tn*10* ends), is bound to the target core of an appropriate "hot-spot". Subsequent to the initial contact, the sequence at the target forms two kinks in the DNA. These kinks seem to improve the contact of the incoming complex, allowing an additional region of the transposase, in the transposome, to bind with the target DNA. Kink formation may be stabilized by the binding of the bivalent ion to the DNA. It is also possible that the kink is required to create a high metal affinity and to activate the scissile phosphate in the target DNA for the nucleophilic attack by the transposome. As for the role of the divalent ion, it was revealed that in the presence of Mn^{2+} (rather than the Mg^{2+}) there is a relaxation of target sequence-specificity.

The target-capture efficiency is also affected by the DNA sequence at the terminals of the incoming Tn*10*; the exact nature of this effect awaits further clarification. However, it is thought that the energy derived from the correct, sequence-specific binding is used to drive the conformational changes in the protein-DNA complex that is required for catalysis. The conformational changes are essential for the correct positioning of the substrate with the various active-site components, in order to accomplish the chemical reactions.

15.5. The Hairpin Cleavage

It was indicated above that in the transposition mode used by Tn*10* (which is also

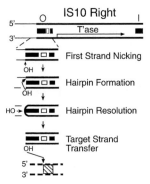

Figure 29. IS10 right and the chemical steps in Tn10/IS10 transposition. The two ends of IS10 right (thick lines), designated "outside" (0) and "inside" (1) ends, contain the binding determinants for transposase (black rectangles). The outside also contains the binding site for the E. coli host factor, IHF (open rectangle). The four chemical steps in Tn10/IS10 transposition carried out by transposase at one end are shown. The striped box indicates the position of a target site. (From Kennedy et al., 2000).

used by some other transposable elements such as the P element of *Drosophila*, the Tc1 element of nematodes and plant T.E. such as Tam1/Tam3), the two donor strands (the "transferred" and the non-transferred) are cleaved *before* strand transfer is initiated. In the other mode, which will not be handled here, the strand transfer starts without prior cleavage of the non-transferred strand. In the first mode, there is a typical donor cleavage that includes the hairpin formation. This is schematically shown in Figure 29, and was mentioned above with respect to Tn5. Briefly, the following stages can be defined. First, one nick is formed in each of the two strands of the Tn10 (or IS10) (first strand nicking). These nicks result in a 3'-OH at each of the ends. The 3'-OH then attacks the other strand so that both ends of the transposon are now "closed" in the form of a hairpin (hairpin formation). In the next stage the ends are opened again (hairpin resolution) and the two 3'-OH are again available for the strand transfer of the resolved end.

The formation of the first strand nicking lasts about 15 minutes and the lifespan of the "closed" hairpin is about 30 minutes; thus these are rather slow processes when compared to other catalytic processes. During the hairpin formation the IHF is joined to its appropriate binding sites at the Tn10 ends. But it is removed before the transfer of the resolved end to the target.

15.6. The Tn10 Transposase

The Tn10 transposase is actually the transposase of IS10R. If we count the bp of Tn10 from left to right, the last (most right) nucleotide is no. 9147. The transcription is driven by a promoter at the outside (right) end of IS10R (see Figure 27) and the ORF of the transposase start from nucleotide 9040 (i.e. 108 bp away from the right terminal) until nucleotide 7832. These 1206 bp encode the 402 amino acid transposase (46 kD). Hence, there is a single open reading frame. The overall amino acid sequence has similarity to the transposases of the IS4 family; they have

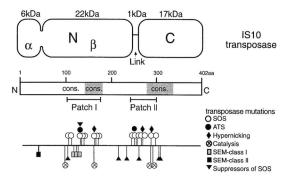

Figure 30. Domain structure of IS10 transposase and location of mutations. Top: domain map determined by limited proteolysis. Nαb and C are the two main domains; the link is a nine-residue protease-sensitive region. Middle: scale representation of the primary sequence of transposase. Areas of sequence conservation with other bacterial transposases are shaded. Patch 1 and Patch II are two regions defined by the mapping of SOS mutations. Bottom: locations of mutations of various classes. Multiple symbols appearing one above another indicate different classes of mutations that result from changes in the same residue, although the specific change is not necessarily identical. (From Kleckner et al., 1996).

the DDE motif that was mentioned above as existing in other transposases. In Tn*10*/IS*10* the locations are the following. There is a D (aspartic acid) 97, a D (aspartic acid) 161 and an E (glutamic acid) 292. Numbers denote amino acid number, counted from the N-terminal of the transposase.

As shown in Figure 30, the transposase can be divided into "domains" and "patches". Limited proteolysis results in two main domains, N and C, that are connected by a short "link" where the proteolysis cleaves the transposase. The N (close to the N-terminal) domain can be divided again into Na and Nb. These two sub-domains are separated by limited trypsin digestion. These domains were separated and purified and then utilized to assay transposase activity. When purified Nab was mixed with purified C, even in the absence of the link, full activity was obtained. Mixing Nb with C did not lead to transposition activity. This means that while the link is not essential for transposition, the Na has an essential role.

The "maps" of Figure 30 also indicate that numerous mutations were found in the transposase and these were located mainly in Patch I and Patch II. Also the "consensus" motif of YREK (tryptophane-arginine-glutamine-lysine) that is located at amino acids 285 to 299 is marked on the map. The similarity of the Tn*10*/IS*10* transposase to the Tn*5* transposase is rather striking. The major cleavage sites in the Tn*5* match well with these sites in the Tn*10*/IS*10* transposase, but Tn*5* has an extension at the C-terminal that does not exist in Tn*10*/IS*10*, although the overall identity of amino acids in these two transposases is only 20%. The chemistry of the transposition and the transposase/DNA interactions were solved in Tn*5* by crystalography.

Due to the similarity of the two transposases the information from Tn*5* can lead to a fuller understanding of the transposition of Tn*10*/IS*10*.

CHAPTER 4

RETROTRANSPOSONS

1. GENERAL CHARACTERISTICS

1.1. Terminology

The terminology of T.E. is problematic. The terms Controlling Elements, Mobile DNA and Mobile Genetic Elements, all describe DNA sequences that can change location in the sequence of genomic DNA or even in plasmids. Moreover, even elements that have lost the capability to change location but apparently had it in the past are included in these terms. The latter elements, which lost transposition capability were nicknamed, in the Introduction of this book, as "skeletons". Because of limitation in the capacity of this book, I shall exclude viruses, such as retroviruses and the bacteriophage Mu, from our handling of T.E.

The main division of T.E. is into two groups. Those T.E. that reinsert into a new DNA site without an RNA intermediate are termed *Class II* (or by some just *Transposons*). The bacterial Insertion Sequences that we handled above, belong to this group. Further in this book we shall handle the Class II T.E. of eukaryotes. The other main group is either termed *Class I* T.E. or *Retrotransposons*. These elements are first transcribed into RNA. The transcript is then retrotranscribed, by an RNA-dependent DNA-polymerase, into DNA. We shall see that molecular-evolution studies suggested that retroviruses are related to a sub-group of retrotransposons. Obviously retrotransposons could not be characterized as undergoing a "reverse-transcription" before the RNA-dependent DNA-polymerase (reverse transcriptase) was discovered. This discovery was published simultaneously by Baltimore (1970) and by Temin and Mizutani (1970), in two "head-to-tail" papers in *Nature*. It is quite amazing that what was then considered a speciality of specific animal viruses is now known as one of the most prevalent "potential" enzymes: it is encoded in retrotransposons and the latter are repeated countless times in the genomes of animals and plants.

The retrotransposons thus all have one common feature: they have a code (or at least had one or have a truncated one) for the *Reverse Transcriptase* (RT); the enzyme was first coined RNA-dependent DNA polymerase. However, they can also be divided into sub-groups. The details of the sub-groups will be presented in our further discussion of the retrotransposons. The division into sub-groups is based on whether or not they have *Long Terminal Repeats* (LTR). Thus, there are *Non-LTR retrotransposons* and *LTR retrotransposons*. The LTR retrotransposons are again divided into two types according to the relative location of the code for the enzyme *integrase* (IN). Those in which the IN coding sequence is at the right end of the coding sequences (as in retroviruses, but in the latter there is also a code for the

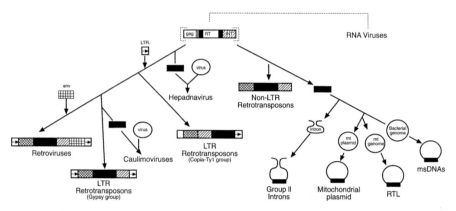

Figure 31. A scheme for the origin of retroelements. Important structural features of each category of RT are presented. Shaded boxes correspond to domains of the ORF of the element, solid shading, RT region; stippled, gag region; diagonal shading, integrase domain; crosshatched, envelope gene. LTRs present in certain elements are diagrammed as an open box with an arrow. Structural features of the ancestral retrotransposable elements (shown in brackets) are assumed based on structures present on both major branches of the tree. For the hepadnaviruses, caulimoviruses, group II introns, the Mauriceville mitochondrial plasmid. RTL sequences and msDNA it is assumed that only a portion of the pol gene, containing the RT domain, entered an already existing element. (From Xiong and Eickbush, 1990).

envelope, downstream of IN) are termed Ty3/*gypsy* group of LTR retrotransposons. Those in which the IN code is upstream of the codes for the reverse transcriptase (RT) code are termed *Ty1/copia* LTR retrotransposons. A scheme of this nomenclature is shown in Figure 31. This figure was presented by Xiong and Eickbush (1990) to show a scheme for the evolution of retroelements. This evolution was updated and detailed in a recent review (Eickbush and Malik, 2002).

The non-LTR retrotransposons constitute a very large and widespread sub-group. The whole sub-group was also called *LINE-like elements, poly (A) retrotransposons* or *retroposons*. This sub-group can again be divided into at least five types. One of these are the *LINEs*, for *long interspersed nucleotide elements*. Another type of the non-LTR retrotransposons are the *SINEs*, for *short interspersed nucleotide elements*. For those readers who can follow the subdivision up to here, there is a further division: types such as *LINE*1 are divided again and there are at lease six *LINE*1 clades: L1, DRE, Cin4, Tx1, Zepp and Zorro (see Eickbush and Malik, 2002 for details and references). We shall leave all the details to experts in this field but keep in mind that there is an almost endless number of clades of retrotransposons. Moreover, these clades exist in almost the entire eukaryotic world. Retrotransposons were found wherever they were looked for – with only one exception.

On the other hand, no retrotransposons were found in prokaryotes. The retrotransposons thus probably emerged with (or soon after) the emergence of eukaryotes; that means about 10^9 YBP. It is less than 25 years since the first report on retrotransposons. The diversity reached by retrotransposons during their 1000 million years of existence is only now being appreciated. Now the research on the diversity and distribution of retrotransposons is vastly excelerated! There is an old

Hebrew saying: "The toil of saints is performed by others". As the genomes of several eukaryotes are now fully sequenced (e.g. yeast, the plant *Arabidopsis*, *Drosophila*, mouse, man) and many partial genome sequences are available, computer programs can be put to work to fish out retrotransposons with only a little effort. The great majority of recent publications deal with the "fishing" of retrotransposons from various organisms and discussion of their evolution. Less publications deal with the chemistry of retrotransposon mobilization. Should we consider the respective investigators as "saints"?

1.2. Two, Rather Different, Modes of Integration

As can be seen in Figure 31, non-LTR retrotransposons have been separated from LTR retrotransposons from a very early stage of evolution. Thus, it is no wonder that these two sub-groups of retrotransposons differ vastly in their respective modes of integration.

This is schematically shown in Figure 32. The non-LTR retrotransposons (Figure 32A) integrate by the "*target-primed reverse transcription*" (TPRT). These elements not only lack (always) the terminal repeats but also differ from the other sub-group by having (almost always) a poly (A) 3' end. The element that is integrated into the target DNA is first transcribed into RNA. This RNA encodes several proteins that are essential for the further steps. One such protein is an endonuclease that cleaves the first strand of the target DNA. The transcribed RNA then goes through a reverse transcription in such a way that a "bridge" of an RNA/DNA double-strand is formed at the target. Another protein, encoded in the transcribed RNA, then removes the RNA strand from the "bridge" and the second strand DNA of the bridge is then synthesized.

The integration of LTR retrotransposons differs basically from the integration of non-LTR retrotransposon. In the former transposons (Figure 32B) the reverse transcription is not primed by the target DNA. The target DNA has no role in the reverse transcription and it enters the scenario only after the RNA transcript of the mobile element is converted (in two steps) into a double-stranded DNA fragment. The target DNA is cleaved by a staggered cut, so that at the last stage of integration there is a filling of gaps by DNA repair, which generates a typical target-site duplication on both sides of the integrated element. The location of the reverse transcriptase (RT) therefore differs substantially in the two types of retrotransposons. In the non-LTR retrotransposons the RT is active in the nucleus, while in the LTR retrotransposons the RT is active in the cytoplasm.

We should recall that we are now handling an eukaryotic system. Thus, the formation of the intermediate double-stranded DNA is synthesized in the cytoplasm. For integration into the target DNA (in the nuclear genome of the cell) this intermediate has to enter the nucleus. The distribution of retrotransposons in animal taxa was surveyed on the basis of their RT sequences.

This has been summarized by Arkhipova and Meselson (2000), from whom Table 2 was derived. Most probably, additional information will accumulate in the coming years, but the data on the listed 46 taxa are already very impressive. The data are not only important for what they show, but also for the "negative" information. Thus, in the *Bdelloidea* group of rotifers there are no retrotransposons with either: "*gypsy*"-type (LTR retrotransposons) RT or with *LINE*-type (non-LTR retrotransposons) RT. This group of rotifers is assumed never to have had sexuality

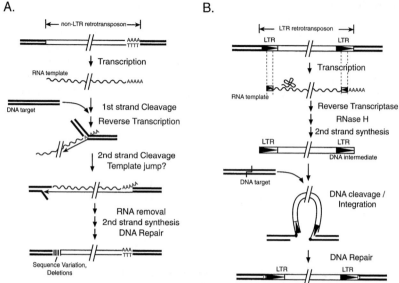

Figure 32. A. Target-primed reverse transcription (TPRT) model for the integration of non-LTR retrotransposons. Most elements end in an A-rich or poly(A) 3' end. Transcription is usually mediated by an internal promoter which initiates RNA synthesis upstream of itself. However, some elements appear to be expressed as co-transcripts with host genes. Most of the information concerning the TPRT reaction is based on in vitro studies with the protein encoded by the R2 element. This TPRT accounts for many of the integration properties of other non-LTR retrotransposons but the degree to which this mechanism reflects that used by these different elements is not known. In the initial step of the integration reaction, an element encoded endonuclease cleaves the first (primer) strand of the target site and uses the released 3' hydroxyl of the terminal nucleotide to prime reverse transcription starting within the poly(A) tail. Cleavage of the second (non-primer) strand occurs after reverse transcription. Thick lines, DNA target sequences; wavy lines, element RNA sequences; thin lines, element DNA sequences. The mechanism by which the 5' end of the newly made cDNA is attached to the upstream target sequences is unclear, but it appears to be functionally equivalent to the reverse transcriptase simply jumping from the RNA template onto the DNA target. If this jump occurs before the reverse transcriptase reaches the end of the RNA template, a 5' truncated copy is generated. The process of 5' attachment generates considerable sequence variation at the junction of the element with the target site even in the absence of 5' truncations. The means by which the RNA is removed and the second strand of the element synthesized is not known. Non-LTR retrotransposons differ in whether they generate a target site duplication and in whether that duplication is of a defined length. B. Retrotransposition mechanism of retroviruses and most LTR retrotransposons. Only the basic steps of this reaction are shown for comparison to that of the non-LTR retrotransposition mechanism (A). Thick lines, flanking chromosomal sequences; thin lines, element sequences for protein translation and reverse transcription begins and ends within the LTR sequences. The sequences within these terminal repeats that are used by the reverse transcriptase to template jump between ends are indicated with a smaller boxed triangle. First strand DNA synthesis is primed by a tRNA annealed near the 5' terminal repeat of the RNA (clover leaf), while second strand synthesis is primed by an RNAse H resistant RNA near the 3' terminal repeat (not shown).

in their evolutionary history. A similar survey of retrotransposons in fungal organisms or in plants is not yet available.

Before the discussion of the two major sub-groups of retrotransposons, I would like to make two general remarks. The first remark concerns the fidelity of the reverse transcription. This process is known to be error-prone as well as recombinogenic (see: Boeke and Sandmeyer, 1991); thus the "same" element of a retrotransposon may be rather heterogenic among individual genetic lines that carry this element. The second remark concerns T.E. in eukaryotes. In publications that handle the integration of retrotransposons into the chromosomal DNA, the re-entry of the double-stranded intermediate of the element into the nucleus is indicated, but one very important issue is ignored. Obviously the nuclear DNA is not a "free" entity, it is coiled and complexed with histones in nucleosomes, forming the chromatin. The latter, as we shall indicate in an appendix in this book, is a very elaborate structure. Thus, integrating a DNA fragment into it may require an elaborate process.

2. THE LTR RETROTRANSPOSONS

We noted above that this is a main sub-group of retrotransposons. The length of these retrotransposons is a few thousand bp (unless truncated); they have direct repeats of several hundred bp on both ends and have several ORFs in their central part. They resemble retroviruses but also differ from retroviruses in several important characteristics. Retroviruses, but not LTR retrotransposons, have an "independent" existence. They have an envelope, can be "free" in the cells and can also survive outside of cells. Retroviruses are also apparently restricted to vertebrates while LTR retrotransposons are widely distributed in many eukaryotes. Actually, they were first discovered in yeast and in the fly *Drosophila melanogaster*. They are also abundant in flowering plants (angiosperms).

Examples of LTR retrotransposons from a wide range of organisms will be provided. No attempt shall be made to list all the known LTR retrotransposons, because "new" elements of this kind are revealed almost daily. Actually, those who have the inclination for it can go out fishing for LTR retroelements, in the ever-increasingly available DNA sequences of various organisms. All that is required are databases of DNA sequences, "fishing gear", meaning computer programs that will identify these elements, and a computer. However, there is a warning! Due to the error-prone RT the "fishes" will not be identical, even in the same species of the screened organism; because of this lack of precise identity the "fishing gear" should also be clever!

Figure 32. Legend continued.
After formation in the cytoplasm of a linear DNA intermediate, the element-encoded integrase binds to termini of the intermediate, migrates to the nucleus, cleaves a target site on chromosomal DNA and inserts the DNA intermediate in a reaction similar to that of DNA transposons. Finally, DNA repair fills in the gaps that had been generated by the staggered cut in the chromosome generating a target site duplication of uniform length for each element (small triangles). (From Eickbush and Malik, 2002).

Table 2. LINE-like and gypsy-like RTase sequences and Mariner/Tc1-like transposase sequences in diverse species.

Phylum	Species	Mariner	Tc1	LINE	Gypsy
Sarcomastigophora	*Giardia lamblia*			+5	
Porifera	*Halichondria bowerbanki*	+		+	
	Spongilla sp.	+		+	+S
Cndaria (L,M)	*Hydra littoralis*	+S	+S	+	−
	Aurelia aurita	+		+	
Ctenophora	*Condylactus* sp.	−		+	+
Platyhelminthes (M)	*Dugesia tigrina*	+S		+S	+S
Rotifera (*Acanthocephala*)	*Moniliformis moniliformis*	−	−	+S	+
Rotifera (*Monogononta*)	*Brachionus plicatilis*	+	−	+S	+S
	Brachionus calyciflorus	−	−	+	
	Sinantherina socialis	−	+	+S	+
	Monostyla sp.	−	−	+S	
Rotifera (*Bdelloidea*)	*Philodina roseola*	+	−	−	−
	Philodina rapida	+	−	−	−
	Habrotrocha constricta	+S	−	−	−
	Adineta vaga	+S	−	−	−
	Macrotrachela quadricornifera	+S	−	−	−
Gastrotricha	*Lepidodermella* sp.	+	+	+S	−
Nemertea	*Lineus* sp.	+	+	+S	+
Priapulida	*Priapulus caudatus*	−		+S	+
Sipuncula	*Themiste alutacea*	−		+S	+
Annelida	*Glycera* sp.	−		+S	+
Echiura	*Lissomyema mellita*	+	−	+S	+S
Mollusca (L)	*Chione cancellata*	+S		+S	+
Brachiopoda	*Glottidea pyramidata*			+S	+
Bryozoa	*Amathia convoluta*	+	+	+S	+S
Phoronida	*Phoronis architecta*	+		+	+
Nematoda (L,G,M,T)	*Caenorhabditis elegans*	+	+	+S	+S
Onychophora	*Euperipatoides rowelli*	+	+S	+S	+S
Arthropoda (L,G,M,T)	*Drosophila melanogaster*	−	+	+S	+S
	Drosophila pseudoobscura	+	+	+	+S
	Drosophila virilis	+	+	+	+S
	Lasius niger			+S	+
	Formica polyctenum			+	+S
	Aphis sp.			+S	+
Tardigrada	*Milnesium* sp.	+		+	+
Chaetognatha	*Sagitta* sp.	+		+S	+
Echinodermata (G)	*Echinometra mathaei*			+	+S
	Strongylocentrotus purpuratus			+	+
Hemichordata	*Saccoglossus kowalevskii*	+		+S	−
Chordata (L,G,M,T)	*Branchiostoma floridae*			+S	−
	Danio rerio			+	+S
	Onchorhynchus keta			+	+
	Xenopus laevis			+	+
	Mus musculus			+	+S
	Bos taurus			+	+

Superfamilies previously reported to be present in representatives of a phylum are indicated in parentheses: L, LINE; G, gypsy; M, Mariner; T, Tc1.
Presence or absence of diagnostic PCR bands is indicated by + or −, respectively; S, verified by sequencing; blank, not done.
(From Arkhipova and Meselson, 2000).

2.1. The Transposable Elements of the Budding Yeast (Saccharomyces cerevisiae)

The yeasts have a long history of applied and fundamental research. In a way they permitted the first "biotechnology" for the production of alcoholic beverages (e.g. wine and beer) and of bread. For our deliberations they have many features that render them excellent subjects for genetic and biochemical studies. Thus, by the late seventies of the 20th century their genetics was well-studied. It was already known by then that in *S. cerevisiae* there are many repeated sequences. Repeated sequences, based on knowledge from bacteria and the fly *Drosophila melanogaster* (see below), could mean transposable elements. The first positive identification of a T.E. in yeast was by Cameron et al. (1979). These authors identified such elements, coined by them Ty1 (for transposon-yeast-1), and described their basic structure. There was actually a family of Ty1 in yeasts. About 35 copies per haploid genome. The full length was given as about 5.6 kbp. The elements had non-inverted repeated terminals of about 0.25 kbp each that were termed "δ". There were also many copies of δ in each yeast genome that lacked all the central part of the full-size Ty1. Notably, in this first publication of Ty1 it was not claimed that the transposition involves retrotranscription. The publication of Cameron et al. (1979) was soon followed by additional reports on this T.E. of yeast, by G.R. Fink and associates who already had a long experience with yeast genetics (e.g. Farabaugh and Fink, 1980; Chaleff and Fink, 1980; Roeder and Fink, 1980; Roeder et al., 1980; Fink et al., 1980) as well as by others (e.g. Elder et al., 1980; Errede et al., 1980a,b). One should recall the dates of these publications. Only in 1975 did Southern (1975) blot hybridization become available and DNA sequencing became an efficient procedure only after the publications of the respective procedure by Maxam and Gilbert (1977) and by Sanger et al. (1977). In one of the early reviews on the T.E. of yeast (Roeder and Fink, 1983), the possibility of an RNA intermediate in Ty transposition was considered, but the proof for it came some years later (Boeke et al., 1985).

There are a number of Ty T.E. that were revealed in yeast. We shall focus on one of these: Ty1. Ty2 is rather similar to Ty1 while Ty3 (*gypsy*-type) differs substantially in its coding arrangement from the Ty1 (*copia*-type) coding. We shall handle Ty3 briefly below. The Ty5 was discovered in *S. cerevisiae* but it is truncated in this yeast species and no functional Ty5 was found. However, a full-length and functional Ty5 was identified in *S. paradoxus*. When the latter was transferred to *S. cerevisiae* it did transpose. Ty4 is closely related to Ty1 and Ty2 and we shall not detail it further.

2.2. The Structure of Ty1

The following description is mainly based on the reviews of Boeke and Sandmeyer (1991), Wilhelm and Wilhelm (2001) and Voytas and Boeke (2002). The scheme of the structure of Ty1 is shown in Figure 33. Ty1 belongs to those retrotransposons in which the coding sequence for RH (RNase H) is at the right end of the ORFs. This is the *copia*-type (also termed *Pseudoviridae*), which differs from the Ty3, *gypsy*-type (also termed *Metaviridae*), in which the coding sequence for IN (integrase) is at the right side of the ORFs. This is not the only difference between Ty1 and Ty3. As indicated above in Figure 31 and elaborated by Xiong and Eickbush (1990) and by Eickbush and Malik (2002), there is a great phylogenetic distance between Ty1 and Ty3.

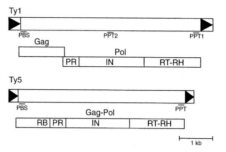

Figure 33. Genomic organization of Ty1 and Ty5. DNA sequences are indicated by the upper open boxes. LTR sequences are shown as solid triangles. The short thin lines denote the location of the PBS and polypurine tracts (PPT), which prime minus- and plus-strand DNA synthesis, respectively. Open reading frames are indicated by the open boxes below each element. Amino acid sequence domains conserved in retroviral proteins include PR, IN, RT, and RH. RNA-binding (RB) domain denotes the finger motif characteristic of nucleocapsid proteins. Ty1 and Ty5 are drawn to scale.
(From Voytas and Boeke, 2002).

Further studies based on exact sequencing showed that the full-length of Ty1 is 5.9 kb. This length includes the two flanking direct-repeats (also termed *delta sequences*) that are 334 bp each. These are the LTR of Ty1. Recombinations can result in *Solo* LTRs, meaning sequences containing only the LTR DNAs. Beyond the LTR, on both ends are conserved dinucleotides: 5'-TG...CA-3'. All the central part of Ty1 is termed *internal domain*. At its 5' end are *cis*-acting sequences that are important for replication, including the *primer binding site* (PBS). This PBS is complementary to 10 bases of the receptor stem of *S. cerevisiae* initiator methionine tRNA. Thus, the PBS serves as the site of initiation of the first, minus-strand, of DNA synthesis. At the right end of the internal domain and in its middle are polypurine tracts (ppt1 & ppt2) that are the priming site for the second, plus-strand, of DNA synthesis. The internal domain also encodes the GAG protein that is required for replication. Further downstream the middle domain encodes POL proteins that are (from left to right): the protease (PR), the integrase (IN), the reverse-transcriptase (RT) and the associated RNase H (RH). The coding of GAG overlaps by one frame the code of POL.

Most of the research on Ty1 was done with a specific strain of *S. cerevisiae* (S288c) that contains 32 intact Ty1 elements, as well as numerous Solo (LTR only) sequences. As noted above, in other stains there are different numbers of Ty1 and the base sequence of Ty1 in different strains is not exactly the same. Moreover, in order to be useful for studies such as the mechanism of transposition, the Ty1 used for experimentation was modified. The "regular" transposition of Ty1 is only about 10^{-6} per element per generation. To vastly increase transposition, a GAL promoter was introduced into the Ty1. When the yeast carrying such a modified Ty1 was cultured in a medium that contained galactose, the rate of transposition increased up to 0.05 per generation. One of the obvious results of retrotransposition is that when an intron is placed in this retrotransposon, the splicing during the reverse transcription will eliminate this intron. This elimination of introns from Ty1 during transposition was used in genetic studies with Ty1.

The transcript of Ty1, which is rather abundant in yeast cells that harbor Ty1, spans almost the whole length of the internal domain. The first and the last nucleotides of this RNA are retrovirus-like domains that are termed R (for repeated terminus of the transcript); there are U5 (for unique at 5') and U3 (for unique at 3'). This transcribed RNA probably has a dual function: it is the template for retrotranscription into DNA and can serve as mRNA for protein synthesis.

The most abundant Ty1 transcript has a length of 5.7 kb. The transcript is amazingly stable, much more stable than regular mRNA. The reason for this was given by their inclusion in virus-like particles (*VLP*). These VLP may either be a "memory" of LTR retrotransposons from their retroviral way of life, or, conversely, this is a stage from where retroviruses proceeded, during evolution to become "free".

There are two, partially overlapping ORFs, termed TYA1 and TYB1, respectively. The codon lengths are 440 and 1,328 (or 1,329) nucleotides, respectively for TYA1 and TYB1. The initiator codon for TYA1 is preceded by an adenine residue at position-3 (i.e. ..A..ATG...). This is the consensus nucleotide at the initiation of polypeptide translation of eukaryotes.

2.3. Transcription and Transcriptional Regulation in Ty1

The dual-purpose RNA is transcribed from the DNA of the Ty1 element by RNA polymerase II; this is the polymerase that transcribes the regular mRNAs in eukaryotes. The major transcripts start at position 241 (counted from the left). This means it starts within the 5' LTR. Upstream of the initiation of transcription there is an essential domain that is AT-rich. When this domain is mutated there is no transcription. There are mechanisms for the termination of the transcription that are still being clarified.

There are several means by which the Ty1 activity can be regulated (see: Voytas and Boeke, 2002 for details and references). An important type of regulation is associated with the cell-type regulation. Haploid and diploid cells differ in the production of regulatory proteins. In diploid cells a protein complex is produced that represses the haploid-specific genes. This affects also Ty1 so that its transcription in diploid yeast cells is only one tenth of the transcription in haploid cells. The change from growth as cells to filamentous growth, caused by starvation, also affects the rate of transcription of Ty1. While the control of integration is a rather complicated issue and shall not be detailed here, it should be noted that Ty1 integrates preferentially into specific chromatin domains that are "silent". Such domains may also include those that are upstream of RNA polymerase III transcription initiation sites (tRNAs and 5S rDNA). When Ty1 elements are inserted into the tandem rDNA repeats, they are silenced.

2.4. The Transcript and the VLP of Ty1

The transcript of Ty1 is transported from the nucleus to the cytoplasm. In the cytoplasm the further stages of the transposition take place; proteins that are active in this process are translated. These exert their activities and the VLPs are formed. These VLPs contain also tRNA, which is essential for the further process of retrotransposition. I shall briefly describe these steps.

As indicated above, there is a frame shift in translation: from the GAG transcript to the POL (PR, IN RT-RH) transcript. For this frame shifting Ty1 uses a +1 ribosomal shift to produce the GAG-POL fusion protein. This occurs in a sequence of seven bases of the coding region: CTT-AGG-C (or CUU-AGG-C at the mRNA). For the shift there probably should be a slowdown (or pause) of translation as well as a special peptidyl tRNA that can undergo a "slip". The AGG triplet is a "rare" codon for arginine and causes a pause. The special peptidyl tRNA is an unusual leucine isoacceptor, tRNA-Leu (UAG). It can either decode the CUU codon, or in about 5% of the mRNAs, it will slip forward one base to the UUA codon. The same shift system is also used by Ty2 while a somewhat different +1 frame shift is used by Ty3. Because this frame shift is essential for the production of the protein arsenal that is required for transposition, any interference with this shift will adversely affect transposition.

The translation of Ty1 results in two proteins Gag-p49 and Gal-Pol-p199. The Gag-p49 is processed to produce Gag-p45. The Gag-Pol-p199 is processed to PR (Pol-p20), IN (Pol-p71) and RT (Pol-p63). The Gag-p45 is a capsid protein that is the main component of the VLP. The processing of Gag-Pol-p199 is essential because it generates the PR (protease) and the latter cleaves the polyprotein into the respective essential enzymes (IN and RT).

As indicated above, the virus-like particles (VLPs) are composed mainly of the GAG protein but also contain the mRNA of Ty1 (in dimeric form) and the other GAG-POL proteins, as well as the initiator methionine primer tRNA and probably also host-encoded factors. In these particles the actual reverse transcription takes place. The VLP particles can be seen in electron micrographs, provided the yeast stain is over-expressing Ty1. It is probable that GAG can self-assemble into these spiny particles that measure 30 to 60 nm in diameter.

2.5. Reverse Transcription in Ty1

The reverse transcription of the mRNA into dsDNA that represents the whole Ty1 element takes place inside the VLP. It is a rather elaborate process that starts with the synthesis of a short fragment of the minus-strand DNA that is initiated from the 3'-OH of a tRNA that is paired to the primer binding site (PBS) and the synthesis of the ssDNA is first towards the left (see Figure 34). Because of the terminal redundance of the mRNA, the newly formed ssDNA can shift to the right and its synthesis again proceeds towards the left. Only after three more stages the dsDNA, representing the full-length Ty1, is finally produced. One process (stage E to F in Figure 34) involves the activity of the RNase H activity of the RT. In this process the RNA strand is removed from most of the RNA/DNA duplex. Only a short polypurine tract of this RNA strand is retained. This tract primes the synthesis of the second DNA strand. While it is one protein, the RT of Ty1 (and of retroviral RTs) performs two main biochemical activities. It acts as a DNA polymerase, using either RNA or DNA as templates, and it has also an RNase activity, removing RNA from the RNA/DNA duplex. The two activities are actually harbored in two domains of the RT protein. The RNase H domain is very conserved phylogenetically and can be aligned with the *E. coli* RH sequence.

Moreover, there is also similarity to the parallel human immunodeficiency virus (HIV) enzyme. In HIV the subdomains were coined as "hand" components: "fingers", "thumb" and "palm". The similarity between Ty1 and other systems of

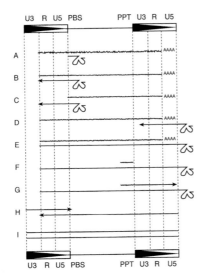

Figure 34. The mechanism of reverse transcription. The template mRNA is depicted as a squiggly line. Solid lines indicate DNA; solid lines and arrowheads indicate DNA strands in the process of being extended by reverse transcriptase.
(From Voytas and Boeke, 2002).

reverse transcription is very great. Thus, hybrid proteins, such as Ty1/HIV and Ty1/HBV (hepatitis-B virus) are fully functional.

One more note is due for the priming of the synthesis of the first (minus) DNA strand inside the VLP. The PBS site of Ty1 has a sequence that is complementary to 10 nucleotides at the 3' end of the "stem" of the initiator methionine tRNA. The latter tRNA thus binds to the Ty1 transcript and primes the synthesis of the first (minus) DNA strand (stages A and B in Figure 34). Obviously, in the process of the synthesis of the dsDNA, the tRNA (as well as the template RNA) should be removed.

As noted above, the reverse transcription has a low fidelity in Ty1 as well as in other RT systems. In Ty1 the rate of nucleotide substitution is about 2.5×10^{-5} substitutions per replication cycle. This results in variability that may have an advantage to retroviruses: new strains could avoid the immune system. However, one has to consider that this lack of fidelity has much earlier phylogenetic roots than the facing of retroviruses with the elaborate immune systems of higher vertebrates.

2.6. Recombination in Ty1

The DNA produced in the reverse transcription (in the VLPs) can assist in the retrotransposition into new targets in the yeast geome, but it is also available for recombination with existing (Ty1) elements. It seems there is a competition between integration into novel sites and recombination. The recipient Ty1 may gain a mutation from an incoming retroelement donor. The recombination products are generated either by reciprocal recombination or by gene conversion (non-reciprocal).

Recombination may also result in tandem or multimeric elements. In Ty1 elements that are integration-defective, the tandem elements and other complex rearrangements may reach 70% of the recombination products.

Another type of recombination is also possible, although this type is probably rare. The machinery of reverse transcription may use ("regular") cellular mRNA rather than the Ty mRNA and reverse-transcribe it. Obviously, the cellular mRNA does not have the specific features of Ty1 mRNA. This may explain the rarity of this reverse transcription. The cDNA produced from the cellular mRNA can then be involved in processes that include recombination. This recombination probably requires host replication proteins, such as proteins involved in chromosome-recombination repair (e.g. Rad52p, Rad1p).

Interestingly, the cDNA generated by Ty1 can also repair double-stranded DNA breaks. Although, the rate of such repairs is low: only about 1% of such breaks are repaired by this cDNA. It is presumed that the repair is mediated by bridging with the ssDNA. The repair requires the RT but it does not require one of the RAD proteins (Rad52p).

The ratio between the actual integration of a Ty1 into a new site and the amount of available cDNA of this Ty1 is extremely low: only one integration for about 14,000 molecules of transcript. Thus, it was suspected that yeast has mechanisms to down-regulate the integration. Indeed a hunt for interfering factors was initiated and several such factors were identified. Mutation in interfering factors (e.g. SSL2, RAD3) increased Ty1 transposition many-fold.

We noted above that the initiation-methionine tRNA is a host factor that is active in the VLP as primer. There is however, an additional host factor that is involved in the posttranscription process of Ty1 retrotransposition: the yeast debranching enzyme (Dbr1p). This enzyme has a hydrolyzing activity on the 2'-5' phosphodiester linkage at the branchpoint of excised intron loops. Mutation of the respective gene (*dbr1*) increased Ty1 retrotransposition. The mechanism for this is not yet know.

2.7. The Reintegration of Ty1

The overall scheme for Ty1 reintegration into the yeast chromosome involves a not yet fully characterized *preintegration complex* (PIC). The PIC should contain the cDNA of Ty1 as well as the main protein that performs the integration: the IN (integrase). IN activates the chemistry of integration and also has a role in target preference (sites that are upstream of RNA polymerase III transcription). Although the IN probably does not sense the preferred insertion domain directly, but rather through certain chromosomal proteins.

A scheme of the integration reaction is shown in Figure 35. Basically, this integration has similarity to the integration of bacterial Insertion Sequences, which were handled above. The IN binds to the two ends of the cDNA of Ty1, thus forming the core of the PIC. The complex is then entering the nucleus. While still complexing the cDNA ends, the IN cleaves the target DNA and creates in it a 5-base staggered nick and joins the 3'-OH ends of the cDNA to the 5' ends of the nicked target DNA. The 5-base gaps, on either side of the integration are then repaired by the host machinery.

The IN has several domains but one of them is the DD35E domain that we met while dealing with the transposases of bacterial Insertion Sequences. This IN

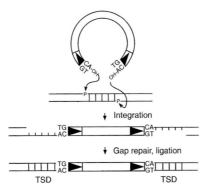

Figure 35. The integration reaction. The conserved LTR end-sequences are shown. Details of the reaction are provided in the text. TSD, target site duplication. (From Voytas and Boeke, 2002).

domain, the bacterial transposases and the respective retroviral enzyme have probably also great three-dimensional similarity. In all the three types the two asparates (the DD of the DD35E) coordinate a divalent metal cation (e.g. Mg^{2+}). While the DD35E domain of the IN of Ty1 is at the left (N-end side) of this protein, the right side (the C-end side) is probably involved in nuclear localization (in Ty5 this domain is involved in target specificity).

In vitro retrotransposition was developed for Ty1 (Eichinger and Boeke, 1988). The macromolecular components required for this *in vitro* system are only two: VLPs and a target DNA. There are two interesting notes to this system. One is that when in addition to the "natural" cDNA of Ty1, exogenic DNA fragments were also added and these had the Ty1 end-sequences, the latter fragments were also integrated by the VLPs into the target DNA. Another point is that the *in vitro* system differed from the *in vivo* situation in insertion site preference. While in *in vitro* integration the insertion is (almost) random, the *in vivo* integration has a site preference. Of the various proteins inside the VLPs it seems that only the IN is essential for *in vitro* integration.

Investigators engaged in nuclear import and export frequently tend to ignore the fact that in almost all eukaryotes there is no nuclear membrane (and thus no coherent nucleus) during mitosis. Therefore, in dividing cells there is a "window", during the cell-division cycle, in which there is no nuclear-membrane barrier. However, fungal organisms are an exception. In fungi there is commonly an intact nucleus also during mitosis. There must thus be a mechanism for the entry of the cytoplasmic-located VLPs into the nucleus. Actually it is not essential that the whole VLP enters the nucleus – only the PIC is required. Thus, the IN of the PIC can be exposed and if it has a nuclear localization domain, it can enter the nucleus. It seems that indeed such a domain exists at the C-end of IN. This was termed *nuclear localization signal* (NLS). The ability of NLS to lead the entry into the yeast nucleus was shown by a fusion protein in which green fluorescent protein (GFP) was fused to NLS and the pigment was introduced into the fungal nuclei. The target site preference of the various Ty elements differ. While Ty1 has a "window" of about 750 bp upstream of the RNA Pol III attachment, the preference of Ty5 is different. However, the target preference of Ty1 does not mean that Ty1 cannot integrate into other sites. With the complete sequencing of the yeast (*S. cerevisiae*) genome, it was found that 90% of the Ty1 integrations are in the above noted "window". While there are hot-spots for integration, cold-spots were also revealed.

There are indications that chromatin configurations also affect the integration of Ty1. These indications were revealed especially when insertions occurred upstream of promoters for regular genes that are transcribed by RNA Pol II. Mutations that affect histones and nucleosome assembly also affect target selection. Much investigation was devoted to the role of RAD6 in Ty1 target preference. This protein has multiple cellular roles, including effects on histones (H2A, H2B, H3). It was found that mutations in the gene *rad6* strongly affected the integration into specific sites. It was suggested that the RAD6 effect is mediated through its capability to change chromatin structure, rendering it, in specific cases, more accessible to the incoming Ty1.

After the complex between the incoming PIC (with its IN and cDNA) and the target DNA is established, the 3'-OH of the cDNA "attacks" the phosphatediester bond on the opposite strands of the target DNA (Figure 35). A gapped intermediate is established and this gap has to be filled. This filling is probably accomplished by the host machinery. The yeast protein Ku, that has a role in double-strand break repair of yeast DNA, seems to be involved in the gap filling.

2.8. Other Ty Elements in Yeast

Before providing information on Ty elements, other than Ty1, that reside in yeast, we should look at the general situation of transposable elements in yeast. This is now possible because the whole sequence of nuclear DNA in yeast is available. Moreover, Kim et al. (1998) did the comprehensive search for T.E. in *Saccharomyces cerevisiae*. Sandmeyer (1998) rightfully indicated that the knowledge of which types of T.E. are *not* found in the yeast genome, is no less important than the knowledge of which types of elements *are* found in this genome. Briefly, the T.E. that are not found are all the elements that do not have an RNA intermediate during transposition. Yeasts lack the bacterial Insertion Sequences as well as the terminal inverse-repeats of eukaryotic T.E. In other words, yeasts have only retrotransposons. However, of this latter group yeasts do not contain the non-LTR retrotransposons. Only the LTR retrotransposons were found in yeast. Of the LTR retrotransposons, the great majority of sequences that were found were not the full-size but rather the LTRs as shown in the following list (the LTRs are coined by Greek lower-case letters).

Element name	Number per genome		
	Full size	LTRs	Total
Ty1	32	δ217	249
Ty2	13	δ32	45
Ty3	2	σ41	43
Ty4	3	τ32	34
Ty5	1	ω7	7

In general terms, Ty1, Ty2, Ty4 and Ty5 are similar and all the four are termed as *copia*-type. Ty3, as already indicated above, has a different order of codes for the POL proteins. On the other hand, with respect to target preference there is a similarity between Ty1, Ty2, Ty3 and Ty4. All have a preference to insert upstream of RNA polymerase III-transcribed genes. Ty5 has a preference to insert into silent regions of the yeast genome. The various Ty elements are also different with respect to the variability within each type of element. Ty1 is more variable and therefore was considered to be the oldest Ty.

2.9. The Main Features of Ty3

The Ty3 has been thoroughly reviewed recently (Sandmeyer et al., 2002), thus the following information will be mainly based on this review. One obvious structural character of Ty3 is the order of its POL genes: PR, RT-RH, IN, rather than the order: PR, IN, RT-RH in the other Ty elements of yeast. Consequently, Ty3 is termed *gypsy*-like while Ty1, Ty2, Ty4 and Ty5 are termed *copia*-like (see Table 3).

The length of Ty3 is 5,351 bp. It has two highly conserved 340 bp LTRs (also termed σ) that flank the middle (or internal) domain. The terminal ends of the LTRs are inverted repeats of 8 bp (TGTTGTAT and ATACAACA, on the left and right ends, respectively). These terminal sequences have the conserved TG/CA ends of other T.E. As in Ty1, genomic insertions of Ty3 are flanked by 5-bp repeats generated from host sequences upon integration.

Laboratory stains of *S. cerevisiae* contain one to five copies of full-length Ty3. The fully sequenced strain of *S. cerevisiae* has two Ty3 (Ty3-1 and Ty3-2) but only one of these (Ty3-1) is active in retrotransposition. The general organization of Ty3 is similar to that described for Ty1, with one exception: in the POL ORF (termed in Ty3 as *POL3*) the order of the protein codes is different; in Ty3 the code for IN is downstream of RT while in the other Ty element RT is downstream of IN. Also the general life-cycle of Ty3 is similar to that of Ty1 although there are notable differences. For example, as in Ty1, there is also in Ty3 a +1 frame shift, but the details of these shifts are different in these two elements. Both Gag3p and Gag3-Pol3g are polyproteins that are processed later into the respective proteins (e.g. capsid, nucleocapsid, PR, RT and IN). These were detected in the Ty3 VLPs. Here a note is in place for the "perplexed". POL, Pol and pol are abbreviations that stand for two very different entities. POL (or Pol) may be used as an acronym for polyprotein (such as a polyprotein that is processed to RT and IN). But pol can also mean polymerase as in RNA pol III.

The RT-RNaseH of Ty3 is also similar to that of Ty1 although, again, with some notable differences in detail. For example, in Ty3 the catalystic centers of RT and RNaseH are more separated from one another than in other retroelements. Also, by several experimental approaches it was concluded that in Ty3 IN has an effect on the reverse transcription. The possibility was thus raised that in Ty3 the RT is a heterodimer composed of RT and RT-IN fusion subunits.

As indicated for Ty1, Ty3 also has a preference to integrate near pol III transcribed targets. In Ty3 it was found that while the tRNA genes are preferred, integration also occurred adjacent to 5S genes. No or very few integrations are associated with pol II or pol I promoters. The careful conclusion with respect to integration preference of Ty3 (Sandmeyer et al., 2002) was: " ... Ty3 insertion is highly specific for pol III-transcribed genes". The determinants of this specific

Table 3. Properties of selected retrotransposons.

Organism Element (class)[1]	Element/LTR length (bp)	Inverted repeat[2]	Duplication of target site (bp)	Strand primer[3]	Frame shift/site[4]	Refs.
Saccharomyces cerevisiae						
Ty1 (*copia*)	5918/334–338	TG...CA	5	GGGUGGUA1	+1/CUU AGG C	1
Ty2 (*copia*)	5961/332	TG...CA	5	GGGUGGUA1	-1/CUU AGG C	2
Ty3 (*gypsy*)	5351/340	TGTTGTATATACAACA	5	GAGAGAGAGGAAGA1	+1/GCG AGU U	3
Ty4 (*copia*)	6226/371	TGTTG...CAACA	5	AGGGAGCA1	+1/CUU AGG C	4
Ty5 (*copia*)	5375/251	TGTTGA TCAACA	5	GGGGGGA1	NA	5
Schizosaccharomyces pombe						
Tf1 (*gypsy*)	4941/358	TGTtAGC GCTaACA	5	GGGGAGGGCAA1	NA	6
Drosophila/*gypsy*	7469/482	AgTTA...TAAtT	4	GAGGGGGGAGU1	-1/AAU UUU UUA GGG	7
17.6 (*gypsy*)	7439/512	AgTgaCA...TgcaAtT	4	AAGGGAAGGA1	-1/GAA AAU UUU CAG	8
297 (*gypsy*)	6995/414	AGTgA...TtACI	4	AAGGGAAGGGG1	-1/GAA AAU UUU CGG	9
Tom (*gypsy*)	7060/474	AGTgA...TtACT	4 or 5	AAGGGGGGAGG1	-1/GAA AAU UUU CAG	10
412 (*gypsy*)	7440/481	TGTAgT...AtTACA	4	AAAAGGAGGGAGA1	-1	11
Ulysses (*gypsy*)	10653/2136	TGTT...AACA	4	GGUGA1	+1	12
Micropia (*gypsy*)	5487/505	TGTCG...CGACA	4	GAAATGTCAGAATGGCCG1	NA	13
Copia	5146/276	TGTTG...CAACA	5	GAGGGGGCG1	NA	14

[1]Class indicates order of pol gene products: copia, IN-RT/RH; gypsy, RT/RH-IN, where IN is integrase and RT/RH is reverse transcriptase/RNAse H. [2]Refers to sequence at ends of LTRs. Uppercase denotes part of repeat structure; lower case denotes not part of repeat structure. [3]Presumed plus strand primer = polypurine tract adjacent to 3' LTR in element's genomic RNA. Indicates border with 3' LTR sequence. [4]+1 and −1 indicate the number of nucleotides that RF2 is shifted relative to ORF1. Site is the sequence of the mRNA where the frame shift occurs. Codons indicated represent o frame. NA, not applicable because the element has a single ORF.
(From Sandmeyer and Menees, 1996, where references were provided).

location of integration seem to be the TF IIIB and TF IIIC that bind near the initiation of pol III transcription.

When yeast cells are exposed to various stress conditions, such as elevated temperature or chemical stress, the transposition of Ty3 is affected: only low levels of the Ty3 polyproteins are produced under stress and there is no formation of VLPs; moreover, previously formed VLPs are degraded under stress. Stress elevates the ubiquitin-mediated proteolytic activity causing Ty3 protein to undergo an enhanced turnover.

For those not familiar with yeast reproduction, the knowledge that scent plays a vital role in the sex (i.e. mating) of yeast may come as a surprise. Yeast mating types send scents, called *pheromones*, during sexual communication. These pheromones also enhance the activation of Ty3. As in Ty1, Ty3 activity is suppressed in diploid yeast, and enhanced in haploid cells. Hence, Ty3 transposition could be induced in haploid cells that undergo mating. Such an activation of transposition in germ cells is a property of several investigated retroelements.

Is the Ty3 of any gross advantage or disadvantage to yeast? There is no definite answer to this general question but there are specific bits of information. For example, it was shown that tRNA genes in yeast are repressive to pol II promoters, even at a considerable distance (Hull et al., 1994). Thus, Ty3 insertion adjacent to tRNA genes could insulate this deleterious effect. On the other hand, there is no evidence that elevated transcription of Ty3 (as can be caused in yeast cells that harbor high copy plasmids with Ty3) causes a gross change in pol II activated transcription. It seems thus that the expression of Ty3 has no gross stress effect on yeast.

2.10. LTR Retrotransposons in Other Fungal Organisms

In recent years numerous transposable elements have been reported in fission yeast (*Schizosaccharomyces pombe*), pathogenic yeast (*Candida albicans*) and several filamentous yeasts. A part of these T.E. belong to the LTR retrotransposon. The latter were reviewed by Levin (2002). Most of them are very similar to either the Ty1/*copia* retrotransposons or to the Ty3/*gypsy* retrotransposons. Therefore, they will not be detailed here.

2.10.1. LTR Retrotransposable Elements of the Fission Yeast (S. pombe)

The two yeasts, *S. cerevisiae* and *S. pombe*, appear to "outsiders" as two related types of yeasts. This is wrong! They are separated evolutionarily by about one billion years. Thus, if the two yeast types share retrotransposons of the same family and assuming that no horizontal transfer of these T.E. took place, it is reasonable to assume that the common elements are also "old". Levin et al. (1990) revealed two Ty3/*gypsy*-like elements in *S. pombe*. They were termed Tf1 and Tf2. Some fission yeast strains have only Tf1 elements while others have Tf1 as well as Tf2. In the latter yeasts the mRNAs of both elements are actively expressed. Tf1 and Tf2 differ from some other LTR retrotransposons by the fact that, in the former, the same ORF serves for Gag as well as for PR, RT and IN. Hence, there is no frame shift. There is a difference in the mobility. Tf1 is several-fold more mobile than Tf2. There is another difference between these two elements. In Tf1 the two enzymes PR and RT are separated, while in Tf2 they are expressed as a PR-RT fusion. The fact that despite a common ORF, Tf1 particles contain higher levels of Gag than IN (by immunoblot evaluation) was explained by a selective degradation process.

An interesting difference between the LTR retroelements mentioned above and Tf1 is in the initiation of the retrotranscription. There is no sequence in the LTR of Tf1 that binds a tRNA and thus tRNA is not a primer for the reverse polymerization of Tf1. The primary binding site (PBS) near the end of the left LTR is complementary to the first 11 nucleotides of the mRNA. There is, therefore, a kind of self-priming. The mode of self-priming was also revealed in a few other Ty3/*gypsy*-type LTR retrotransposons.

2.10.2. LTR Transposons in the Pathogenic Yeast Candida albicans

The *C. albicans* is a human pathogen that has had increasing relevance since the emergence of the human immunodeficiency virus (HIV); this relevance came to light because of the use of immunodepressive treatments. These treatments reduce the defensive potential of the patient against the burst of *C. albicans*. Under regular conditions, the infections range in severity from mild mucocutaneus forms to severe systemic and disseminated forms that may be fatal. Immunodepressive treatment may push the infection to the severe form. *C. albicans* exists in many forms and it is dimorphic: the cellular form and the hyphal form. Chen and Fonzi (1992) inferred that the variability and the frequent mutability of *C. albicans* stems from the existence of T.E. in this fungus. They thus isolated a repetitive element that they termed *alpha*. They then revealed a composite element containing two copies of alpha and termed it *Tca1*. Tca1 had two LTRs of 388 nucleotide each that flanked a sequence of about 5.5 kbp. Additional characteristics of the Tca1 sequence indicated that it is a LTR retrotransposon.

A further search for T.E. in *C. albicans* by Matthews et al. (1997) yielded pCal. This is a Ty1/*copia*-like LTR retrotransposon. One of the outstanding features of pCal is the abundance of the dsDNA of this element in *C. albicans* that harbors it. About 50 to 100 free, linear dsDNA copies of pCal were found in each cell. This is much higher than in any previously studied LTR retroelement. It was suggested that these dsDNA can serve as donors for reintegration promoted by the pCal integrase. There is another peculiarity in pCal. Its Pol enzymes are probably expressed via the "pseudoknot-assisted suppression" of an upstream, in-phase stop codon; a mode of expression found in the Moloney Murine Leukemia virus, as well as in some other mammalian retroviruses.

2.10.3. Additional Ty3/gypsy-like Retrotransposons in Hyphal Fungi

As summarized in Table 4 (from Levin, 2002) there are Ty3/*gypsy*-type retrotransposons in several filamentous fungi. It is a fair guess that with the availability of genomic sequences in additional fungal organisms, more retrotransposons will be revealed. Several transposable elements were revealed in the studies of M.J. Daboussi and associates (see: Daboussi and Langin, 1994; Daboussi, 1996). Most of these were found not to be of the LTR retrotransposon type. The one that probably belongs to this type, *Foret*, was found to be defective. Anaya and Roncero (1995) did find a Ty3/*gypsy*-type retrotransposon in *Fusarium oxysporum*.

Two retrotransposons were found in the phytopathogenic fungus *Magnaporthe grisea*. One of them was characterized by Dobinson et al. (1993) and coined

Table 4.   Type Ty2/gypsy retrotransposons in fungal organisms.

Transposon	Host	Length (kb)	LTRs (bp)	Copy number (estimated)
Afut	*Aspergillus fumigatus*	6.9	282	10
Boty	*Botrytis cinerea*		596	10
CfT-1	*Cladosporium fulvum*	7.0	427	25
Grasshopper	*Magnaporthe grisea*	8.0	198	65
Maggy	*Magnaporthe grisea*	5.6	253	30
Skippy	*Fusarium oxysporum*	7.8	429	20
Tf1	*Schizosaccharomyces pombe*	4.9	358	0–20
Tf2	*Schizosaccharomyces pombe*	4.9	349	15
Ty3	*Saccharomyces cerevisiae*	5.4	340	1–4

(Adapted from Levin, 2002).

grasshopper. It is a Ty3/*gypsy*-type element with two 198 bp LTRs. The other element was discovered by Farman et al. (1996) and was called MAGGY (for MAGnaporthe *gypsy*-type). MAGGY served in studies on the impact of stress conditions on the activation of a retrotransposon (Ikeda et al., 2001). These studies were assisted by the availability of a transformation system. Thus, the stress reactivity could be analyzed also in a heterologous host: *Colletotrichum lagenarium*. It was found that the MAGGY promoter was strongly activated in *M. grisea* by heat shock, copper sulfate (an antifungal agent) and oxidative stress. Some other stress conditions did not affect the MAGGY promoter. The stress effect was also observed when MAGGY was introduced into a *M. grisea* isolate that had no endogeneous MAGGY. Also, the effect was only in *M. grisea* but not when MAGGY was transformed into *C. lagenarium*. In a recent report on *gypsy*-like elements in the oomycete genus *Phytophthora*, Judelson (2002) reviewed the Ty3/*gypsy* retroelements in fungal organisms.

2.11. LTR Retrotransposons of Drosophila

It was known for many years that repetitive DNA sequences inhabit the genome of *Drosophila melanogaster* (fruit fly), but awareness of transposable elements in this species was probably triggered by the phenomenon termed *hybrid dysgenesis*. In the following I shall provide a brief description of hybrid dysgenesis and then focus mainly on one LTR retrotransposon, the *gypsy*. Additional information on hybrid dysgenesis will be provided when dealing with the *P* element of *Drosophila melanogaster*.

2.11.1. Hybrid Dysgenesis
The phenomenon of hybrid dysgenesis in *Drosophila* genetics turned out to be very much involved with several transposable elements. An accurate and concise description of hybrid dysgenesis was provided by the manual of Redei (1998). We shall describe it here and use the *P* element to demonstrate hybrid dysgenesis, although *P* is not a *gypsy*-type LTR retrotransposon. *Drosophila* geneticists

traditionally (ever since the "fly-rooms" of Morgan, mentioned in the Historical Background) looked at the eyes of this fly! Looking at the eyes of *Drosophila* is not a symmetric activity: the investigator looks with one or two eyes and the fly looks back with hundreds of eyes! The eye of the fly is a compound eye. This has genetic relevance because there can be a mosaic pattern: the eye of a fly can be composed of *white* (mutant) and *red* components. Actually white (*w*) eyes is a well-known mutation in chromosome X of *D. melanogaster*. The location of the *w* mutant was known and a DNA clone, of about 25 kbp, that contained this gene, could be cloned and investigated. When this was done, with one such mutation, it was found that the mutated clone contained an insertion of about 1.4 kbp that occurred in this *w* locus. This "extra" DNA fragment was found in other locations of the *Drosophila* genome. Not all strains of *D. melanogaster* harbor this fragment, but those that do, contain about 40 copies of it.

Technically the search for DNA sequences that are homologous to the sequence found in the mutated locus could be performed only after molecular methods, such as cutting DNA by restriction endonucleases, running the digested genomic DNA on gels and Southern blot hybridization, became established. It was then found that the *P* elements may be of different lengths, the longest (and intact) are 2.9 kbp. These latter elements have terminal inverted repeats (as we met in the bacterial IS) of 31 nucleotides. It was found that the sequences of the *P* elements are very conserved but they are located at different sites in different lines of flies. Moreover, the mutated white-eyed gene could revert to red-eyed and with this reversion the *P* insert disappeared from the locus. In short, intensive studies (reviewed by Bregliano and Kidwell, 1983; Rio, 1991; Engels, 1989) clearly established that *P* is a transposable element.

P has a preferred sequence of integration: 5'-GGCCAGAC; its integration is accompanied by the formation of an 8 bp duplication. When *P* is excised from a locus it leaves footprints, showing where it was inserted. For autonomous mobility in the genome the *P* element has to be intact (full length).

Up to now we are sailing on calm seas. The storm starts when female flies of one strain are mated with male flies of another specific strain. As noted above, some strains of *D. melanogaster* are devoid of *P* elements, these are *M* strains. Others do contain *P* elements and are *P* strains. The latter are frequent among wild-type flies. When *M* female flies were mated with *P* males the "storm" was created. Notably there is no "storm" in *M* females because they do not harbor *P* elements. There is also no "storm" in *P* males, nor in the progeny of their mating with *P* females, probably because *P* males and their female sibs of *P* strains, also harbor suppressing factors that reduce transposition to a very low level. When the sperm of *P* males enters *M* female eggs, the suppressing factors do not join and the "chaos" starts. The *P* elements are excised from one location and inserted into new locations, causing a lot of mutations. Some of them end in sterility. The rate of insertion can be as high as one per chromosome in each generation of flies. This genetic peculiarity, termed *hybrid dysgenesis*, also has other genetic consequences. While there is usually no crossing-over in male flies, males that are progenies of hybrid dysgenesis do show crossing-over. The phenomenon of this high rate of transposition occurs only in germ-line cells of the flies. A possible reason is that effective transposition requires splicing of the mRNA of the *P* element and this can occur only in the germ-line cells of the female flies. Once the damage (or the curing) was done, the progeny were affected.

Recently, Simmons et al. (2002) found that the splicing of the third intron of the transposase encoded by *P* is involved in the *pecular inheritance* of hybrid dysgenesis. Their experimental results suggested a unique characteristic of the pre-mRNA of the *P* transposase. The third intron of this pre-mRNA, in the oocyst, can be either spliced, causing the formation of a functional transposase, or, it is not spliced, thus no functional transposase is produced. It appears that there is an *exclusion* mechanism that excludes the fully spliced mRNA from the region of the oocyst that will later form the soma. The fully spliced mRNA does exist in the region that will form the primordial germ-line in the embryo, after fertilization.

Drosophila has a characteristic that renders the location of mutations (such as those caused by the insertion of a *P* element) easily visible. The flies have polytenic chromosomes in their salivary-gland cells. In these chromosomes the DNA is duplicated (lengthwise) about 1000 times and they appear as banded structures in cytological preparations. The bands represent sites along the chromosomes. When a radioactive DNA or RNA clone is hybridized ("*in situ*") to polytenic chromosomes, the site of homologous hybridization becomes radioactive and can be visualized by a proper photographic film covering the cytological preparation. By this method, the "exact" location of a *P* element could be assured. Moreover, when a *P* element was assumed (by genetic studies) to be excised from a mutated locus, this could be analyzed by *in situ* hybridization with a radioactive *P* element probe. Incidentally, Russian molecular geneticists, who returned to *Drosophila* and studied retrotransposons, made ample use of *in situ* hybridization of polytenic chromosomes (Ilyin et al., 1980a,b,c; Gerasimova et al., 1984a,b,c; Evgen'ev et al., 1997, 2000).

At the end of this description of *P*-mediated hybrid dysgenesis, the author of this book apologises: the *P* element will be handled again in a later part of the book when Class II elements of insects will be handled.

As we shall see below, the LTR retrotransposons of the *gypsy*-type are also involved in hybrid dysgenesis, but before handling these elements we shall divert once again to another group of retrotransposons that can cause this phenomenon: the I transposable elements. The I elements of *Drosophila* belong to the non-LTR retroelements. This is a large group of elements that will be detailed in a subsequent section. They are represented in many kinds of plants and animals. In mammals they are represented by *LINE*s and a specific kind of the latter are the L1 elements of primates (e.g. Fanning and Singer, 1987a). L1 elements can reach 100,000 insertions per genome and thus may cause a wide range of maladies in humans.

The *I-R* induced hybrid dysgenesis has some similarities with the *P* element induced phenomenon that carries the same name. Namely, it appears after a cross is made between two different strains of flies. Matings within each of these strains do not result in *I-R* type hybrid dysgenesis. Also, the phenomenon appears only in females and in the germ-line cells of these females.

But there are major differences. *P* and *I* are very different elements. While the *I* has similarity to non-LTR retrotransposons, the *P* element has similarity to bacterial IS. The transposition of *I* is probably of the "replicative" type, meaning that when the *I* transposes to a new site, it leaves behind the same *I* element. Thus, excision of an *I* element from its insertion site was not observed. While there is a sequence diversity among I elements, they all have some common features. In almost all of them the 5' end sequence is CAGTA (or with only a minor modification of this sequence). The 3' end is also rather preserved. They are TCA (TAA)n where n can be 4 to 8. The total length is several thousand bp and they transpose by reverse transcription. They contain two ORFs but seem to lack an IN.

The *I-R* hybrid dysgenesis is observed only in F_1 females (SF females) that result from crosses between "reactive" females and "inducer" males. SF females have strongly reduced fertility and their germ-lines have many mutations, chromosomal rearrangements and non-disjunction. Actually SF females are sterile because their eggs fail to hatch. While their whole genetic relations are rather complicated (e.g. there are several levels of "reactivity" in females, affected by chromosomal and cytoplasmic factors), it is clear that the *I* elements are involved. The *I* elements from the males are causing the hybrid dysgenetic in females, while in the male parents these *I* elements are not mobile, consequently no hybrid dysgenesis happens in them. Obviously, since the hybrid dysgenesis is restricted to the female germ-line, causing the damage in eggs and very early embryos – males are excluded by definition.

2.12. The gypsy LTR Retrotransposon

2.12.1. Basic Features of gypsy

The notion that there are transposable elements in *Drosophila melanogaster* was brought up by two review publications. One of these is by Finnegan et al. (1977) and the other is by Green (1977). Finnegan et al. analyzed the many previous reports on repeated gene families such as *Copia* and *412*. They also provided data that was based on the available methods (restriction maps and *in situ* hybridization of polytenic chromosomes). One section of this review actually speculates: "How terminal redundancies of dispersed families can yield transposable elements that maintain sequence homology". Thus, the mobility of the repeated gene families was already obvious in 1977. The publication of Green (1977) deals with mutations, mostly eye-color mutations, and their reversion to wt (wild-type) or to nearly wt pigmentation. After reviewing such changes in eye-pigmentation and especially changes that are facilitated by X-ray irradiation, Green comes to the conclusion that: "... no authentic case of the transposition of a specific DNA element from one genetic site to another has been reported in *D. melanogaster*." However, his summary ends with the suggestion that *Drosophila* insertion mutants are analogous to those described for proved insertion mutants of *E. coli*. He obviously meant the bacterial Insertion Sequences that we handled above. A later review, by Rubin (1983), cited numerous publications that provided evidence for transposable elements in *D. melanogaster*. One of the convincing arguments was that the very same genetic elements appeared at different locations in different strains. This means they were able to move. Actually, as early as 1973 Ising and Ramel (1973) published an abstract on: "The behavior of a transposing element in *Drosophila melanogaster*".

In a series of publications Russian geneticists reported on their studies on "*mobile dispersed* genetic elements" in *D. melanogaster*. They termed them MDGs (Bayev et al., 1980; Ilyin et al., 1980a,b,c; Kulguskin et al., 1981). Thus, Mdg1 and Mdg3 had all the characteristics of mobile elements. A few years later, this team (Bayev et al., 1984) named another MDG as Mdg4. By then there was already an earlier publication on this element. Modolell et al. (1983) found the very same element but gave it the name *gypsy*. In this way the fashion for fancy names for transposable elements in *Drosophila* started.

Penelope stirs the mobility of *Ulysses* in a phenomenon in which, among others,

Helena and *Paris* are also involved. While this set of names sounds as if taken from a Helenic myth, this phenomenon is *hybrid dysgenesis*; a genetic peculiarity that was handled above. Evidence for the role of *Penelope*, a retrotransposon, in hybrid dysgenesis of *D. viridis*, was provided by a sober publication of Evgen'ev et al. (1997) and Zelentsova et al. (1999). *Helena* (Zelentsova, one of the authors) is of the Russian Academy of Science, in *Vavilov* Street in Moscow. There is some irony in all this. Nicolai Ivanovich Vavilov was a world-famous Russian botanist/geneticist who probably died in 1943 in a Soviet concentration camp after encounters with Trofim Lysenko, the then favorite Soviet plant breeder. In 1948 Lysenko headed an "All-Union" meeting of Soviet scholars engaged in agricultural sciences. There Lysenko scolded the Soviet geneticists, blaming them that while, in World War II, Soviet patriots sacrificed their lives to defend Stalingrad, they were experimenting with flies (*Drosophila*). The geneticists were given the choice: to admit their grave mistake and to become adhererents of Lysenko, or to be sent to Siberia. Times were changing: most of the dissidents of 1948 perished in Siberia, Lysenko fell out of favor in the Soviet Union and "disappeared", but the "fly-rooms" flourish again in Vavilov Street in Moscow. *Ulysses*, *Helena* and *Paris* are retrotransposon elements in *Drosophila*, as are numerous other elements of this kind. They are much more ancient than their mythological namesakes; they probably emerged soon after the first eukaryotic organisms. However, back to Vavilov Street. This was not named after Nicolai Vavilov but rather after his brother Sergei Ivanov, who was promoted in Stalin's time to become President of the Soviet Academy of Science.

The three early studies on *gypsy* (Modollel et al., 1983; Freund and Meselson, 1984; Bayev et al., 1984) where on different *D. melanogaster* strains but provided the same basic information. It became clear that *gypsy* is a transposable element of an estimated total length of 7.3 kbp (the Messelson teams estimate) or 7.5 kbp (Bayev et al. 1984) and that this element has long terminal repeats (LTRs) of 479 bp (Bayev et al., 1984) or 482 bp (Freund and Meselson, 1984). The insertion and excision could be followed by *in situ* hybridization of radioactive *gypsy* probes to salivary chromosomes. It was also found that there is a preferred insertion into the sequence TACATA. Furthermore, *gypsy* signals for RNA polyadenylation (AATAAA) and transcription initiation (TATATAA), were also revealed. The *in situ* hybridization was very useful in the subsequent studies of Gerasimova and her associates (Gerasimova et al., 1984a,b,c) who analyzed the results of hybrid dysgenesis, especially in one unstable *D. melanogaster* strain, ct^{MR2}. In this line the investigators witnessed an "explosion" of excision and reinsertion of several transposable elements, including *mdg4* (meaning *gypsy*). The investigators focused on events in chromosome X. The sudden mobilization involved actually *mdg1*, *mdg2*, *mdg4*, *copia*, *Foldback* (*FB*) and i elements. It was assumed that the "explosions" could be traced back, in each case, to one and the same germ-line cell and they happened simultaneously, although the process could be repeated several times. Moreover, during this extensive mobilization transposable elements tend to come back and reinsert into loci from where they were previously excised (reverse transposition). Why are the elements coming back to the locations from where they were excised? The authors (Gerasimova et al., 1984c) suggested a kind of "memory" in which a part of the element was retained when it was excised, and this part then led the element again to the same location. In all the studies mentioned up to now, on *gypsy*, the issue of reverse transcription has not raised.

On the other hand, the impact of the "transposition explosion" on evolution was elaborated. It was argued that the insertion of mobile elements can lead to several genetic modifications: gene inactivation, down mutations, enhancement of gene activity, change in gene regulation, etc. Therefore, the "explosion" may be an important factor in causing genomic diversity. Novel complex features could emerge and those with selective advantages could lead to an evolutionary optimization of novel taxa. In other words, the transposition explosions may lead to "evolutionary jumps".

The events that took place in the Soviet Union, during the All-Union meeting organized by Lysenko, were mentioned above. Well, there is more to it! In a recent publication of Lyubomirskaya et al. (2001) that deals with variants of the *gypsy* LTR retrotransposons, it was noted that among the many strains of *D. melanogaster* used in this study, were some old strains kept in the collection of the Moscow State University for more than 50 years! This means that the flies survived the assault of Lysenko! We should note that in order to keep a fly-strain someone has to tend the flies continuously. The building of the Moscow State University should thus not be considered merely an Ivory Tower – it actually served as a Granite Castle. Moreover, any scientist caught with *Drosophila* research, during the Lysenko period, was in severe danger. Rather than staying in the Sergei Vavilov Street, he or she would have been forced to follow the path of Nicolai Vavilov in Siberia. One should never underestimate the willpower and determination of fly researchers!

2.12.2. The Effect of gypsy on Gene Expression
Further studies focused on the question of what is the impact of *gypsy* insertion on the expression of a specific gene. One such study was performed by Parkhurs and Corces (1985). These investigators focused on the *forked* locus of *Drosophila*. In this locus the insertion of *gypsy* does not cause the appearance of new RNAs. But the insertion, that probably occurs in an intron, causes the disappearance of a 4.3 kb transcript and the lowering of 2.4 and 1.9 kb transcripts. On the other hand, there is a slight increase of a 1.4 kb message. These authors suggested that there is some interference by the regulatory elements of *gypsy*. They pointed out that the *gypsy* sequence has in its LTRs a TATA box that is part of a promoter and there is also a polyadenylation signal in one of the LTRs of *gypsy*. These are under developmental regulation. They are activated maximally in 2–3 days old pupae. This peak of activity is also the peak of the activity of *forked*. Thus this parallel in the temporal control of the expression of *forked* and *gypsy* is not a mere coincidence but explains the mutational effect of *gypsy*. Possibly the enhancer-sequences located in the LTR of *gypsy* can act at a distance and override the normal transcription control of the *forked* locus.

A further publication of the team of Corces (Marlor et al., 1986) appeared after it was already evident that a similar element, Ty3, is a LTR retrotransposon (see above and Roeder and Fink, 1983). Marlor et al. made the effort of obtaining the complete nucleotide sequence of *gypsy*. They obtained the sequence from the mutated *forked* locus. This sequencing revealed a striking homology between *gypsy* and retroviruses of vertebrate animals. Thus, *gypsy* emerged as more similar to the retroviruses than to other LTR retrotransposons such as *copia* of *Drosophila* or Ty1 of yeast. The *gypsy* element had a total length of 7,469 bp and encoded three putative protein-products (or three ORFs). The latter authors verified the lengths of the LTRs as 482 bp (as originally reported by Freund and Meselson, 1984). The sequencing also revealed the transcription initiation and termination signals as well as sequences that are homologous to the polypurine tract and the tRNA^{Lys} primer binding-site that are essential for the reverse transcription of retroviruses. Several sequences in the *forked* derived *gypsy* as well as information from another *gypsy* element indicated that the *gypsy* element is evolutionarily closer to another T.E. of *Drosophila* (17.6) and to vertebrate retroviruses than to the *copia* element of *Drosophila*.

How exactly the element *gypsy* exerts its effect on the expression of host genes is obviously complex. Corces and Geyer (1991) returned to this subject and tried to provide some explanations. They stressed the role of the suppressor of Hairy-wing (*su(Hw)*) in the *gypsy* effect. The protein of *suHw* is a regulatory protein with a DNA-binding capability and thus it is active in the regulation of some fly genes. These authors suggested that when a *gypsy* insertion becomes evident, the impact of the *suHw* protein on the regulation of gene expression is altered. The whole issue is rather complicated and even the language used for explanation is enigmatic. Here is a sentence from the conclusion of Corces and Geyer that shows the problem: "Furthermore, when *gypsy* insertion takes place in an intron of the gene, the effect on transcription rate is compounded with an effect on the stability of the precursor transposable element-mutant gene hybrid RNA". A further publication of the Corces laboratory (Smith and Corces, 1995) added information and clarification. The general picture that emerges from the role of *gypsy* in affecting various *Drosophila* genes (as *cut, forked, yellow* and *scute*) is that when *gypsy* inserts, the result is a mutant phenotype. This phenotype can be reversed by a second-site mutation in the *su(Hw)* gene. The insertion of *gypsy* that causes the change in phenotype is between the *cis*-regulatory sequences (of the affected gene) and the promoter of the gene. The protein produced by *su(Hw)* has "zinc-fingers" that bind to DNA. The binding occurs in the 5' transcribed, but not translated, portion of the *gypsy* sequence, thus regulating the expression of *gypsy*. Probably, the wild-type *su(Hw)* protein is a positive regulator and is also involved in the correct temporal and spatial expression of *gypsy*. A mutation in the *su(Hw)* gene will alter the protein product and thus interfere with the effect of the inserted *gypsy*. Under "normal" conditions there is a typical temporal and spatial expression of *gypsy*. In the embryo of flies, *gypsy* is expressed in the gonads, fat body and salivary glands precursors. The expression in the salivary glands is then gradually reduced. In adult female flies the expression of *gypsy* is only in the fat body and in the nurse and follicle cells during oogenesis. Hence, there is a rather elaborate pattern of expression levels. It appears that some domains of the *su(Hw)* protein are involved in this regulation. It also appears that the *su(Hw)* protein interacts with other fly proteins to exert its regulatory activity.

2.12.3. The Similarity of gypsy to Retroviruses
The similarity of *gypsy* to retroviruses became evident after its full nucleotide sequence was revealed (e.g. Marlor et al., 1986). It became clear that *gypsy* has the three ORFs found in retroviruses, as well as other features that brought *gypsy* very close to retroviruses. How close? This was analyzed in a series of studies by Eickbush and associates (Xiong and Eickbush, 1990; Malik and Eickbush, 1999; and see review of Eickbush and Malik, 2002). While the similarity between the nucleotide sequence of *gypsy*-like retroelements and retroviruses are striking (and shall be handled again below) the question remains whether or not *gypsy* can be infectious. Kim et al. (1994) of Gif-sur-Yvette, France, approached this question. These authors noted that *Drosophila* flies usually contain 20–30 defective *gypsy*

elements that are located in pericentromeric heterochromoatin. These elements do not transpose. In addition, some fly strains also contain potentially active *gypsy* elements. Among those are strains having 30–40 very active elements and one, "Mutator" (MS) strain, shows high rates of transposition of *gypsy* not only in the germ-line cells but even in somatic cells. Consequently, Kim et al. injected egg-plasm (devoid of cells and nuclei) of a line of MS flies into embryos of SS flies. The latter have no *gypsy* elements, but genetic tests showed that they are receptive to such elements. The injection caused gene reversions of *gypsy* as expected. Moreover, the injected *gypsy* elements were inserted into the SS chromosomes. Injection was not the only means to infect receptive larvae. When such larvae were raised on food that was mixed with homogenized MS pupae – the daughters of the infected flies did show that *gypsy* was integrated. From this and other studies, the French team that investigated *gypsy* considered this element as a retrovirus.

Syomin et al. (1993) were also asking whether or not *gypsy* can be considered a kind of retrovirus. They acknowledged that the third ORF of *gypsy* has a code for ENV (envelope) proteins, meaning components required for the formation of the Virus-Like Particles (VLP). The question was, whether such isolated VLP can be preserved in an extracellular form and then used to "infect" flies. If this were so there should be a way for *gypsy* to be transmitted "horizontally" (not meaning from parents to offspring, that is considered "vertical" transmission). *Drosophila* cells can be cultured *in vitro*. *In vitro* cultivation of such fly cells actually enhances the expression of *gypsy*. Therefore, the medium in which the *gypsy*-containing cells are cultured will contain VLP that are released from the cells. Syomin et al. collected such media and were able to obtain a fraction of VLP. By various criteria the particles concentrated from the culture medium were identical to authentic intracellular VLP of *gypsy* or at least some of these extracellular particles had the composition and shape of intracellular VLP. Here this study stopped. Are there extracellular VLP infections? From the experiments of Kim et al. (1994), mentioned above, it can be deduced that there are infections.

2.12.4. Flamenco and gypsy Mobility

We shall now turn to *flamenco*. *Flamenco* is not another transposable element, but rather a gene that controls the mobility of *gypsy*. We owe this fancy name to the French investigators in Gif-sur-Yvette (Prud'homme et al., 1995). These investigators collected the accumulating information that in most *Drosophila* strains there are numerous *gypsy* sequences, mostly located in the pericentric heterochromatin and thus not active. There also exist a few *gypsy* elements in chromosomal arms (euchromatin) but even those are mostly repressed. In relatively rare cases, such as in the stock MS, the *gypsy* elements "come to life". Prud'homme et al. found a gene, on chromosome X, that suppressed the mobility of *gypsy*, and called it *flamenco* (*flam*). When *flam* is mutated and female flies are homozygous for this recessive mutation, the transposition of *gypsy* in their germ-line cells becomes very active.

How does one score the mobility of *gypsy*? For that *Drosophila* has a gene that is termed *ovo*. When a mutant of this gene, ovo^{D1}, is expressed in female flies they become sterile. But when ovo^{D1} is mutated, as can happen by insertion of *gypsy* into ovo^{D1}, fertility is restored. Actually the rate of changes from sterility to fertility monitors the mobility ("dancing capability") of *gypsy*. Thus, by evaluating the restoration of fertility it became evident that *flamenco* is involved in this "dancing":

in the progeny of double-mutant *flam* females *gypsy* is actively transposing. It happens that *ovo* is a hot-spot for *gypsy* integration (Dej et al., 1998). We may even claim that the *ovo* locus is a "sink" for mutating agents. As early as in 1968, Lifschytz and Falk (of the Hebrew University in Jerusalem) found that the location, in chromosome X of *Drosophila*, where *flam* was located, is a "hot spot" in this chromosome, that is liable to breaks by mutagenesis. This obviously reminds us of the *Ds* induced breaks in chromosome 9 of maize that led Barbara McClintock to the discovery of transposable elements. The story of *gypsy*, *flamenco* and *ovo* is actually even more complicated (thus more interesting for those who continue to follow the story). It was suggested that for full activity of *gypsy* this element has to adopt the lifestyle of a retrovirus. That means it has to express the envelope (ENV) proteins into which it is packaged and become infective. These proteins include a signal peptide, a transmembrane anchor and a cleavage site to produce transmembrane and glycosylated surface proteins. The claim was that ENV expression is facilitated in the ovaries of *flam¹/flam¹* females, where a respective splicing activity is manifested and the initiation of translation of the third ORF of *gypsy*, that encodes the ENV proteins, is probably facilitated. It was considered (Bucheton, 1995) that *gypsy* transcripts do not accumulate in the germ-line cells themselves (that are the oocyte and the nurse cells) but in the somatic follicle cells that surround the oocyte. Thus, the ENV proteins are also produced in the follicle cells. The infectious enveloped viral particles of *gypsy* may then enter the permissible (*flam¹/flam¹*) oocytes and there start their enhanced integration. However, the story became complicated. In a further study, this French team (Chalvet al., 1999) presented data that suggested that the transmission of *gypsy* from soma cells to germ-line cells does not require expression of the *env* gene of *gypsy*. The possibility thus remains that non-enveloped particles enter oocytes by other means, such as endocytosis.

The relation between *flamenco* and *gypsy* requires further clarification. The permissiveness of *gypsy* expression that is regulated by *flamenco* is determined in the soma of the female: when the follicle cells have mutated *flamenco* (*flam¹/flam¹*) in their genome, the *gypsy* activity is permitted. The *gypsy* elements are then entering the germ-line and are thus transmitted vertically. What is the chemistry of the regulation-control by *flamenco* on *gypsy*? To answer this question the first step would be to isolate the gene and then to characterize the translation product of this gene. Robert et al. (2001) started to do just that. The isolation of the coding sequence of *flamenco* should have been a simple task because the total genome sequence of *D. melanogaster* was already published (Adams et al., 2000; Myers et al., 2000). In practice it turned out to be a formidable task, because the location of *flamenco* in the X chromosome is in a region of β-heterochromatin, that is extremely difficult to sequence. The *flamenco* is actually located in a *gap* of the published *D. melanogaster* sequence: this location is so filled with repeated sequences that neither Adams et al., nor Myers et al. detailed it. Robert et al. (2001) came close to it but the gene is still elusive.

A final "twist" is noteworthy. While in *D. melanogaster* the *env* gene of *gypsy* is functional and ENV proteins are produced, the *gypsy* elements in some other *Drosophila* species (e.g. *D. subobscura*, *D. virilis*) have no functional *env* genes. Can it be that *gypsy* plays a "double lifestyle": it can be transmitted vertically and for that a functional *env* is not essential, and it can also be transmitted horizontally, and for this transmission a virus-like envelope is essential.

2.13. A Brief Overview of Additional Retrotransposons of Drosophila

A detailed survey of the retrotransposons (active and "skeletons") of *Drosophila* is way beyond the scope of this book. Moreover, as indicated above, with the accumulation of sequenced genomes it is expected that many more such transposons will be revealed. As noted above, the nucleotide sequence of *D. melanogaster* has been reported and it is expected that the genomes of other *Drosophila* species will be fully sequenced. These additional sequences will probably provide further information on LTR retrotransposons in this genus. In addition to studies on LTR retrotransposons of *Drosophila* that have already been mentioned, a lot of investigations were performed on other LTR retrotransposons in *Drosophila* such as *297* (Inouye et al., 1986), *Foldback* (FB, Potter, 1982), *R1* and *R2* (Burke et al., 1993), *412* (Will et al., 1981) that were reviewed by Bingham and Zachar (1989). Another, relatively recently identified LTR retroelement is *tom* (Tanda et al., 1994). Steinemann and Steinemann (1997) of the Heinrich-Heine University of Düsseldorf, Germany looked at the *neo*-Y chromosome of *Drosophila miranda*. This species is a relative of *D. pseudoobscura*. The *neo*-Y emerged by the fusion of an autosome with a Y chromosome. These investigators revealed a massive accumulation of a novel LTR retrotransposon, termed *TRAM* in the *neo*-Y chromosome. They attributed the cytological evolution of *neo*-Y to the presence of *TRAM* elements. *TRAM* is a 3,452 bp element with direct LTRs of 372 bp each.

An interesting relationship was studied in *D. virilis* that actually comprises a group of species. In *D. virilis* five families of T.E. were investigated by Russian molecular geneticists (Evgen'ev et al., 1997, 2000; Zelentsova et al., 1999) and were given fancy names: *Ulysses*, *Penelope*, *Paris*, *Helena* and *Telemac*. The last name probably stands for Telemachus (the son of Penelope in Greek mythology). All these transposable elements are involved in hybrid dysgenesis but those studied in great detail are *Penelope* and *Ulysses*. The Russian authors were interested to investigate the possibility that the great karyotype variation in the *D. virilis* group members resulted from the ample hybrid dysgenesis that occurs in this species. The elements *Penelope* and *Ulysses* co-mobilize in dysgenic crosses. Moreover, *Penelope* was able to induce the transposition of *Ulysses* when it was infected in the early embryos of *D. virilis* flies that did not contain *Penelope* elements. These investigators (Evgen'ev et al., 2000) actually found that many (hybrid dysgenesis derived) rearrangement break points coincided with the chromosomal locations of *Penelope* and *Ulysses* insertions in the parental strains and with break points of inversions previously established for other species of the group. The authors concluded that their results "are consistent with the possibility that these elements play an important role in the evolution of the *virilis* species group". The authors did not reveal their considerations in choosing the fancy names of these retroelements of *D. virilis*. There are two mythological versions of Penelope. According to one, she was a faithful wife, but according to the other, she was promiscuous. As for Ulysses (that is the Latin name for Odysseus) there are also two versions of his character. Thus with respect to myths we are left in the mist, but in the *D. virilis* reality *Penelope* causes the mobility of *Ulysses*.

2.14. A Survey of LTR Retrotransposons in Metazoa

In the previous section we detailed the LTR retrotransposons of *Drosophila*. As this fly genus was a subject of intensive genetic investigation, their T.E. study was very prolific. There are obviously T.E. in other insects such as the genus *Bombyx* that shall be mentioned below. Also, T.E. were found in additional insects such as mosquitoes and it is reasonable to expect that T.E., including LTR retrotransposons, will be found in all the 30 (or 31) orders of insects. These revelations are expected to happen almost "automatically" with the availability of genomic sequences in taxa of these orders. Reports on T.E., including LTR retrotransposons, are thus likewise expected in other metazoan phyla. However, there seem to be notable exceptions. The human and the mouse genomes are now fully sequenced and there is an almost complete sequence of other mammalian genomes. No active LTR retrotransposons were found in mammalian species. Birds were less subject to genomic sequencing but up to now no active LTR retrotransposons have been revealed in birds. We assume that if indeed mammals and birds are found to be "clean" of LTR retrotransposons, there is a reason why they are missing.

In the following sections we shall survey a few representatives of LTR retrotransposons of metazoa according to phyla. We shall first mention insects (of the Arthropoda), then look at representatives of worms, Echinodermata and several taxa of Chordata. In the latter we shall start with the urochordate *Ciona* (sea squirt) and go through amphibia and reptiles to fishes. The last group, fishes, will be detailed to some extent as they have been subjected to more detailed research.

2.14.1. Insects other than Drosophila

The silkworm species (*Bombyx mori*) was a source of a LTR retrotransposon termed *Pao* (Xiong et al., 1993). In this case the authors indicated that the name *Pao* stands for "running" in Chinese. The isolated, full-length, element was detected in one of the spaces between the repeated codes for ribosomal RNA. The *Pao* element has 4,791 bp that include two terminal direct repeats (LTRs) of 634 bp each. The element has one ORF coding for 1,158 amino acids. Among these, 200 amino acids were similar to reverse transcriptase. In addition to this full-length element, several other (truncated) elements were found in *B. mori* that diverged by about 5% from the above-mentioned full-length *Pao*. The LTRs of *Pao* and other similar LTR retrotransposons of *B. mori* have a unique sequence. They contained several tandem repeats of a 46 bp motif. However, the *Pao* that was isolated (as well as similar truncated elements) had no codes for RNaseH and IN. Thus *Pao* could be a "skeleton", rather than an autonomous T.E. Now comes the question of "drawers". According to the sequence of its RT *Pao* does not fit into either the Ty1/*copia* or the Ty3/*gypsy* "drawer". According to the RT sequence, *Pao* is similar to an *Ascaris* T.E (*TAS*). However, by other criteria *Pao* has similarities to Ty3/*gypsy*.

A large team of Japanese investigators (Abe et al., 2001) were "running" after silkworm T.E. and indeed discovered at least two elements that had common structural features to *Pao* ("running"). This time fancy Japanese names were given: *Kamikaze* and *Yamato*. Both of these elements had codes for RH and IN. The order of the enzymes is as in Ty3/*gypsy* but there were differences between the latter element and the two *B. mori* elements. Moreover, Abe et al. (2001) went back to *Pao* and found in *B. mori* the codes for RH and IN in this element that was missing from the isolate of Xiong et al. (1993). Finally, Abe et al. (2001) agreed with the suggestion of Xiong et al. (1993) that *Pao*, *Kamikaze*, *Yamato*, and possibly some additional insect and nematode LTR retrotransposons should be given a "drawer" of their own and belong neither to the Ty3/*gypsy* nor to the Ty1/*copia* groups. So we may have a new category of LTR transposons but whether or not *Pao* (and its two cousins *Kamikaze* and *Yamato*) are really "running" – meaning actively mobile, is still not certain.

The mosquito *Anopheles gambiae* is the vector for African malaria. It attracted the attention of molecular geneticists and cytogeneticists. One team of investigators (Rohr et al., 2002) focused on the Y chromosome of this mosquito where they found repetitive DNA regions. They screened the DNA of male mosquitoes and the sequencing revealed a DNA region that encoded proteins with putative LTR retrotransposon amino acid sequences. Such screening indeed revealed an intact element that, by DNA sequence, had the characteristics of a typical *copia*-like LTR retrotransposon. Its total length was 4,284 bp and it had two LTRs of 119 bp each. The terminal bp of the LTRs were 5'-TG...CA-3'. No "classical" transcription initiation signal was revealed but there was a putative termination signal: AATAAA. Near the 5' LTR a sequence complementary to 14 bp of tRNA^{Met} was found – a possible primer location for the reverse transcription (PBS). The middle region of this putative element had two long overlapping ORFs. The order of the putative Pol genes indicated that the element belongs to the *copia* superfamily of LTR retrotransposon; i.e. the order of the genes is IN, RT, RNaseH. Two sequences that represent the full-size putative element were then revealed in the DNA of the Y chromosomes. The authors named the element *mtanga*-Y and assumed that it could be transcribed and was capable of mobility. Further search suggested that in total the Y chromosome of *A. gambiae* contains about 12 intact elements.

There are, in addition, elements that are not intact. *Mtanga*-Y was found in the Y chromosomes of *A. gambiae* from several African countries, but not in females, nor in other *Anopheles* species. Less stringent screening indicated the existence of a family of *mtanga* elements, some of them in female mosquitoes. Actually the *mtanga*-Y elements in the male mosquitoes were found to be clustered in one region of the Y chromosome. The two LTRs of *mtanga*-Y differ only in two positions. Thus it appears to the authors that this element inserted only recently into the Y chromosome of the mosquitoes. They also suggested that *mtanga*-Y was not eliminated from its location on the Y chromosome because in this location the element does not cause genetic disturbance. Within a certain range on the Y chromosome, the *mtanga* could also be mobile without inflicting any damage because this region does not contain essential genes. Interestingly, the authors also suggested that one or two intact and active *mtanga*-Y located in the Y chromosome could cause the mobility of other, *mtanga*-like elements that are defective and have lost their autonomous mobility.

2.14.2. LTR Retrotransposons of Worms

With worms we move to a completely different branch of metazoa. The first LTR retrotransposon of worms was identified by Felder et al. (1994) of Switzerland. These investigators identified *Tas* (*transposon-like element from Ascaris*). The element was found in about 50 copies per haploid of *Ascaris lumbricoides*, a parasitic nematode. The length of *Tas* is 7,627 bp and it was completely sequenced. The internal domain of *Tas* is 7,128 bp and it is flanked with two identical sequences of 236 bp (LTRs). Both LTRs are bounded by the nucleotides TG and CA. At 6 bp downstream of the left LTR there is a 19 bp sequence that is identical to a sequence of tRNA^{Arg} of the nematode *Caenorhabditis elegans*. This probably facilitates the priming in the reverse transcription. Three ORFs were detected in *Tas*.

The ORF1 encodes the *Gag* protein (631 amino acids), with an RNA-binding "zinc-finger" motif. There was a loss of RT function within the ORF2 of most genomic *Tas* elements. The authors inferred that this could be a means for this nematode to regulate *Tas* transposition. The ORF3 included a coding region for *env* although this sequence is very different from other *env* coding regions in retroviruses and other LTR retrotransposons.

Because of several features such as variable LTR sequences in the different *Tas* elements of individual worms, the authors assume that *Tas* is mobile. The defective RT code found in ORF2 of most (or all) *Tas* elements of *A. lumbricoides* is suggested by Felder et al. (1994) to be complemented by some intact RT in other *Tas* elements, or even in other LTR retrotransposons that exist in this worm.

"Ancient" families of retroviral-like elements were more recently revealed in the much studied nematode *Caenorhabditis elegans* (Bowen and McDonald, 1999). The complete nucleotide sequence of the genome of *C. elegans* was previously available. Thus, these authors could perform a detailed computer search and analyze the retrotransposons. They found 12 families of a LTR retrotransposon type and named the family members *Cer1* to *Cer12* (*C. elegans* retrotransposon). The elements all had some basic features in common but also differences; this justified giving them a "family" status, although the authors found only one or two elements for each "family". The division into families was based on the sequences of the amino acids of their respective RT. When two *Cer* elements had a lower than 90% amino acid identity, they were separated into different families. Families were separated into different elements, when they were less than 99.9% similar in their nucleotide sequences. Thus, *Cer1* could be separated into *Cer1*-1 and *Cer1*-2.

Identity and divergence between the 5' LTR and the 3' LTR of a given LTR retrotransposon has very relevant consequences for the evolution of T.E. This identity/divergence provides a clue for the "history" of the LTR retrotransposon. It is assumed that if in an LTR retroelement the two LTRs are identical (or very close to identical), it is mobile. When an element is "dead" (or a "skeleton"), mutations can occur independently in its two LTRs so that they divert extensively. Thus, when two LTRs of the same element differ extensively from one another, it is assumed that this element has not moved for a long time. This approach is based on the assumption that a transposing element is 100% symmetric: both LTR have the same nucleotide sequence. Is there good evidence for this? According to this criterion only *Cer1* is an "old" and static element, other *Cer* elements have been recently transposed. The *Cer* elements were found to have a unique NC (nucleo capsid) domain in their Gag region: this unique CCHC motif ($Cys-X_2-Cys-X_4-His-Cys$) is thought to bind RNA. The CCHC of several *Cer* elements diverges from the norm. Also, the sequence of the RT of *Cer* is rather unique. Because the RT is rather conserved in all LTR retrotransposons, it was given a prime position in the building of phylogenetic trees. Analyzing the RT of *Cer* revealed a surprise: some *Cer* elements (*Cer1–6*) group with Ty3 and *gypsy* while others (*Cer7–12*) group in a different cluster of LTR retrotransposons to which *Tas* and *Pao* belong. Finally, there is another noteworthy feature that points towards an interesting evolutionary possibility. The *Cer7* has a very unique *env* code and protein structure. Its structure is very close to specific vertebrate retroviruses. This, in the authors' opinion, indicates that vertebrate retroviruses evolved very early in metazoan evolution.

An LTR retrotransposon, named *Gulliver* was reported by Laha et al. (2001) in the genome of the human blood fluke *Schistosoma japonicum*. Its full length is

4,788 bp and it is flanked by LTRs of 259 bp. It has coding sequences for Gag and Pol proteins and by various criteria seems to be close to the Ty3/*gypsy* LTR retrotransposons. As there may be many copies of *Gulliver* in each genome of *S. japonicum* and additional LTR and non-LTR retrotransposons are probably inhabiting this blood fluke, the authors believe that retrotransposons occupy a considerable fraction of the genome of this organism. Being aware of what happened to our own genome (see below when we deal with non-LTR retrotransposons of man), we should not be surprised. The increase of T.E. numbers in a given genome can be compared to a snow avalanche. The increase has a tendency to be exponential. The assumed reason is that in a very "condensed" genome such as the relatively small genome of the plant *Arabidopsis*, the genes are packed densely to each other. In such a situation an incoming T.E. has a great chance to cause a fatal mutation. If the genome is huge as in the genome of maize or the genomes of a member of the *Lily* genus, the genes are separated from each other and thus an incoming T.E. has only a small chance of causing a fatal mutation. The more T.E. enter into the genome the more the genes are further spaced. This is obviously a simplification of the actual situation because T.E. may have a preference of integration into certain sites. On the other hand, some T.E. integrate into existing T.E. sequences and by that promote the "avalanche" trend. We shall come back to this issue when we shall deal with non-LTR retrotransposons, as these elements seem to be responsible for huge differences in genome size between genomes of related organisms.

The human liver fluke *Chlonorchis sinensis* is a human parasitic trematode of the *Platyhelminthes* phylum. This liver fluke is an important parasite in East Asia, including Korea. A team of investigators in Korea (Bae et al., 2001) searched for LTR retrotransposons in *C. sinensis* ... and found an active one: *CsRn1* (*Chlonorchis sinensis* retrotransposon 1). There were several, very similar *CsRn1* elements in each genome of the investigated flukes and they were found to have the "classical" basic structure. The length of *CsRn1* is 5,026 bp and its LTRs are 471 bp. It is bounded by direct repeats of four nucleotides (target-site duplications, TSDs). The LTRs are flanked by short inverted repeats of 3 bp (5'-TGT...ACA-3'). A poly (A) signal was found in the right LTR but no TATA box could be identified. A likely primer binding site (with a sequence as in tRNA^{Try}) was also revealed. Also, a priming site for the second cDNA strand (a poly-purine tract, PPT) was identified upstream of the 3' LTR. A single ORF was revealed and the deduced amino acids indicated the Ty3/*gypsy* order of Pol proteins: PR, RT, RH and IN. No code for an *env* protein was found. The derived Gag protein has an unusual CCHC motif (CHCC). There were about 100 *CsRn1* elements per haploid genome that seemed to be randomly distributed. The investigators isolated 14 full-size elements and found that they were similar but not identical. Using changes in individual nucleotides, a phylogenetic tree could be constructed to indicate which element was the progenitor of the others. Although the levels of mRNA of *CsRn1* was rather low, it was positively identified and indicated that *CsRn1* is an active group of T.E. elements. When phylogeny was based on Pol amino acid sequences it was found that *CsRN1* closest LTR retrotransposon is an element called *Kabuki* from *Bombyx mori*.

2.14.3. LTR Retrotransposons in Echinoids
R.J. Britten, E.H. Davidson, M.S. Springer and their associates of California (USA) were interested in T.E. of sea urchins for many years (see: Springer et al.,

1995, for literature). They focused on one group of LTR retroviral-like elements and termed them *SURL* (*sea urchin retrotransposon*) elements. Basically the *SURL*s of sea urchins show considerable intraspecific diversity. Springer et al. (1995) thus divided them, somewhat arbitrarily, into groups. *SURL*s that diverted, by the sequence of their derived amino acids of RT, by 0–10% were put into one group. Each sea urchin species contains *SURL* elements of different groups. The data collected served to investigate phylogenetic questions. The phylogeny of Euechinoidea was worked out in detail, especially in *Echinacea*. It was estimated that the *Echinacea* started to divert about 250 million YBP. Thus, the genomes of species representing different super-orders, orders, families and genera could be included in the study. Practically, the authors studied 10 representative species and looked at the diversions of the species genomes relative to the diversity of the *SURL* elements that are found in these genomes.

A further study of the same kind was performed by Gonzalez and Lession (1999) who used 40, rather than 10 sea urchin species. One of the conclusions from these studies, which were based on almost identical *SURL*s in two unrelated sea urchin taxa, is that horizontal transfer of *SURL* occurred. Gonzalez and Lession (1999) suggested that although *SURL*s are mostly transmitted vertically, there are four cases of horizontal transfer between sea urchin species. In one case, *SURL* elements with identical RT sequences were found in sea urchin species that have been separated for at least 3 million YBP.

2.14.4. The Emergence of the Tail and a Rich Repertoire of T.E. in the Urochordate Ciona intestinalis

For those who remember their basic zoology lessons it will not come as a surprise that we humans are bound by our tails to the very early evolution of chordates. In other words, the dorsal nervous system and the tail, including the a notochord, appeared at the junction when Urochordates were separated from Cephalochordates and Vertebrates. The Urochordates consist of several orders, among them the *Ascidians*. The latter gained much attention as representatives of Chordates that are not Vertebrates (Figure 36). M.W. Simmen, A. Bird and their associates looked at the T.E. of the ascidian *Ciona intestinalis* (a sea squirt). In a survey of several T.E. (Simmen et al., 1999), they already found "treasure". In this small urochordate that contains only about 15,000 genes there are about 75 copies of an LTR retrotransposon; about 17,000 copies of miniature terminal-inverted-repeat (*SINE*) elements, and 40,000 copies of a composite tRNA-derived *SINE*.

In a further study of this team (Simmen and Bird, 2000) the investigations focused on specific T.E. One of these was the Ty3/*gypsy*-type LTR retrotransposon, that was termed *Cigr*. Here were no surprises: the Cigr elements have an amazing similarity to *gypsy* and Ty3. the *Cigr* had a total length of 4,226 bp and the two terminal repeats (LTRs) consisted of 245 bp. There is an ORF encoding 1,209 putative amino acids that indicates *gag* and Pol genes (but no genes for *env* proteins). The order of the putative Pol enzymes is PR, RT, RH and IN; thus the same order as in Ty3/*gypsy*-type elements. Although the mobility of *Cigr* elements was not traced individually, the authors found indications that *Cigr* elements are mobile or were mobile until recently. One indication was that the location of the very same *Cigr* element, within the genomes of individual *C. intestinalis* organisms, is different. The estimate of the number of copies of one type (*Cigr-1*) was 75 per genome.

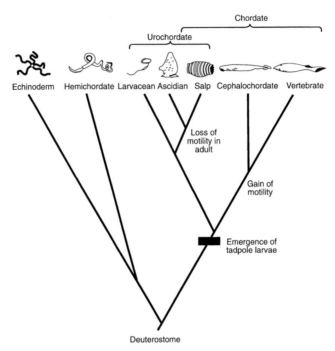

Figure 36. Deuterostome phylogeny inferred from cladistic and molecular analysis. The tree suggests that deuterostomes are a monophyletic group and that urochordates are the sister group of the cephalochordate/vertebrate line. (From Satoh and Jeffrey, 1995).

However, there are additional *Cigr* elements in *C. intestinalis*. Phylogenetic studies were conducted to obtain an answer to the question of whether the Ty3/*gypsy*-type element was transmitted horizontally. The authors could not provide a clear-cut answer but their results suggested two possibilities. One is that the putative horizontal transmission event took place *after* the divergence of the ascidians from the protovertebrates. The other possibility is that a *suchi*-like element does exist in the *Ciona* genome but has not yet been revealed.

2.14.5. LTR Retrotransposons in Amphibia
This will be a short paragraph because relatively little information on LTR retrotransposons exists on this group of animals. One study is noteworthy (Marracci et al., 1996). Marracci et al. focused on the terrestrial salamander *Hydromantes*. They intended to obtain an answer to the huge genome sizes of some amphibian organisms. Could this have resulted from extensive multiplication of T.E. elements? The choice of *Hydromantes* seems to be appropriate because it contains the largest known genome among all terrestrial vertebrates. One long repetitive fragment (*Hsr1*) of 5 kbp was revealed in several species of *Hydromantes*. The derived amino acid sequences of a putative RT protein, as well as some other features of *Hsr1*, indicated that it can be grouped with the Ty3/*gypsy*-type

retrotransposons. About one million copies of this *Hsr1* were found per *Hydromantes* genome. This is the highest number of copies so far (until 1996) discovered for retrotransposon-like elements in eukaryotic organisms. As we shall see below, the very large *Lily* genome also contains a very large number of a specific T.E. While such correlations between number of T.E. copies and genome sizes are impressive, they are not yet proof of a causal relation.

2.15. Fishing for LTR Retrotransposons in Fishes

The idea of studying the genome structure of fishes was initiated by S. Brenner and associates (Brenner et al., 1993; Elgar et al., 1996). These authors looked for an organism that would have a much smaller genome size than the human genome but with a similar number of genes. They therefore focused on the tetraodontoid fish *Fugu rubripes*, that is the pufferfish, nicknamed *Fugu*. They estimated *Fugu* to have a genome of about 400 Mbp of which more than 90% should be "unique" sequences.

The authors indicated that *Fugu* has a similar number of genes as humans, and the intron arrays in their genes are similar or identical to comparable arrays in human genes. But, the genome of *Fugu* is 7.5 times smaller than the human genome. *Fugu* should therefore be a very good model for the discovery of human genes. Well, only seven to eight years have passed since this claim by the investigators of the MRC Molecular Genetics Unit in Cambridge, UK; and by now, due to the hard work of investigators in several laboratories, the human genome has been totally sequenced. The leaders of the Human Genome Project, who were encouraged by S. Brenner himself, preferred the human genome itself over a fishy model.

2.15.1. The Sushi Family of LTR Retrotransposons from the Fugu Fish
As indicated above, S. Brenner and associates (Brenner et al., 1993; Elgar et al., 1996) chose the *Fugu* fish for comparative genome analysis. These authors gave several good reasons for their choice of the *Fugu* fish but did not stress the fact that this fish is a symbolic kind of a fish in Japan because it serves for the production of that Japanese delicacy – *Sushi*. Two investigators in New Zealand (Poulter and Butler, 1998) received cosmids from the *Fugu* project in Cambridge as well as sequence data and started fishing for *Fugu* T.E. Their first screenings and comparisons with full-length and corrupted retrotransposons from other organisms indicated that *Fugu* DNA contained sequences that were more similar to the *MAGGY* LTR retroelement (from the fungus *Magnaporte grisea*) than to another LTR retrotransposon, the *Mag* element (from *Bombyx mori*). After some further screening, the authors focused on a genomic region in *Fugu* that contains the IgM immunoglobulin heavy-chain variable domain. As this region was a subject of intensive study in the MRC laboratory in Cambridge, the relevant sequences of this region were readily available.

Poulter and Butler detected two similar, but not identical, retrotransposon sequences in the IgM region. They termed them *sushi-ichi* and *sushi-ni*. It was then found that *sushi-ni* is not an intact retrotransposon in the *Fugu* genome. Further work was focused on *sushi-ichi*, and ultimately the complete sequence of this retrotransposon was obtained. The *sushi-ichi* element emerged as a 5,645 bp LTR retrotransposon. Its LTRs start with TG and end with CA. The derived internal sequence contained an ATG initiation triplet, just distal of the left LTR and a TAA stop triplet just proximal to the right LTR. There are two long ORFs. The ORF1 (*gag*) includes the CCHC motif (mentioned above) that is assumed to bind RNA. There is a –1 phase shift between ORF1 and ORF2. ORF2 is uninterrupted and encodes the usual pol proteins (PR, RT, RNaseH, IN) in the order known from the Ty3/*gypsy*-like elements (and retroviruses). No code for *env* was revealed. The *sushi-ichi* that was sequenced cannot move autonomously because it contains a stop in the *gag* motif. There may be active elements of the *sushi*-type that could mobilize *sushi-ichi*. The *Fugu* fish has additional surprises. Dalle Nogare et al. (2002) fished from its genome another, novel-type element that they termed *Xena*. It has unique LTRs, with some similarities to *Penelope* of *Drosophila*. A phylogenetic analysis put *Xena* on a separate branch – away from *copia*, *gypsy*, hepadnaviruses, caulimoviruses and retroviruses.

2.15.2. Fishing for LTR Retrotransposons in Herring, Salmon and Xiphophorus

The detection of LTR retrotransposons in diverse fishes was achieved by "fishing". Thus Britten et al. (1995) looked for Ty3/*gypsy*-type retrotransposons in various marine animals, mostly invertebrates but also in the herring (*Clupea pallasi*). Their strategy was to rely on a conserved region of the sequence encoding the RT of Ty3/*gypsy*-type retrotransposon. Using degenerate primers and templates that were genomic DNA from the studied organisms, they obtained DNA fragments that were sequenced and compared to known or to other sequences of this RT region. Such "fishing" could identify probable retroelements. Thus, the authors did identify a probable Ty3/*gypsy* element in herring and termed it *Cpr1*. Hence, there is a name but the complete structure was not provided. Incidentally, "fishing" for retrotransposons in herring is especially easy. One has to have the appropriate PCR-generated probe and utilizing this, look at hybridization in herring DNA. Herring DNA can be obtained "off the shelf" as herring sperm DNA is readily obtained commercially. In fact, as was noted above in the Historical Background, fish-sperm was originally a source of the nuclein of Friedrich Miescher of Basel, in the 19th century.

A similar approach, of using amplified fragments of fish DNA, to isolate certain sequences of probable LTR retrotransposons, was employed by Tristem et al. (1995) of the Imperial College and Cambridge University, UK. The latter authors used a somewhat different strategy to build the primers for PCR amplification than the strategy used by Britten et al. (1995). Tristem et al. looked at the genomic DNA of several salmonid genera (*Oncorhynchus*, *Salmo* and *Salvelinus*). In each of these genera two to six species were screened. The authors revealed a LTR retrotransposon group that had similarity to the Ty3/*gypsy*-type elements and termed this group *Easel*. It was suggested that *Easel* is a phylogenetically basal member of the Ty3/*gypsy*-type LTR retrotransposons. *Easel* occurred in some of the same fish species from which retroviruses have previously been isolated. This means that the Salmonidae fish may be the "first" organisms known to harbor the retroviral branch elements and the *gypsy* LTR retrotransposons branch elements. This group of fishes may thus be subjected to further studies to investigate the validity of the claim of H.M. Temin (a discoverer of the reverse transcriptase) that retroviruses were derived from retrotransposons. The methods used by Tristem et al. (1995) provide convincing evidence for the existence of the *Easel* element in Salmonidae and clearly relate this element to *gypsy*; these authors did not detail the structure of *Easel* or its mobility.

After *Sushi* was isolated and characterized by Poulter and Butler (1998) as noted above, a German team from the University of Würzburg (Volff et al., 2001a) looked for Ty3/*gypsy*-like elements from a different fish. Their fish was *Xiphophorus maculatus* (Teleostei; Poecillidae). The choice of these investigators was because the genus *Xiphophorus* is an established model for cancer research, for the analysis of sex determination and many questions of evolutionary biology. Probably, this genus was readily available in the laboratory of these authors. It should be noted that in this genus, overexpression of the receptor tyrosine kinase oncogene *Xmrk* is responsible for hereditary melanoma. The *Xmrk* and the oncogene are located in the subtelomeric region of the X and Y sex chromosomes in the platyfish *Xiphophorus maculatus*. Several repetitive DNA sequences are also nearby, so these regions were screened for the LTR retrotransposons. Indeed, Volff et al. (2001a) isolated a complete LTR retrotransposon from the platyfish and termed it *Jule*. The element has a length of 4.8 kbp and it is flanked by two LTRs of 202 bp. The element encodes Gag proteins and Pol proteins but not *env* proteins. The Pol protein order is as in *gypsy*: PR, RT, RNaseH and IN. There were three to four *Jule* elements per haploid genome and two of them were located in a subtelomeric region of the sex chromosomes and associated with *Xmrk*. *Jule*-related elements were detected in several other fishes. *Jule* seemed to be more closely related to the *Mag* elements of silkworms, than to other elements (e.g. *Maggy*, *Sushi*, *Cer1*, *Ulysses*) but it is quite close to the SURL elements of sea urchins.

The same team of investigators from the University of Würzburg (Volff et al., 2001b,c) also turned to other fishes, such as the medakafish (*Orysias latipes*) and to pufferfishes, to look for additional LTR retrotransposons. They were searching for *Penelope*-like elements (*Penelope* was originally isolated from the fly *Drosophila virilis*). Two such elements were revealed and the authors selected two names: *Neptune* and *Poseidon*, from Roman and Greek mythology, respectively. Actually, they represent the same sea-god and as such probably seemed appropriate to the authors as names for mobile elements in fishes. Volff et al. (2001b) recalled that there are many fishes in the waters: more than half of the 48,000 vertebrate species are fishes. Thus, they went fishing for an insect LTR retrotransposon in fishes. One should also recall that there are about 40 times more insect species than fish species. The *Penelope* element has similarity to the Ty3/*gypsy* element but has also sufficient differences (in its RT) so that regular screening did not detect *Penelope* in other *Drosophila* species (e.g. *D. melanogaster*). A partial element found in *O. latipes* belongs to the *Poseidon* group. While Evgen'ev et al. (1997) identified the classic DD35E domain in the IN of *Penelope*, Volff et al. (2001b) did not find this motif in either *Penelope* or the *Penelope*-like fish elements (*Poseidon* and *Neptune*).

Instead they found similarity between the derived amino acids, downstream from the RT, in the *Penelope*-like elements and an endonuclease type termed *Uri*. The *Uri* GIY-YIG domain was claimed to induce DNA double-strand breaks and it was previously found to be encoded in some group I introns. On the basis of available derived amino acid sequences of the RT, a phylogenetic tree was constructed. In this tree the *Penelope*-like elements form a separate branch and *Poseidon* is closer to *Penelope* than to *Neptune* (Figure 37). The authors also suggested that if no horizontal transfer occurred, *Poseidon* and *Neptune* lineages might have diverged before the separation of the arthropod and the vertebrate lineages. That means more

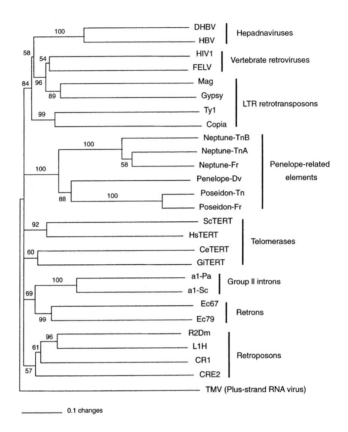

*Figure 37. Phylogenetic position of Penelope-related retroposons.
Phylogenetic analysis was performed using the conserved domains of the reverse
transcriptase according to Xiong and Eickbush (1990). The tree has been rooted
in the Tobacco Mosaic Virus sequence (TMV) plus-strand RNA virus according
to Xiong and Eickbush (1990). (From Volff et al., 2001b).*

than 600 million YBP. Well, some screening of insect species could substantiate (or refute) this suggestion. Actually, both *Poseidon* and *Neptune* are not yet fully sequenced. Moreover, many of the elements that were detected in the DNA of fish genomes were truncated at their 5' end, and the truncation varied considerably. Are *Poseidon* and *Neptune* really intact elements and even if they are, are they still mobile? There are as yet no answers to these questions.

2.16. LTR Retrotransposons in Plants

We detailed above the term LTR retrotransposons but what is our definition of plants? For this section the definition is simple and broad: plants will include algae, liverworts (Bryophyta), ferns (Pteridophyta) gymnosperms and angiosperms. An actual list of plant species in which one type of LTR retrotransposons (the Ty1/*gypsy*-type) has been reported, was already provided by Flavel et al. (1992a,b,

1994). The first reported LTR retrotransposon from plants was probably *Cin1* (Shepherd, 1984) that was identified in maize; actually in the same laboratory that started the molecular genetic work on the Class II T.E. in maize (at the Max-Planck-Institute in Köln, Germany). Thus, only 8 years later Flavel et al. came with their list of LTR retrotransposons that included 56 plant species, all attributed to the Ty1/*copia*-type of T.E. The number of species does not represent the full wealth of elements because in some plant species, such as in maize and in *Arabidopsis thaliana*, numerous LTR retrotransposon families have been revealed. As was already noted above, revealing T.E, and especially LTR retrotransposons, is a simple procedure. Hence, with the additional genomic sequences that will become available to the public, additional T.E. will also be revealed. In short, the number of LTR retrotransposons and even LTR retrotransposon families is already so huge that it would be senseless to record them in this book. It is a fair guess that either or both Ty1/*copia*- and Ty3/*gypsy*-type elements exist in every angiosperm species. They probably also exist in many or even in all lower plant species. They have already been reported in algae (e.g. in *Volvox carteri*, Lindauer et al., 1993), although some of them may divert from the standard type of LTR retrotransposons (e.g. in *Chlamydomonas reinhardtii*, Day et al., 1998). We shall therefore focus on a limited number of plant species in which interesting information on LTR retrotransposons is available. We shall also handle certain specific issues of LTR retrotransposons in plants.

2.16.1. LTR Retrotransposons in Arabidopsis thaliana

A few years after the first Ty1/*copia* elements were revealed in maize, came the first report of such elements in the model plant *A. thaliana* (Voytas and Ausubel, 1988). These researchers already had in their hands sequenced fragments of the genome of *A. thaliana*. They started their search by using appropriate primers and the RFLP analysis to screen for Ty1/*copia*-type elements. After finding some truncated elements they also revealed full-size elements. Two such elements were given the name Ta1-3 (for transposable element of *Arabidopsis*). These elements had a total length of 5.2 kbp and an internal domain of 4,190 bp that was bounded by LTRs of 514 bp each. The insertion was flanked by 5 bp direct repeats and the terminal of the LTRs were 5'-TG...CA-3'. The Ta1-3 also had the typical priming sequence, adjacent to the 5' LTR, complementary to a sequence in the tRNA^{Met}. All these characteristics were indicative of a typical LTR retrotransposon. The order of IN-RT was as in the Ty1/*copia* elements. From their finding Voytas and Ausubel deduced that such elements will be found throughout the *A. thaliana* genome and that transposable element activity may be a major source of genetic variability in this species. We should recall that *A. thaliana* has the smallest (i.e. most condensed) genome among angiosperm species – it is less than a tenth, in size, of the maize genome. A further study by this team (Voytas et al., 1990) reported that six Ta1 element copies were found in three geographically diverse *A. thaliana* races. These six elements were rather similar to each other in their sequences and occupied distinct sites in the genome of this species. However, they could not be defined as "mobile elements" because they have probably lost the transposition capability. Their coding sequences were altered so that the proteins required for transposition were rendered non-functional. The sequencing of these elements suggested that Ta1-1 predated the global dispersal of *A. thaliana*. Only after the spread throughout the world were the two elements Ta1-2 and Ta1-3 added.

From families of Ta we move to a "superfamily". Further studies of Ta in *A. thaliana* revealed (Konieczny et al., 1991) two more types of elements, Ta2 and Ta3. By the derived amino acids of their RT, the three elements (Ta1, Ta2 and Ta3) have about 75% identity. Seven more Ta were then recovered (Ta4 to Ta10). The authors calculated that the members of this Ta superfamily comprise about 0.1% of the total *A. thaliana* genome.

Further studies (e.g. Peleman et al., 1991; Pelissier et al., 1995, 1996) indicated that additional, truncated retroelements exist in *A. thaliana*. Each of these elements was found in moderate numbers, unlike in other plant species where the number of certain retroelements exceeds 100. Thus, the element *Athila* is flanked by typical LTRs but its middle region does not encode the Gag or the Pol proteins that are required for autonomous transposition. Why this element, which does not have a mobility capability, was given the name *Athila* was not explained by the authors.

The Tat1 is a small element that is also lacking coding sequences. The latter is suggested by Wright and Voytas (1998) as a kind of *solo* LTR element; that means it contains the left LTR but is missing much of the right side of a regular LTR retrotransposon. Such a solo LTR can arise when the two LTRs of an integrated retrotransposon recombine, deleting (all or part of) the internal region and leaving behind a single LTR that is flanked by a target-site duplication. By analyzing in detail several Tat elements in *A. thaliana* the authors suggested that the Tat elements are close to the Ty3/*gypsy*-type LTR retrotransposons. Since some elements of this type also encode *env* proteins, the Tat elements were considered retrovirus-like. This points to the possibility that in the past *A. thaliana* harbored infectious LTR retrotransposons.

The suggestion that Ty3/*gypsy*-type elements did exist, at least in the past, and were integrated in *A. thaliana* is supported by a previous study of Knoop et al. (1996). The latter investigators studied the mitochondrial genome rather than the nuclear genome of *A. thaliana*. These investigators had much experience with mitochondrial molecular genetics and, specifically, much of the DNA sequence of the mitochondrial genome of *A. thaliana* was in their possession. The transport of nuclear sequences into the mitochondrial genome is well documented. Therefore, Knoop et al. (1996) could handle the question of whether retroelements were transferred from the nuclear genome into the mitochondrial genomes. They used the appropriate tools to look for three kinds of elements: Ty1/*copia*-like, Ty3/*gypsy*-like and non-LTR retrotransposons. They did find "remnants" of all these three types of elements. Nine retrotransposon-derived fragments were revealed. These ranged from 67 to 420 codons. Sequences with similarities to the three types of nuclear retroelements were scattered on the mitochondrial genome, and were found in intergeneric regions at various distances from coding sequences. Five out of the nine locations on the mitochondrial genome indicated remnants having similarity to the Ty1/*copia*-type elements and the other 3 locations had sequences indicative of remnants of Ty3/*gypsy*-type elements. The details are quite surprising. One of the nine regions, RT5, showed highest similarity to the Ty3/*gypsy* of yeast while another region, RT6, had greater similarity to the Ty3/*gypsy* of *Drosophila*. The third region, RT9, had the greatest similarity to a LTR retroelement of the fungus *Fusarium oxysporum*. Moreover, one of the mitochondrium-located (truncated) elements had greater similarity (as evaluated by the derived amino acid sequence) to an element of maize than to any of the known nuclear elements of *A. thaliana*. In total the authors estimated that about 5% of the mitochondrial genome of *A.*

thaliana consists of "skeletons" of retrotransposons. It is suggested by the authors that numerous independent transfer-insertion events occurred in the evolutionary history of the *A. thaliana* mitochondrial genome. Is that specific to *A. thaliana* mitochondria or is the situation similar in the mitochondria of other angiosperms? As yet, this is an open question.

In a more recent investigation Wright and Voytas (2001) studied the *Athila* (probably referring to Attila, the King of the Huns) retrotransposons in detail. They detected what seemed to be intact members of the *Athila* family in *A. thaliana* that are 12–14 kbp in length and have the coding regions of retrovirus-like ORFs, including the *env*-like ORF. Transcription was analyzed in an *A. thaliana* strain with decreased DNA methylation. The expression of *env* protein could render *Athila* a potential invader of other species and cause horizontal transfer of this retrotransposon. The association between *Athila* and the King of the Huns thus becomes rather obvious. *A. thaliana* had a more effective defense than the Eastern and the Western Roman Empires and after initial invasions into the genome of *A. thaliana*, *Athila* was stopped before causing too much damage.

Since the early years of regenerating functional plants from *in vitro* cultures of explants, there have been reports that this culture causes variability in the derived plants. In some reports the variability was restricted to the first generation of the regenerated plants, while other reports claimed that this variability was not sexually inherited. This variability was of special concern to commercial propagation of (mostly ornamental and forest) plants where a uniform product was required. The dispute was going on specifically in cereals such as wheat. A group of investigators insisted that they had evidence for variation of wheat regenerated from tissue culture and even termed this phenomenon *Somaclonal Variation*. The author of this book was involved in this dispute, claiming that in his laboratory regenerated wheat plants had no somaclonal variation (Breiman et al., 1989). We used embryoids that developed from the scutellum of immature grains for regenerating wheat plants.

In later studies, by many investigators, it appeared that when plants were regenerated from *in vitro* cultured calli these plants showed somaclonal variation. What has this "story" to do with LRT retrotransposons in *A. thaliana*? A recent investigation shows such a connection! It was already known that tissue culture (and other stresses) strongly enhances the transcription of LTR retrotransposons (e.g. Mhiri et al., 1997, and see the review of Grandbastien, 1998). A team of investigators (Courtial et al., 2001) exploited the fact that genetic transformation of *A. thaliana* can be performed by two very different procedures. In one procedure root explants of this plant are co-cultivated with *Agrobacterium tumefaciens*. The agrobacteria contain a binary vector for genetic transformation; this means they contain a helper plasmid (encoding *vir* genes) and another plasmid that contains the engineered T-DNA that includes the transgene (the transformation vector).

Courtial et al. (2001) used agrobacteria that harbored a vector that contained the LTR retrotransposon *Tnt1* of tobacco. This is an element that is active in tobacco. When the transformation was done by the root-cocultivation procedure, the authors regenerated 35 plants from independent explants. The transgenic *A. thaliana* plants that were regenerated showed numerous *Tnt1* elements; up to 25 such elements in one genome. Moreover, the transcription of *Tnt1* was already apparent in the tissue culture two days after the start of *in vitro* culture.

The other procedure to transform *A. thaliana* is termed the *in planta* method. In this method there is no tissue culture stage. The inflorescence of the plant is dipped into a suspension of agrobacteria and by operating a vacuum the suspension is forced into the inflorescence. The plants are then grown to maturity and the resulting seeds (from self-pollination) are replanted. Due to marker genes, the transgenic plants can be recognized. When agrobacteria containing the plasmid with *Tnt1* were used for *in planta* transformation of *A. thaliana*, the resulting transgenic plants did not show *Tnt1* activity.

These results provide a clear indication that tissue culture activates *Tnt1* and that without a tissue culture phase there is no activation of this element. Whether or not this is a general phenomenon and thus somaclonal variation is caused by the activation of the LTR retrotransposon, seems to be a possibility that requires further evidence.

Hirochika et al. (2000) exploited the features of *A. thaliana* as a model for genetic manipulation and searched for the activation and inactivation of a tobacco retrotransposon, *Tto1*, that was introduced by genetic transformation into the *A. thaliana* genome. They reported that the transposition and transcription of *Tto1* was affected by the state of DNA methylation. They changed this state by introducing the *ddm1* gene. This gene, when homozygous, will reduce DNA methylation. This potentially active tobacco retrotransposon (*Tto1*) was recommended by the same team of investigators (Okamoto and Hirochika, 2000) to serve in insertion mutagenesis because it seems that *Tto1* has no preferred insertion sites and can insert in various *A. thaliana* genes.

2.16.2. LTR Retrotransposons of Maize
The same laboratory that was active in the early molecular studies of the Class II T.E. elements of maize (e.g. *Ac/Ds, Spm*) started to investigate the "middle-repetitive" sequences in this plant (Shepherd et al., 1982, 1984; Gupta et al., 1983). These authors found that a DNA sequence was present in some maize lines but absent in others. As their colleagues in Moscow, these authors chose a mythological name for the repetitive sequence. They termed it *Cin1* after the young maize god Cinteotl. The *Cin1* was absent from a wild relative of maize: *Teosinte guerrero*. The ends of the *Cin1* element (5'-TGTTGG...CCAACA-3') indicated repeated terminal ends, similar to those earlier found in the LTR retrotransposons. The authors also found other indications that *Cin1* is a truncated LTR retrotransposon; it contained a signal for the initiation of RNA transcription and a signal for the termination of RNA transcription. In those maize lines where *Cin1* was detected it was found in many copies, appearing as a repetitive sequence. Still these investigators did not reveal a full-size sequence of *Cin1* that was similar to LTR retrotransposons found previously in yeast. Such full-size elements were, soon thereafter, found in maize by another team (Johns et al., 1985) in the USA.

The "history" of the latter element, *Bs1*, is very different from that of *Cin1*. It was revealed in a progeny of a maize plant that was infected with the RNA of barley stripe mosaic virus (BSMV). Such an infection was previously reported to be mutagenic. Indeed *Bs1* appeared as an insertion mutation in the *Adh1* gene. This gene should be remembered! The *Bs1* appeared only twice in each plant. When *Bs1* was sequenced, direct repeats of 304 bp were found flanking a central region and the whole element had similarities to the *Ty1* of yeast and to *copia* of *Drosophila*. Incidentally, once *Bs1* is inserted into the *Adh1* gene, the pollen of this plant is resistant to allyl-alcohol. This procedure to select for pollen grains with *Adh* mutations was invented years ago by Drew Schwartz. There are not many such efficient selection procedures of mutations by the application of chemicals that are toxic only in the wild-type.

Another procedure is the application of chlorate for the selection of nitrate-reductase mutations. This latter procedure was developed for the fungal genus *Aspergillus*, but also works in plants. It seemed that the first inserted *Bs1* underwent replicative transposition so that an additional *Bs1* element was inserted in the *Adh1* region. We should not be surprised because we know that twenty or more retrotransposons can be found in the *Adh1* region of maize (SanMiguel and Bennetzen, 1998) by a "nested" insertion process that shall be dealt with below. Contrary to *Cin1*, all analyzed maize lines had one to five *Bs1* elements and all elements were intact rather than truncated. What was the role of BSMV? There was no definite answer to this question but it could be that the infection induced a stress that activated *Bs1*. Interestingly, when the research on *Bs1* was published (Johns et al., 1985) the authors could still state: "*Bs1* is unique among plant transposable elements in being bounded by long direct repeats". Because in 1985 all other known plant T.E. were of Class II, thus with inverted repeats. Soon a "shower" of LTR retrotransposons of plants started to appear in the literature.

Maize genes seem to be a treasure trove of T.E. Hence, S.R. Wessler and her associates (Varagona et al., 1992) were focusing for several years on the *waxy* (*Wx*) gene-locus of maize. *Wx* has advantages for the investigation of mutants. The wild-type allele produces ADP-glucose-glucosyl transferase that is required for the synthesis of amylose. The presence of the active gene can easily be detected in maize grains – by observing whether or not the mature grains are waxy. Also the endosperm and even the pollen grains of maize can be stained by a solution of I/KI. In the presence of amylose a blue-black color is visible. Because amylose is required only at specific times of the development of the maize plant and only in specific tissues (e.g. endosperm, pollen grains) *Wx* gene expression is normally under spatial and temporal regulatory control. Any diversion from this control can be swiftly recognized. Moreover, even if the gene is completely inactivated, this does not lead to lethality.

Mutations in the *Wx* gene were analyzed by Wessler and her associates and three of these mutations had to do with LTR retrotransposons: *Stonor*, *G* and *B5*. Two of these, *B5* and *G* were found in introns of the *Wx* gene (in intron 2 and 8, respectively). The third element, *Stonor* was inserted just at the border between exon 6 and its upstream intron (intron 5). As can be seen in Figure 38, the insertion of an LTR retrotransposon into an intron of *Wx* causes a vast increase in the distance between the exons, and consequently, the splicing may be affected. The consequences of the insertions of the three LTR retrotransposons were analyzed in detail and a phenomenon termed *Alternative Splicing* was revealed. The transcript size of the *Wx* gene seemed to be altered (extended) only after the insertion of *Stonor*, while after the insertion of *B5* or *G* normal-size transcripts were found among the alternatively spliced transcripts. The inserted elements had the features of LTR retrotransposons. Another element was revealed by Wessler and associates (Purugganan and Wessler, 1994) and it was called *Magellan*. It was found in the *Wx* locus of a maize plant and had the characteristics of a Ty3/*gypsy*-type element. The element that was first detected had the 3' LTR (of 341 bp) but it lacked the 5' LTR. Thus, this specific *Magellan* was "anchored" to the *Wx* locus and could not "sail" anywhere. The sequence of this *Magellan* revealed other *Magellans* that were intact and located in other sites in the maize genome.

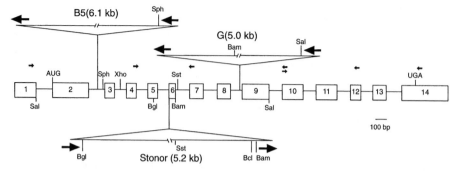

Figure 38. The position and orientation of the retrotransposon with respect to the exons and introns of the wx gene. The limits of the Wx protein are indicated by the translation start and stop codons. The positions of relevant restriction sites and oligonucleotide primers used for PCR amplification (small horizontal arrows) are also noted. Exons, open boxes; introns, connecting lines. (From Varagona et al., 1992).

Interestingly, *Magellan* was found not only in *Zea mays* but also in the related *Tripsacum* genus, but there only in *T. andersonii*. There was a considerable variability between the sequences of individual *Magellan* elements. The variability in *Magellan* sequences was considerably greater than variability in genomic sequences (of *Zea* and *T. andersonii* genomes). This should not come as a surprise to us since we already mentioned that the number of "mistakes" during reverse transcription is much higher than during DNA replication. Thus, with each transposition (that involves reverse transcription) there is a chance for an increase in variability.

A few hundred miles north of the laboratory of S.R. Wessler, in Athens, Georgia, the search was going on for LTR retrotransposons of maize in the laboratory of J. Messing, at the Waksman Institute of Rutgers, the New Jersey State University (Hu et al., 1995). Hu et al. were analyzing a tandemly duplicated, 27 kDa maize storage-protein locus. This locus can undergo rearrangement and in this rearrangement an element was found that had two long terminal repeats, a primer binding site, a polypurine tract and a *gag*-related ORF. In short – an LTR retroelement. The element, termed *Zeon-1*, was then found to be a middle-repetitive LTR retrotransposon of 7,313 bp, hence a member of a family of *Zeon* elements. The *Zeon* elements were detected in maize and teosinte but not in other species of the Gramineae family such as wheat, barley, sorghum and rye. Paradoxally, while wheat is considered the oldest cultivated crop, *Triticum* is one of the youngest genera; wheat branched off from barley only about 10 million YBP. While *Zeon-1* has an ORF for *gag*, it lacks the codes for the *pol* proteins, thus it is not autonomous. From the sequences of *Zeon-1* that were available, this element seemed to be of the Ty3/*gypsy*-type. Is *Zeon-1* still mobile and if it is, what mobilizes it? These are still open questions.

When we move further north we reach the State University of New York at Albany. There J.P. Mascarenhas and associates have much experience with molecular-genetic studies of pollen grains. Focusing on pollen, Turcich et al. (1996) of the Mascarenhas laboratory identified a retroelement of maize that is specifically expressed in the early microspores but not in any other maize tissue. They called it *PREM-2* (pollen retroelement, maize-2). *PREM-2* was found to be of very respectable size: its total length is 9,439 bp, it has two LTRs of 1,307 bp each and its internal region is 6,825 bp. The internal region encodes all the proteins of a Ty1/*copia*-type retroelement (*gag*, proteinase, IN, RT and RNaseH). This laboratory previously revealed another LTR retroelement, which they named *PREM-1* (Turcich and Mascarenhas, 1994). While *PREM-1* also produced its transcript mainly during the early stage of microspore development, the transcript did not appear to code for any specific protein because the ORFs of *PREM-1* contained stop codons that abolished normal translation. This defective element was very abundant in the maize genome: the authors estimated that about 10,000 to 40,000 copies of *PREM-1* exist in the haploid genome. *PREM-2* was located upstream of the polygalacturonase genes. This gene family has 12 or 13 genes in maize. The authors (Turcich et al., 1996) could build a phylogenetic tree of these genes and then ask in which branches of this tree there is the *PREM-2* insert. In this way, they could estimate when in the evolution of maize the *PREM-2* was inserted near the polygalacturonase gene. The estimate was that it occurred before about 2–3 million YBP but not earlier than about 5 million YBP. Here we can see the advantage of LTR retrotransposons for phylogenetic studies, over other T.E. Many of the Class II elements move by a "cut and paste" process, this means that when they transpose the element leaves its previous insertion site and leaves behind only footprints (short, duplicated sequences). On the other hand, when LTR-retrotansposons are transposed they first produce a duplicate by the reverse transcription process, and the duplicate inserts into a new site. The original element stays in its original location.

A specially interesting situation of LTR retrotransposons in maize emerged from the studies of J.L. Bennetzen of the Purdue University in Indiana, a place that is surrounded by maize ("corn") fields (e.g. Bennetzen et al., 1993; Springer et al., 1994; Avramova et al., 1995; SanMiguel et al., 1996; 1998; SanMiguel and Bennetzen, 1998). These investigators focused on the *Adh1* gene locus. Their earlier interest was on *matrix attachment regions* (MARs) in this locus (Avramova et al., 1995, and see Avramova, 2002 for review). MARs are attributed with essential roles in attaching the nuclear DNA to the nuclear matrix and by that affect the control of transcription and other nuclear phenomena (reviewed in a previous book: Galun and Breiman, 1997). However, as the whole fragment of 280 kbp was cloned, and subsequently much of it was sequenced, very interesting features emerged. It became clear that the region contained many repetitive DNA sequences. It was then revealed that some repetitions appear as pairs of identical sequences that are spaced from each other.

Further analysis showed that these "pairs" are actually LTRs of potential retrotransposons. The complete sequencing of one such "pair" with the DNA between the pair revealed one complete LTR retrotransposon that was called *Opie-2*. *Opie-2* has a length of 8,987 bp with two LTRs of, respectively, 1,271 and 1,292 bp. It also has the other features of an LTR retrotransposon. Many other such elements were revealed in this region and these all received their respective names (e.g. *Huck, Kake, Fourf, Milt, Reina, Victim, Ji, Grande, Cinful*). Most of these were actually members of families. Thus there were *Opie-1, Opie-2, Opie-3* and *Opie-4*. The twenty elements revealed in this region accounted for 150 kbp out of the 240 kbp that were analyzed – meaning over 60 percent of the analyzed region. Most interestingly, the 20 elements were not dispersed linearly but were inserted

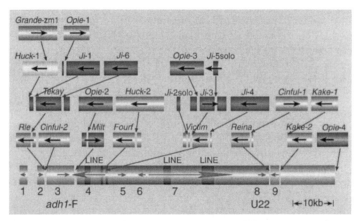

Figure 39. Nested insertions retrotransposons in the adh1-F region of the maize genome. (Courtesy of J.L. Bennetzen).

one into another. For example, into the *Victim* element the *Ji-2* element was inserted (as well as a number of additional full and truncated elements). Further, the element *Opie-3* was inserted into *Ji-3*. Thus, an elaborate nested integration emerged as shown in Figure 39. This is not an isolated case! This nested retroelement structure is apparently the standard organization of intergeneric regions in maize.

The number of repetitions in the whole maize genome could be huge. Thus, the members of the *Opie* element family have about 30,000 copies, constituting more than 10% of the 2,400 Mbp of the maize genome. When calculated, after taking into account other highly repetitive and middle repetitive elements, the retroelements constituted more than 50% of the maize genome (SanMiguel et al., 1996). Obviously, when elements are inserted into each other rather than in the coding sequences of essential genes, such integrations cause little (or no) damage to the plant.

The "filling" of the genome of at least some organisms (such as maize) with T.E operates in an apparently strange manner: the more T.E. elements are added to it the more the capacity of the genome to accept additional elements is increased. This is reminiscent of the "law of human memory", suggested with a smile, by Amiel Ben David Halevy (the pseudonym of the present author). This law says that if you add information to the human memory you make additional space for more information. This law refutes the claim of Sherlock Holmes (in a book by Sir Arthur Conan Doyle), who was told by Dr Watson that the planetary system is heliocentric (rather than the earth being in the center). The reaction of Sherlock Holmes was that this information is irrelevant to him and he should throw it out of his head right away, in order to avoid overfilling his brain with irrelevant information.

What happened in the maize genome was a kind of avalanche. The phenomenon is especially peculiar when we look at the genome of a related genus, the *Sorghum*. There is no such nested integration in the sorghum genome, which is less than a third of the maize genome in size. No such clustering was found in a region of *Sorghum* that is orthologous to the maize *Adh1* (SanMiguel and Bennetzen, 1998). The generae *Zea* and *Sorghum* separated approximately 16 million years ago. Examining the phylogeny of several LTR retrotransposons of maize, SanMiguel et al. (1998) estimated that they all were inserted within the last 6 million years; most of them actually only 3 or less million years ago. This "explosion" of retrotransposition probably caused the maize genome to increase in size from about 1,200 to 2,400 Mbp.

2.16.3. LTR Retrotransposons in Wheat, Barley and Oats

The term "*chromosome walking*" is used to describe a procedure to reach the coding sequence of a target gene after its genetic location has been roughly determined. A fragment that includes this location is isolated and the investigator gradually approaches the target sequence. "Walking" on chromosomes of a condensed genome, as in *Arabidopsis thaliana*, is not relatively tiring but walking on wheat chromosomes is obviously tiresome. The genome size of wheat is 1.6×10^{10} bp. Even when we divide this length by 21 we still have 800,000,000 bp, on average, in each chromosome. By the methodologies that were available in the early nineties of the 20th century such a search would require 2,000,000 sequencings because each sequence analysis would cover at most 400 nucleotides. This is why R. Flavell and his associates at the John Innes Centre in Norwich (Moore et al., 1991) stated for wheat: "These questions (the nature of the repetitive elements and their organization) will have to be addressed before the technology of "chromosomal walking" can be applied to these genomes (barley and wheats)". The reader may ask why should one handle at all such outstanding huge genomes as those of wheats? Some questions, and this question included, have only a historical, rather than a rational answer. These investigators were "mobilized" from the Plant Breeding Institute (PBI) near Cambridge, UK, to Colney Lane near Norwich. In the PBI, wheat and barley were the major crops under investigation. Indeed, major discoveries in wheat cytogenetics and great achievements in the breeding of wheat and barley came from the PBI. Thus the "problems" were mobilized with the investigators to the John Innes Centre near Norwich. Incidentally, the investigators grew their wheat plants in John Innes Compost. This compost was developed at the John Innes Horticultural Institute. During the 20th century "John Innes" wandered from place to place and changed names, however, the recipe of the compost was carried along all the way – unchanged.

In one study of Moore et al. (1991) of Norwich, UK, the investigators focused on a sequence that they termed *Bis1* (for barley insertion sequence 1). The length of *Bis1* was found to be 6.5 kbp and 10,000 *Bis1* repeats were estimated to exist in the barley genome. The authors thus inferred that about 5% of the barley genome is composed of *Bis1*. Other, *Bis1* related elements were also found in the genomes of wheat and rye. Elements similar to *Bis1* were found to be dispersed in the chromosomes of barley rather than clustered. By *in situ* hybridization with barley metaphase chromosomes a *Bis1* probe indicated that the signals were on all chromosomal arms of barley but were absent or reduced from the regions of the centromeres, telomeres and the linkers between the genes that code for rRNAs. This means that the *Bis1* favors euchromatin over heterochromatic chromosomal regions.

The same John Innes Centre team, in collaboration with other laboratories, studied also LTR retrotransposons of wheat (Moore et al., 1991; Lucas et al., 1992; Monte et al., 1995). Much of this work focused on WIS 2-1A. WIS 2-1A was first revealed from the *Glu-1A* locus that encodes a high-molecular-weight storage protein (glutenin) in wheat grains. This element had the features of a *copia*-type retroelement but it had lost its autonomous transposition. Its copy number in the hexaploid bread-wheat was estimated to be 200 (per "haploid"). Related elements were detected in other species of the Pooideae sub-family: barley (*Hordeum vulgare*), rye (*Secale cereale*), oat (*Avena sativa*), and in the genus of a wild relative of wheat, *Aegilops*. However, there was a high level of interspecific variation while the intraspecific variability was low.

From England we move to Japan. As noted above, important cytogenetic discoveries in wheat were obtained in the PBI. These discoveries were, in part, also made in parallel in Japan, where the study of wheat cytogenetics goes back to the twenties of the 20th century, pioneered by H. Kihara. Why study wheat in Japan is another question that has an "historical" answer. It was in Japan where the chromosome numbers of wheats were first correctly determined. Nevertheless, intensive wheat research goes on especially under the leadership of Koichiro Tsunewaki in Fukui, Japan. These investigators also studied LTR retrotransposon families in wheat (Matsuoka and Tsunewaki, 1996, 1999). They accumulated Ty1/*copia*-type elements from several grass species and analyzed 177 of them. The analysis was based on the sequences of the RT. On the basis of these sequence analyses they evaluated the RT domain divergence of the Ty1/*copia*-type retrotransposon families in grasses (i.e. the tribes Oryzeae, Paniceae, Andropogoneae, Chlorideae, Aveneae and Triticeae). The 177 elements were included in a phylogenetic tree that was based on RT nucleotide sequencing. The results indicated four "families" of elements. "Family" G1 was probably established before the radiation of the grass species, and it remained active until recently. The G2 "family" may have originated before the divergence of monocots and dicots, meaning before 200 million YBP. Members of this G2 "family" do not exist in the Pooideae sub-family of grasses. The G3 "family" was found only in *Oryza* (rice). The G4 "family" was found in various species of different grasses indicating that it was also established before the radiation of the Gramineae species. The authors assume that even in "families", where most elements lost the functionality of the RT, this function could be supplied by *trans* from elements that retained the RT function.

The retroelements of barley and wheat that were mentioned above were mostly of the Ty1/*copia*-type of LTR retrotransposon. Those analyzed were found dispersed on different chromosomal arms. An exception was reported by Presting et al. (1998). They used a DNA clone from the centromeric region of sorghum, *pSau* 3A9, to identify repetitive sequences in wheat, barley and rye. Indeed, they identified such a sequence in all three cereal species. Moreover, what they found were apparently complete retrotransposons of the Ty3/*gypsy*-type that were preferentially localized in the centromeric regions. As most of this study was performed with barley, the element was given the name *cerebra* (for *ce*ntromeric *re*troelement of *ba*rley). The *cerebra* was estimated to be repeated 200 times in the centromeres of barley chromosomes. Did the *cerebra* elements improve the functionality of the centromere of cereals or did they accumulate there because of other reasons (e.g. their integration there caused no harm, consequently there was no negative selection)? Future research may tell us.

FISH in the language of molecular cytologists is not an animal phylum but rather an acronym for *F*luorescent *In Situ H*ybridization. FISH was also developed for studying the metaphase chromosomes of plants (e.g. Heslop-Harrison et al., 1991) and is useful when the metaphase chromosomes are large, can be spread nicely and can be identified individually. Also, an extensive probe (that is marked with a fluorescent stain) is useful in FISH and there is an advantage if the target of the probe is made of clusters of repetitive sequences, as in regions that encode ribosomal RNA or contain repeated retroelements. Under appropriate conditions FISH can locate specific DNA sequences on chromosomal arms and also indicate whether these sequences occur in the centromeres, telomeres, heteromatic regions or on the euchromatic regions of the chromosome. Thus, FISH was used in combination with highly repeated LTR transposons, or part of such transposons, such as fluorescent probes. In studies performed with rye (*Secale cereale*) and with oats (*Avena* spp.) the retroelements were used mainly as tools for cytogenetic analyses (e.g. Katziotis et al., 1996; Pearce et al., 1997; Linares et al., 1999, 2000, 2001) rather than being investigated in depth by a molecular approach. We shall therefore not detail these FISH studies with LTR retrotransposons of rye and oats.

2.16.4. LTR Retrotransposons in Rice
The study on the LTR retrotansposons of cultivated rice (*Oryza sativa*) and its wild relatives (*Oryza* spp.) started relatively late, but once started, intensive investigations of these elements were performed by H. Hirochika and his collaborators in Tsukuba, Ibaraki, Japan. In their first publication on this subject (Hirochika et al., 1992) these investigators used two methods to "fish" these elements from the respective genomes. In one method the investigators used the sequence of the primer for the reverse transcription. As mentioned above, adjacent to the 5' LTR is a sequence that is complementary to the RNA sequence of tRNA. In most plant retrotransposons it is from the tRNA^{Met}. Thus, a sequence of 14 nucleotides was used as a probe to hybridize to colonies of a phage library of rice DNA. The respective plaques were identified and clones were obtained which, after due purification, resulted in putative retroelement sequences. The other method for "fishing" retroelements from the rice genome was by using the conservation of certain regions of the RT proteins in such elements. The investigators constructed two primers for PCR that had degenerating nucleotide sequences encoding the conserved amino acid sequences QMDVKT and YVDDM (i.e. 18 and 15 nucleotides, respectively). In this way, the investigators could amplify fractions of codes for RT in the rice genome. Both methods yielded the expected results and the investigators found 10 families of LTR retrotransposons. These were termed *Tos1* to *Tos10*. Of these, *Tos1*, *Tos2* and *Tos3* were similar (with respect to their RT sequences) while *Tos4* to *Tos10* were in another group of families. More work was then devoted to *Tos1*, *Tos2* and *Tos3*. These elements were found also in wild rice species but not in other plants (maize or tobacco). It was estimated that there are about 1,000 copies of *Tos* elements in a haploid rice genome.

The three *Tos* elements (*Tos1*, *Tos2* and *Tos3*) were found to be efficient tools to finger-print cultivars of rice in order to bestow cultivars with a defined identity (Fukuchi et al., 1993).

Hirochika and associates continued their search for LTR retrotransposons in rice and identified 20 *Tos* families. All were of the Ty1/*copia*-type. These families all (or almost all) appeared as intact, full-size elements. Only three families seemed to be active in transposition. Moreover, transposition was triggered by tissue culture (Hirochika et al., 1996a). Of the three potentially mobile families of elements the investigators focused on *Tos17*, which was the most actively transposing after a period of tissue culture. One *indica* cultivar of rice was followed intensively. It was found that transposition of *Tos17* increased with the increase of time of tissue culture. The effect of long *in vitro* culture periods could be evaluated because the investigators had a cell line (Oc) that was kept in cell-culture for many years. The increase in transposition of *Tos17* was also observed in transgenic rice plants that were regenerated from tissue culture. Here we return to the issue of somaclonal variation: the investigators suggested that the initiation of transposition during tissue culture followed by insertion of retroelements into rice genes is a major cause for mutation, and hence, for tissue culture induced genetic variability. The investigators also suggested the utilization of tissue culture of rice as a useful means for *insertional mutagenesis*. In insertional mutagenesis individual plants that show mutant features are analyzed to detect where the alien DNA was inserted. If the alien DNA is a specific LTR retrotransposon, it can be detected by molecular methods and then the sequence that is flanking the alien DNA is sequenced. This sequenced fragment should encode the mutated gene.

More recently, Hirochika (2001) detailed the use of *Tos17* in "functional genomics" of rice. Actually with the progress in fully sequencing the DNA of the rice genome, insertional mutagenesis became an efficient method to relate DNA fragments to biological functions. Hirochika provided examples of using the "tagging", by insertional mutagenesis, to perform genetic studies. However, there is still the problem that the rate of *Tos17* integration is not high enough, even after tissue culture. Thus, Hirochika suggested to look for additional retroelements as well as additional means (other than tissue culture) to further enhance transposition. The additional retrotransposon should retain one great advantage of *Tos17*; the latter element seems not to have a preferential target site, it will integrate into low-copy-number targets, meaning, also inside sequences that encode genes (Yamazaki et al., 2001). Agrawal et al. (2001) demonstrated a system in rice, where *Tos17* insertion, tissue culture and genetic "tagging" can be combined to investigate a change in morphogenesis: the induction of a *viviparous* state (i.e. direct germination of immature grains, even when still attached to the spike). These authors applied what geneticists call *forward genetics*. They regenerated 30,000 rice lines (you need very devoted investigators to do this!) and looked for viviparous plants. One hundred such lines were detected. These lines were then analyzed for co-segregation of transposed *Tos17* with the viviparous feature. They were able to pinpoint a few lines in which the transposition of this element was causally related with the change to the precautious germination of grains and with changes in enzymatic activity. One such viviparous line had an impaired epoxidation of zeaxanthin.

A search for LTR retrotransposon was also performed in a collaborative Chinese/USA effort (Wang et al., 1997). These investigators found many such putative elements but gave them different names: *Rrts* (rather than *Tos*). No transcripts of these retroelements were detected but the sequencing of the conserved RT domain indicated that all were rather closely related and of the Ty1/*copia*-type. They exist in both the *indica* and in the *japonica* rice cultivars as well as in wild rice species.

It should be noted that RTL retrotransposons are not the only T.E. in rice. Thus, for example, 35 complete sequences of another type of T.E. were revealed in rice (Yang et al., 2001). These were MITEs (for Miniature Inverted-repeat Transposable Elements. The authors coined them *Kiddo*.

Finally, if you like to use LTR retrotransposons for mutational insertion and tagging of rice genes and the endogenous elements (e.g. *Tos17*) seem not to be sufficiently efficient – you may choose to recruit help from an alien LTR retrotransposon. Hirochika et al. (1996b) tested this possibility and came up with a useful suggestion. These investigators chose one of the few active plant LTR retrotransposons: the *Tto1* of tobacco. They sequenced the whole *Tto1* element and then asked if, when introduced (by genetic transformation) into rice, *Tto1* would maintain autonomous transposition. It was found that it would. It could also be tested whether *Tto1* indeed underwent retrotransposition (i.e. by reverse-transcription). The answer was positive and was based on the observation that if an intron was added into the RT gene of *Tto1* before it was introduced into rice, this intron was spliced-out in the process of transposition. The ability of *Tto1* to transpose in rice is interesting. It indicates that the host-factors required for this transposition are conserved between a monocot and a dicot, although it is assumed that these two diverted about 200 million YBP.

2.16.5. LTR Retrotransposons in Tobacco and other Species of Solanaceae

Tobacco has several advantageous features as an experimental plant. It can easily be sexually propagated; each plant produces thousands of seeds, there is no self-incompatibility, thus self-pollinations and cross-pollinations will yield ample progeny and it is amenable to tissue culture techniques. It was the first plant in which isolated protoplasts could be regenerated to functional plants. Plants can swiftly be regenerated from callus cultures and anther culture can lead to haploidization. Tobacco also has drawbacks as an experimental plant for molecular genetics: it has very small chromosomes (as seen in metaphase), it is an amphitetraploid, thus many of the genes that encode housekeeping enzymes are active in its two genomes (i.e. in four alleles) and there is no USA federal support, anymore, for research on tobacco.

M.-A. Grandbastien, A. Spielman and M. Caboche of Versailles, France (Grandbastien et al., 1989), intended to look into the possibility that a T.E. had integrated into a vital tobacco gene (nitrate reductase) leading to NR⁻ plants. This requires some explanation of the background.

One of the genes that are required for the formation of the functional NR is *nia*, which encodes the apoenzyme of NR. Being an amphitertraploid there are two *nia* genes in tobacco: *nia1* and *nia2* (each of these is represented by two alleles; a total of four alleles). Only when both these genes are mutated will the plant show the NR⁻ phenotype. If the tobacco plant is haploidized (by anther culture) there are only 2 alleles, one for each gene, and if these are mutated, the mutations are manifested in the haploidized plants ("haploid" tobacco plants are viable and they frequently double their chromosome number, resulting in homozygous plants). When one starts with a situation that one gene is already mutated while the other gene is heterozygous, the haploid will immediately show NR⁻ when the wild-type allele of this heterozygote is mutated, such as by insertion of a retrotransposon. As indicated above, NR⁻ mutants can be identified easily because they are resistant to chlorate, to which wild-type (NR⁺) cells are very sensitive. Now the advantage of tobacco comes to light.

Grandbastien et al. (1989) took advantage of the procedure for regenerating functional plants from isolated protoplasts that are grown in tissue culture. Protoplast-derived colonies were exposed to chlorate and resistant cells were selected to obtain spontaneous NR⁻ clones that can be regenerated into functional plants. Among such regenerated plants some had an "alien" insertion in the *nia* gene. This insertion of 5,444 bp was identified as an LTR retroelement and was named *Tnt1*. It has two LTRs of 610 bp each. The internal region had an ORF that was found to be similar to the ORF of Ty1/*copia*-like elements. The *Tnt1* was estimated to exist in tobacco in about 100 copies. Molecular hybridization indicated that *Tnt1*-like elements exist in species of other Solanaceae genera (*Petunia hybrida*, *Solanum tuberosum*, *Lycopersicon esculentum*) but no cross-hybridization to *Tnt1* was found with dicots outside of Solanaceae or in certain monocots that were tested.

In further studies of this Versailles group (Pouteau et al., 1991, 1994) it was found that what initiates the burst of transcription in *Tnt1* is the exposure of leaf tissue to protoplast isolation medium (that contained enzymes for cell-wall maceration). The burst of *Tnt1* mRNA appeared 2 hours after exposure to this maceration medium. Interestingly, in their 1991 publication the investigators of Versailles did mention that two commercial hydrolases were used for maceration: Macerozyme R10 and Driselase. However, they did not mention that these are actually rather crude extracts of fungal cultures.

The *Tnt1* continued, for many additional years, to be useful as an experimental tool in INRA at Versailles (Casacuberta et al., 1993; Moreau-Mhiri et al., 1996; Melayah, et al., 2001). Thus, the LTR sequences of *Tnt1* were characterized to detect which factors may activate the transcription of *Tnt1* in the early stages of protoplast isolation from tobacco leaves. In one publication (Casacuberta et al., 1993) the investigators analyzed the transcript of *Tnt1*. They revealed that this mRNA is not homogeneous, but consists of a population of different, though closely related, species of RNA. This frequently results in defective *Tnt1* elements. As we already know, reverse transcription is far from perfect, but another finding is noteworthy: the *Tnt1* found in tobacco roots differed from those of protoplasts.

Moreau-Mhiri et al. (1996) investigated the expression of *Tnt1* when it was transferred by genetic transformation into either *Arabidopsis thaliana* or tomato plants. The investigators were impressed that in tobacco, any microbial factor that elicits the hyper-sensitivity effect for defense – also causes transposition activity of *Tnt1*. The authors thus investigated the effect of such microbial factors on the promoter of *Tnt1* or on the whole *Tnt1*. For studying the effect on the promoter a fusion-construct was made in which the LTR of *Tnt1* was fused to the GUS reporter gene, and this construct was used for *Agrobacterium*-mediated genetic transformation. The promoter of *Tnt1* seem to regulate expression in the transgenic plants in a similar manner to its regulation in tobacco. Although there were quantitative differences: the *Tnt1* promoter seems to cause enhanced induction in aging leaves of the transgenic plants. Microbial factors as extracts from the fungus *Trichoderma viride* and supernatant from the culture of the bacterium *Erwinia chrysanthemi* also caused the induction of expression in the transgenic plants. Further studies of the Versailles group (Melayah et al., 2001) showed that not only expression (of transcript) is enhanced by the protoplast-isolation procedure; the transposition itself is also enhanced. About 25% of the plants regenerated from protoplasts had newly transposed *Tnt1* copies. A low increase in transposition was also observed after regeneration of tobacco plants from callus tissue. Actually, the response of the promoters of *Tnt1* is different in three different *Tnt1* sub-families (Beguiristain et al., 2001). The promoter of each of the three sub-families is responsive to stress but they differ substantially with respect to the type of the stress. The *Tnt1A* sub-family is very responsive to microbial elicitors and to methyl jasmonate, whereas the *Tnt1C* sub-family is more responsive to salicylic acid and to auxins.

Hirochika (1993), the same investigator who pioneered in the rice LTR retrotransposons research, also initiated a study on tobacco LTR retrotransposons. He isolated "his own" tobacco retroelements. His strategy was interesting. He assumed that most LTR retrotransposons of tobacco lost their transposition activity. However, those that are active should produce the respective transcripts. He therefore looked for cDNA of such elements in cell cultures of tobacco (the Japanese BY2 cell lines, one of the two lines used by Hirochika, had already been 20 years in culture at that time and it is still an outstanding culture with respect to uniformity and rate of cell division). By the use of appropriate primers and PCR, Hirochika isolated several elements that he named *Tto1*, *Tto2*, etc. They have similarities to the *Tnt* elements isolated in Versailles, but are not identical to them. The *Tto* elements are also triggered to transpose by the same conditions as the *Tnt1* elements: stress caused by *in vitro* culture.

In a further study of the investigators in Tsukuba, Japan (Takeda et al., 1999) the *cis*-regulatory sequence in *Tto1* that is responsible for the stress response was localized: it is a 13 bp repeated motif in the 5' LTR of *Tto1*. The investigators indicated that the *cis*-regulatory motif of *Tto1* has similarity to the regulatory motif in an asparagus defense gene (*AoPR1*). Finally, from the same laboratory came a study which looked further at this 13 bp motif. The investigators studied the binding of several stress-related proteins to this motif. One protein, NtMYB2, that was found to bind to the 13 bp motif, actually activated the transcription of *Tto1*. It also activates the phenylalanine ammonia lyase gene in tobacco protoplasts. This gene encodes an essential enzyme of secondary metabolism.

The gospel of LTR retrotransposons also reached Australia. A group of investigators (Royo et al., 1996) analyzed the genome of *Nicotiana alata*, because one of these investigators was Adrienne Clarke (a devoted investigator of incompatibility in plants), it was obvious that they should look at the S-locus for incompatibility in this species. They found two elements and one of them was a Ty3/*gypsy*-like element. However, this latter element was not likely to transpose.

We shall now travel to Chile. The Chilean investigators (Yanez et al., 1998) looked for Ty1/*copia*-like elements in *Lycopersicon chilense* (a relative of tomato). By the use of primers that had degenerated sequences homologous to conserved sequences in Ty1/*copia*, the investigators found four groups of elements that were *TLC1* to *TLC4*. These elements were probably active in transposition. Actually, partial sequences of TLR retrotransposons were previously reported in tomato. Yanez et al. (1998) had a unique reason to look for active LTR retrotransposons in *Lycopersicon chilense*. They argued that the habitat of this wild relative of tomato is in high altitude deserts and salty soil. These are considered stress conditions. As stress conditions were correlated with the activation of retrotransposons – such elements should exist in *L. chilense*.

A Ty1/*copia*-type retroelement emerged from a study of a starch-synthesis enzyme in potato (Carmirand et al., 1990). The Canadian investigators were actually interested in the enzymology and molecular biology of a α-glucan phosphorylase in potato. This enzyme is of major importance for starch metabolism in the potato tuber. It has large subunits of 105 kDa and it is encoded by a large gene of 14.4 kbp that is interrupted by 14 introns. The Canadian investigators revealed that the fifth intron is exceptionally long. They looked into it and found there a sequence that could represent an LTR retrotransposon. The putative retroelement was given a name: *Tst1*. *Tst1* has two LTRs of 285 and 283 bp, respectively and an internal region of 4,492 bp. The latter contains 4 ORFs. Its sequence had similarities to Ty1/*copia* retrotransposons. The sequence downstream of the 5' LTR showed 18 bp with homology to a tRNA^{Met} of plants and was therefore considered as the primer for the reverse-transcription of these elements for the synthesis of the second-strand DNA synthesis. Did *Tst1* behave as a transposable element or was it just sitting there? It is probably not mobile because of a stop codon in the RT code.

2.16.6. LTR Retrotransposons in Legumes
Legumes were relatively neglected as a source of LTR retroelements. Investigators looked for such elements mainly in genera that were subject to intense molecular-genetic studies (e.g. *Arabidopsis*, *Zea*, *Triticum*) or in plants that constitute major crops (e.g. rice, wheat, maize, potato). Although it is a good guess that such elements will be found in all the species of Phanerogams.

Soybean (*Glycine max*) is an exception to this neglect. It is a major crop plant and an important source of protein and oil for human consumption. The existence of *copia*-like retroelements in soybean was already mentioned by Voytas et al. (1992) but a more detailed study of such elements in soybean was started by Laten and Morris (1993). An element termed *SIRE-1* was revealed by PCR from the genomic DNA of this plant. It was estimated to exist in the soybean in 500 copies per haploid genome. Its length was estimated to be more than 10.6 kbp. Partial sequencing indicated similarity between the *Ta1* of *A. thaliana* and *SIRE-1*.

In a subsequent study, Bi and Laten (1996) further investigated the *SIRE-1* family of LTR retrotransposons. These investigators found that *SIRE-1* is a family of several hundred, non-tandem and relatively homogeneous elements. Two ORFs were detected in the internal region that apparently encode *gag* and *pol* proteins as well as highly conserved motifs for retroelements nucleocapsids (NC). The derived NC of the *SIRE* sequence has motifs that are very similar to the same NC motifs in other retroelements, such as the CCHC nucleic acid binding box, but the NC of *SIRE-1* has also very unique features. The same team of investigators (Laten et al., 1998) went on with their investigations on *SIRE-1* of soybean. They revealed an unusual situation. LTR retrotransposons differ from retrovirus by the fact that the retroviral genome also encodes an envelope (*env*) protein that mediates virion export from a cell and its entry into a host cell. The regular LTR retrotransposons lack the code for the *env* protein. Up to the study of *SIRE-1* by Laten and collaborators all plant retroviral-like elements were Ty3/*gypsy*-like, but *SIRE-1* does encode at least a part of the *env* protein, albeit it is obviously Ty1/*copia*-like. It should be noted that *SIRE*s were not the first T.E. revealed in soybean. A family of *Tgm* Class II elements was reported to reside in soybean in earlier years (e.g. Vodkin et al., 1983; Rhoades and Vodkin, 1988).

In studying the retrotransposons of the genus *Vicia*, Pearce et al. (1996) had mainly one purpose in mind: to investigate whether or not there is a correlation between genome size and the number of element copies per genome. For such research three species of this genus were very appropriate. One species, the broad bean *Vicia faba*, has an exceptionally large genome of 13.3 pg DNA per haploid genome. The genome of *Vicia melanops* is a littler smaller: 11.5 pg DNA. A third species *Vicia sativa* has a much smaller genome of 2.3 pg DNA per haploid genome. When the number of Ty1/*copia*-like elements were estimated the number of elements per haploid genome were 1,000,000, 1,000 and 5,000 for *V. faba*, *V. melanops* and *V. sativa*, respectively. Hence, there seems to be no genome-size/element-number correlation. As could be expected, *Vicia faba*, which contained a large number of Ty1/*copia* elements (most of them truncated) was also the richest in variability of these elements. In all three species the retroelements constituted a major component of the genome, but what percentage of the genome is actually retroelements could not yet be determined. In *V. faba*, where the locations of the retroelements were surveyed by *in situ* hybridization, the Ty1/*copia*-like elements were distributed in the euchromatic regions of the chromosomes.

A family of Ty1/*copia*-type retrotransposons was relatively recently revealed in beans (*Phaseolus vulgaris*). Each genome seem to contain about 40 copies of the elements that were called *Tpv2*. The elements were found mainly inside actively transcribed regions (i.e. genes). None of the isolated elements were sufficiently intact to indicate that they maintained transposition capability (Garber et al., 1999).

For our final example for a family of LTR retrotransposons, we come back to Greek mythology: the name of this family is *Cyclops*. The name was probably given because it is a "giant" retroelement. It was not found in Sicily, near the shores of Catania, where it had an encounter with Odysseus, but rather by a team of investigators from Germany, France and Italy – in pea (*Pisum sativum*). The total length of *Cyclops-2* is 12,314 bp and it contains two LTRs of 1,504 and 1,594 bp, respectively. *Cyclops-2* is an intact element and belongs to the Ty1/*gypsy*-type retrotransposons. It has all the regular features of such elements but an unusual primer-binding-site (PBS) sequence: it has homology to a sequence in tRNA^{Glu}, rather the more common tRNA^{Met}. Although being "intact", the *Cyclops-2* that was sequenced showed that its *pol* ORF was disrupted by several mutations. Thus, this sequenced element had probably lost its capability to transpose. The sequencing of the region that encodes the RT of this element indicated that it is not closely related to any of the known Ty3/*gypsy*-like elements from plants. *Cyclops-2* is rather abundant in pea: about 5,000 copies per haploid genome. By Southern blot hybridization the authors concluded that *Cyclops* elements are abundant in other legume crops such as bean, mung bean, broad bean, soybean and peanut. As a conclusion to the section on LTR retroelements of legumes we may note that people consuming legumes not only enjoy a protein-rich diet, but also swallow a lot of retroelements.

3. THE *DIRS1* RETROTRANSPOSONS

After handling the typical LTR retrotransposons and before discussing non-LTR retrotransposons, here is a brief description of a special group of retrotransposons: the *DIRS1* group of transposable elements. Although the *DIRS1* retrotransposons have similarities to the LTR retrotransposons they differ by several important features. The *DIRS1*s do not have two direct repeats, their termini are *inverted* rather than directly repeated. Moreover, the two "repeats" are far from being identical: the right "repeat" has an additional 27 bp sequence (*re*). The "repeats" may be much shorter than those of LTR retrotransposons: only about 50 bp in the element termed *Prt1* (see below). The coding sequences in the internal region of *DIRS1* are also "unusual". For example, the two enzymes RT and RNaseH are encoded in the third ORF and no DDE-type IN was found in *DIRS1*s. Contrary to other mobile elements the *DIRS1*s integrate in their target without creating a target-site duplication. On the other hand, the RT of *DIRS1* is similar to the RT of LTR retrotransposons. A recent review of *DIRS1* was provided by Goodwin and Poulter (2001).

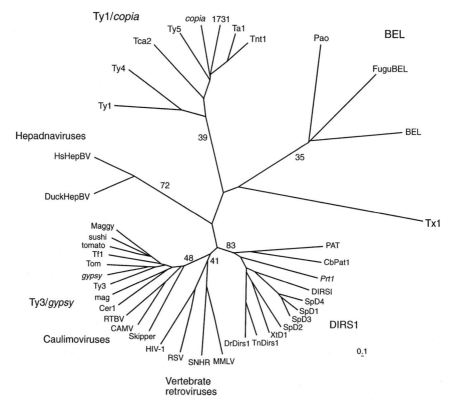

Figure 40. The relationships among long terminal repeat (LTR) retrotransposons and related elements. The tree is based on an alignment of the seven highly conserved domains of reverse transcriptase described by Xiong and Eickbush (1990). The seven major groups of elements are indicated. A non-LTR retrotransposon (Tx1) was included as an outgroup. (From Goodwin and Poulter, 2001).

Earlier studies revealed members of the *DIRS1* group in three distinct organisms: (1) in the cellular slime mold *Dictyostelium discoideum*; (2) in the nematode *Panagrellus redivivus*; (3) in the zygomyceteous *Phycomyces blakesleeanus* (the latter was, in the past, considered to be a fungus, but recent molecular taxonomy puts it closer to some filamentous algae). With the availability of DNA sequences of the genomes of many organisms and with *DIRS1* sequences as probes, it became possible to "fish" for additional *DIRS1* elements. Consequently Goodwin and Poulter (2001) started their "fishing" in zebrafish (*Danio rerio*). They found the *DrDirs1*. This fish-*DIRS1* is very similar to the original *DIRS1*. The total length is 6.1 kbp and its right "repeat" contains a 97 bp internal complementary region (ICR) and a 26-29 *re* sequence. The first ORF encodes *gag* proteins. Another ORF encodes RT and RNaseH proteins. Another type *DIRS1* element was revealed in the genome of the pufferfish *Tetraodon nigroviridis* and was termed *TnDirs1*.

Using the nematode *DIRS1*-like element *PAT* as queries, the genome of the nematode *C. briggsae* was screened, and consequently, an additional element,

CbPat1, was identified. Moreover, when conserved *DIRS1* sequences from the region that encodes the RNaseH protein were used to search for *DIRS1*-like elements in other organisms, several such new elements were revealed in frogs (*Xenopus* spp.) and in sea urchins (*Strongylocentrotus purpuratus*). Additional searches in the genomes of various organisms, including the human genome, clearly indicated that *DIRS1*-like elements exist in many metazoa. The phylogenetic relationship between these *DIRS1*-like elements and other retrotransposons was presented in a rootless tree (Figure 40). Finally, on the basis of sequence similarity Goodwin and Poulter (2001) suggested that the *DIRS1*-like retroelements integrate by λ-recombinase-like enzymes rather than by the IN of regular LTR retrotransposons.

4. MOBILE INTRONS

Introns, and especially Group II introns, can be considered, functionally, as retrotransposons. Eickbush and Malik (2002) reviewed the origin and evolution of the retrotransposons. They suggested that the non-LTR retrotransposons, which are the "oldest" group of retrotransposons, were derived from Group II introns. The awareness of the retrotransposon-like character of Group II introns has emerged only in recent years. The phenomenon of *mobile introns* was indeed revealed in yeast mitochondria, chloroplast genomes, cyanobacteria, proteobacteria and gram-positive bacteria. The subject was thoroughly reviewed recently by M. Bellfort, A.M. Lambowitz and associates (Bellfort et al., 2002). The transposition of Group II introns is an elaborate process that is not yet fully understood. We shall not handle the question whether or not Group II introns should be considered as "regular" types of transposable elements. However, a summary of the mobility of Group II introns will follow.

Most information on mobile introns came from yeast mitochondria. The Group II introns have a conserved secondary structure, forming six stem-loop domains that protrude from a central "wheel". The ORFs of these introns are located in domain IV, with very few exceptions. A bulged adenosine nucleotide, located in the stem-loop of domain VI, has an important role in transposition. The 2'-OH of the bulged adenosine initiates an "attack" on the 5' splice site of the target and a lariad structure is formed by the intron, as well as a 2'-5' linkage.

There is then an additional "attack" on the 3' splice site of the target. The lariad is released and the exons ends are spliced. The ORF of these introns (at least, some of them such as those of yeast mitochondria) code for a few proteins such as RT, *maturase* (for the splicing) and an endonuclease. There is also a code for a Z protein (of unknown function). Of these, the maturase was found in all analyzed Group II introns. The RT, when found, is closely related to the RT of non-LTR retrotransposons. Obviously, some Group II introns do not encode all these proteins, they may lack RT and/or endonuclease. How such introns cope with their transposition is still not clear. The respective functions may be supplied by genes outside of the Group II introns. At least in yeasts there is the formation of a ribonucleo-protein particle (RNP). The RNPs of yeast have the activity of the maturase, the RT and the endonuclease. RNP of yeast introns was used in *in vitro* experiments. It is assumed that the maturase binds to a specific location on domain IV and then makes contact with other regions of the intron RNAs, thus causing correct folding of the catalytic core.

A probably essential feature of Group II intron mobility is its targeting into specific DNA sequences. Defined sequences in the intron dictate this specificity by pairing with homologous sequences in the target.

The ability of Group II introns to move to heterologous sites was first revealed in yeast mitochondria, where ectopic insertions were observed. In these ectopic insertions the intron can move into a site in the mitochondrial genome that is collinear with its original location. This mobility of the intron into a "new" location in the mitochondrial genome, which is collinear with the original location of this intron is termed *homing* or *retrohoming*. The retrotransposition obviously involves, as the term implies, a step of reverse-transcription. It may occur via reverse splicing into RNA, followed by the synthesis of a cDNA and recombination into an ectopic site. While, as mentioned, the chemistry of Group II transposition is still under investigation, the reader can refer to Figure 32a, where the retrotransposition of non-LTR retroelements is schematically presented. However, one should keep in mind that the non-LTR retrotransposon transposition is only a conceptual example how Group II introns may transpose – the two transpositions are not identical; moreover, different members of Group II introns may differ in the details of this process.

As for evolution, it seems rather clear that all Group II introns have a common origin: in all of these introns the ORFs are in domain IV. How they evolved is a matter of speculation. The whole entity may be a remnant of the "RNA world" (that according to some evolutionists preceded the "DNA world"). Group II introns may have evolved from a kind of pre-existing autocatalytic introns that acquired an ORF with a choice of encoded proteins. The high affinity binding site for the RT is always located in a different domain, domain VI. This could mean that the RT was a later addition to the Group II introns. Finally, there is a suggestion that the Group II introns emerged first and from them the non-LTR retroelements have evolved.

A final note on the Group II introns concerns their utilization in genetic engineering. This utilization is based on the very specific targeting of the Group II introns. The target-site recognition involves base pairing between the RNA of the intron and at least a sequence of 14 bp in the DNA target. This allows precise "homing" of the intron into a specific DNA sequence. The recognition on the RNA of the intron could be changed, and thus targeting the intron into respectively different DNA sequences could be achieved. The idea was thus engendered that by this approach a genomic site of a pathogen could be invaded by the engineered intron and cause gene disruption. Thus, this system could be developed for gene therapy against certain pathogens.

5. NON-LTR RETROTRANSPOSONS

The realization that there are unique groups of repetitive DNA sequences in the nuclear genome of mammals emerged in the early eighties of the last century. I had a personal "encounter" with the very early research on what later was termed "non-LTR retrotransposons". In the summer of 1980, I was visiting the USDA at Beltsville, MD, to write a review on the physiology of plant protoplasts for the *Ann. Rev. of Plant Physiology*. I went with my wife to visit a friend, Dr Maxine Singer, at the NIH in Bethesda, MD. The extremely crowded laboratories at the NIH impressed me. But I was even more impressed by the enthusiasm of Maxine Singer, who told me about her recent research on repetitive DNA sequences in mammalian genomes (African green monkeys). She insisted that, contrary to the previous notion that these sequences are DNA "junk", she has indications that there is a biological/genetic meaning for these highly repetitive DNA sequences. I must confess: at the time I did not care much about this subject. My interests were focused on plant-protoplast fusion for the production of hybrids and cybrids in order to study nuclear/organelle interaction in plants. Over 20 years passed until I was searching the literature on the early studies of non-LTR retrotransposons for this book. I found no publication on this subject before 1980. The first report on this subject appeared in 1980 (Adams et al., 1980). Then came a number of publications of Maxine Singer and her colleagues (e.g. Grimaldi et al., 1981; Singer, 1982b; Lerman et al., 1983; Grimaldi et al., 1984; Skowronski and Singer, 1984; Fanning and Singer, 1987b). These publications substantiated the vision Maxine Singer had when she hosted me in 1980.

The term *non-LTR retrotransposons* may not be the most appropriate one to define the group of T.E. retroelements that we shall handle below. However, we need "drawers" for sorting and retrieving information. Thus, this will be our term for a variety of retroelements, such as long interspersed nuclear repetitive elements (*LINE*s), short interspersed nuclear repetitive elements (*SINE*s) and *Alu*. The term *Alu* is an historical remnant – the early *LINE*s and *SINE*s were first termed by the restriction endonucleases used for their isolation; there were *Bam*H1, *Kpn*I and *Alu* elements.

As noted above, the first non-LTR retroelements were reported from mammals in the early eighties of the last century. At about that time some investigators even jumped to the conclusion (without any real basis) that while mammals lack Class II transposable elements they are unique in having non-LTR retrotransposons. However, starting in the mid-eighties *LINE*-related and *SINE*-related elements were reported in many other eukaryotic organisms. We shall first discuss these elements in mammals but then devote some space to the occurrence of these retroelements in plants and in other organisms such as *Trypanosoma*, and insects.

5.1. The Early Studies on Non-LTR Retrotransposons of Mammals

The early studies on the non-LTR retrotransposons, during 1980 to 1987, should be perceived on the background of the methodologies that were becoming available during this period. The efficient sequencing of DNA became available only in 1977 and the Southern blot hybridization only two years earlier. Thus, when Catherine Houck and her associates (Houck et al., 1979) identified the *Alu* repeated DNA family in the human genome (a collaboration between Davis, California and the Virgin Islands), their analytical procedures were mainly ^{125}I and ^{3}H-thymidine labeled DNA, density gradient centrifugations and renaturation under thermal gradients. What is amazing is the amount of relevant information Houck, Rinehart and Schmid could derive from renaturation profiles! The repeated sequence of DNA had a site for *Alu*I cleavage and the authors could estimate that this family of *Alu* elements comprised at least 3% of the human genome, or, repeated in this genome at least 250,000 times. They could also suggest that *Alu* is interspaced with single-copy DNA sequences (i.e. genes). The *Alu* elements were then sequenced by Jelinek et al. (1980).

A further study that probably opened the way for intensive research on *LINE*s in mammalian genomes was that of Adams et al. (1980) that was mentioned above. Before the study of Adams et al. the reported repeated DNA sequences in mammals that were found adjacent to genes were shorter than 1.5 kbp. Adams et al. found a repeated DNA sequence 3,000 bp downstream of the human β globin gene. The authors made a phage library (with DNA fragments of 15–20 kb) to isolate the DNA sequences of interest. With due Southern blot hybridization, restriction endonuclease digested DNA and the globin gene as probe, the authors identified several clones that contained *EcoR*1-fragments, which represented repeated DNA fragments. Restriction-endonuclease maps were constructed and heteroduplex hybridizations finally indicated that the full length of the repeated fragment is 6.4 kb and that it is reiterated 3,000 to 4,800 times in the human genome. This means it comprises about 1% of the human genome. There was some variation between the copies of this element. The element was apparently not transcribed into RNA (or the transcript was below the level of detection). Does this repeated DNA have a function? The authors could give no definite answer but one suggestion was that it "... is carried in the genome in an essentially parasitic fashion ...", thus, it could be a self-duplicating, transposable element.

The M. Singer laboratory of NIH in Bethesda, MD, was engaged intensively with the SV40 virus and the African green monkey genome. They (Grimaldi et al., 1981) found that the family of *Alu* elements, detected previously in the human genome, is also represented in the genome of the African green monkey (AGM). These *Alu* units were separated from each other in the monkey genome by about 8 kbp of unrelated DNA sequences. Thus, the dispersion of *Alu* and their copy number are similar in the two primates. This should not surprise us; only about 16 million YBP the ancestors of these primates were sitting at the junction of the two (phylogenetic) branches that subsequently separated them. Actually, similarities in sequences of non-coding DNA fragments goes much further down the phylogenetic tree – to the emergence of eukaryotic organisms (Miesfeld et al., 1981). This emergence occurred, presumably, a billion or more YBP.

The year 1981 was a great year for repetitive and dispersed elements in mammals and in their brief review on this subject Jagadeeswaren et al. (1981) indicated that the *Alu* elements are the most abundant elements in the human genome (they claimed that there are 300,000 copies in the genome) and that there is an 80% homology between *Alu* members. Experimental results indicated that most *Alu* elements are flanked by tandem, direct repeats; also the DNA of most *Alu* elements served as templates for transcription by RNA Pol III. These authors came to the conclusion that *Alu* elements were frequently transposed. Thus, they suggested that *Alu* DNA sequences are representatives of a type of short, repetitive DNA elements that may share a mechanism of *transposition by self-primed reverse transcription*. In two reviews, Singer (1982a,b) summarized the findings on *SINE*s (including *Alu* elements) and *LINE*s in mammalian genomes. The border lines were set by defining *SINE*s as elements of a length of 500 bp or less. "Highly repetitive" were considered elements that appeared 10^4 or more times in the genome. Evidence has been accumulated that *SINE*s of primates and rodents are related but far from being identical. *SINE*s of rodents and primates contain *Alu* sequences but in different compositions. Also, Singer indicated that: "it is possible that some *Alu*-like *SINE*s are representatives of a new class of eukaryotic *mobile elements*". Singer compared the *SINE*s to the bacterial Insertion Sequences and the (now termed) Class II T.E. (i.e. those found in maize) and indicated that the *SINE*s are different in three respects: (1) the *SINE*s are shorter than the aforementioned T.E.; (2) the *SINE*s lack internal terminal repeats; (3) the lengths of the direct repeats flanking *SINE*s vary from one to another family. On the other hand, *SINE*s have a poly (A) sequence at their 3' end, similar to that found in some *Drosophila* T.E. Hence, a cDNA may be formed from *SINE*s and serve as an intermediate for transposition. *LINE*s, that were revealed only one or two years before the reviews of Singer (1982a,b), were defined as elements that are over 5 kbp in length and are repeated about 10^4 times in the mammalian genome. The *LINE*s were revealed not only in primates but also in rodents and were found to be composed of families and sub-families with a certain level of similarity between the family members. The *LINE*s of primates and rodents share some features but they do not cross-hybridize (at the DNA level). While in 1982 the *SINE*s were already considered as possible retrotransposable elements, this was not yet assumed for *LINE*s.

Just as the name *Alu* was based on the restriction endonuclease used for the isolation of this element, the names given to *LINE*-like elements also were based on endonucleases.   Hence, *Bam*HI elements of mice and *Kpn*I, as well as *Hind*III element-families of man (e.g. Manuelidis, 1982; Soriano et al., 1983) were revealed and investigated.

A significant step in the understanding of *SINE*s and *LINE*s of mammals was the nucleotide sequencing of these elements.   Once the investigator has the full sequence of such an element, she or he can obtain answers to major questions. For example, it could be asked if these elements encode known enzymes, especially enzymes that are known to play a role in regular or retrotransposition. Then, because cDNAs of mammalian cells from defined organs and developmental stages were available, clones from such cDNAs could be used to detect the sequences in the *SINE* and *LINE* elements that encode these clones. The actual rate of expression of these elements could thus be followed. The first question that could be answered was whether or not these elements are still expressed – or they are "dead" remnants of ancient transpositions. Such an approach was taken, for example, by Lerman et al. (1983) and Grimaldi and Singer (1983). These NIH investigators were engaged with *LINE* elements from AGM that were termed *Kpn*I elements. The *Kpn*I elements are polymorphic in both total lengths and in their DNA sequences. The longest ones are about 6 kbp. Already the restriction maps of these elements provided meaningful information such as the fact that the various members are related but not identical and that at least some of the elements were recently inserted into the nuclear genome. Southern blot hybridizations also indicated a close relationship between the AGM *Kpn*I elements and the *Kpn*I elements of the human genome. More information was derived from sequencing. It became evident that the various *Kpn*I elements of AGM share extended sequences but not all are collinear with each other ("scrambled arrangements"). Moreover, it became clear that at least one of the AGM elements (the *Kpn*I-LS1, of about 2 kbp) is transcribed; it hybridized to the RNAs of AGM and of man. At least some of the *Kpn*I elements contain a poly (A)-rich segment at their 3' end. Although the authors of the above-mentioned reports did not mention it explicitly, the information provided was compatible with the notion that *Kpn*I are non-LTR-like retrotransposons.

A similar situation to that of *Kpn*I (LINE) elements in primates was found for *LINE*-like elements in rodents, where the elements were termed the *Bam*H1 family of *LINE*s (Fanning, 1983). The *Bam*H1 family members were approximately 7 kbp long, were repeated about 10^4 times in the genome, were represented as about 0.5% of the mouse genome and contained a poly (A) track at their 3' end. At least in some cases, the *Bam*H1 elements in the rodent genomes were flanked by short direct repeats of the "host" genome. The characterization of the rodent *Bam*H1 families suggested also that they originally arose, and possibly still replicate, by a process that involved an RNA intermediate; meaning by reverse transcription.

Equipped with additional full-length *Kpn*I family members (i.e. *LINE*s) of primates, Grimaldi et al. (1984) could initiate a detailed study of the 5' and 3' ends of such elements. First, by sequencing these ends it became evident that these elements do not contain LTRs. Further, several family members had a polyadenylation signal at their 3' terminus.

The intensive studies, during only a few years, with mammalian *LINE* elements, were summarized by Singer and Skowronski (1985). In addition to what was mentioned by us in the previous paragraphs, it became evident that each mammalian species that was investigated had only one family of *LINE*s. Thus, order could be made in nomenclature. The term *LINE* was abbreviated to *L1* and this abbreviation was followed by the initials of the species. Hence, the *LINE* of mouse (*Mus domesticus*) was termed *L1Md*. The overall structure of the *L1* elements in all mammal species is rather similar. They may be detected in different lengths but those that are full-length are about 6 kbp long. The shorter elements represent truncations at the 5' ends. Actually, one element that was believed to be a mouse *SINE* (R element) of about 500 bp, turned out to be the right end of a *L1* element. It also became clear that while the ends of the primate *L1*s and those of rodent *L1*s differed substantially, the middle regions of these elements were similar.

These and other observations indicated that the *L1*s of mammals code for one or more proteins that are shared by all these *L1*s. Evidence then suggested that all *L1*s are amplified and dispersed through reverse transcription and that, at least under some conditions, the transcripts of *L1* elements might be efficient templates in this reverse transcription.

A question that was not answered at the time (1985) was how important is the role of certain *L1*s in modulating the activity of genes, especially genes that are in proximity to *L1*s. Whatever their "role" in the mammalian genome, the *L1* elements were joined to the "Mobile Elements".

A further adoption of the concept that *L1* elements are a kind of non-autonomous retroviruses appeared in a publication by Fanning and Singer (1987a). These investigators analyzed the *L1* elements of representatives from four orders of mammals. The elements were partially sequenced: from primates (man and monkey), from carnivores (cat), from rodents (mouse), and from lagomorphs (rabbit). First, the apparent coding sequence for a reverse transcriptase (RI) was found to be rather conserved in the *L1*s of all four orders. In another region, closer to the 3' end of the elements, there was a 90% homology with respect to derived amino acid sequences, in a coding region for a nucleic acid binding protein that exists also in retroviral nucleic acids (the *gag* region).

As to possible functions of *L1* in mammals that contribute to the vitality of mammals, the authors (Fanning and Singer, 1987a) only stated that this is an interesting possibility. This publication of Fanning and Singer (1987a) marked the end of the early studies on *LINE*s and *SINE*s. Before 1987 these elements were reported only in mammals. It was subsequently found, as noted in a later publication in the same year (Fanning and Singer, 1987b), that these elements are much more spread in eukaryotic organisms. For an update on these elements in animals we shall separate the *SINE*s from the *LINE*s and shall devote a special section for plant elements.

5.2. SINE Elements in Animals

As indicated above, by 1987 it became evident that *SINE*s are not confined to mammals. Already in relatively early reviews (Deininger, 1989; Okada, 1991), a difference between the *SINE*s was noted in several groups of non-mammalian animals such as newt, tortoise and fish. The *Alu SINE*s of primates and rodents are derived from 7SL RNA (an integral component of the ribosomal particle), while the *SINE*s from non-mammals were probably derived from tRNAs.

While the discovery of Class I and Class II elements in bacteria, maize, yeast and *Drosophila* resulted from genetic studies, the discovery of non-LTR retrotransposons in mammals and other vertebrates was due to molecular studies – mainly sequencing of the nuclear DNA.

As briefly indicated in the survey of early research on mammals, the non-LTR retrotransposon, *Alu*, was one of the earliest elements that was revealed in primates. While earlier estimates were that there were 300,000 *Alu* elements in the human (haploid) genome, later estimates brought it up to 500,000, or even more than 600,000. The length is about 300 bp; it was therefore calculated to compose 5 to 10% of the total human genome. While the general structure of *Alu* is the same in all these elements, they do differ from each other in their specific nucleotide sequence. The difference between two randomly selected *Alu* elements is about 10%. The *Alu* are composed of tandem repeats of two sequences that are separated by an A-rich region. The tandem repeats have homology to the 7SL RNA, but the right monomer has also an insert that does not exist in the left monomer of *Alu*. The 3' ends of *Alu* elements have an A-rich tail that is reminiscent of mRNAs. Flanking the *Alu* elements are short, direct repeats, that suggest the duplication that is created by the insertion of *Alu* into a new host DNA sequence. The *Alu* elements also contain internal promoter boxes for RNA polymerase III (Pol III) transcription. The transcription starts at the consensus 5' end of the *Alu* elements.

Much attention was paid to the evolution of human *Alu* elements and their relation to other *SINE*s in monkeys and other mammals (Schmid, 1996). Those who studied the phylogeny of *Alu* took advantage not only of the diversity in the individual *Alu* elements but also of the transposition-manner of non-LTR retroelements. By this transposition the original element stays at its site and only a reverse-transcript of it moves to a new insertion site. One can thus derive information that will suggest which are the more ancient sites of *Alu* and which sites are more novel. Zietkiewicz et al. (1998) traced the origin of primate *Alu* elements. They concluded that at 60 MYBP, free left and right monomers formed the *Alu* heterodimer that is connected by a 19-nucleotide-long linker. Actually, some human *Alu* elements are still actively transposing. This gives tools to anthropologists to build a phylogeny of *Homo sapiens*. For example, the mobile *Alu* elements can be analyzed with respect to the frequency of their insertion into genomes of: (1) Australian aborigine; (2) Asian; (3) Amerindian; (4) Alaska native; (4) Caucasian; (5) Nigerian people.

The observation that about 80% of the sequenced *Alu* elements are flanked by direct repeats, indicated that the changed locations occurred by transposition rather than by processes such as unequal homologous crossing-over. This does not mean that *Alu* elements are not involved at all in cytogenetic scrambling, but that such scramblings are relatively rare. Actually, a specific case that resulted from *Alu-Alu* cross-over is the establishment of X-X maleness that was caused by unequal crossing-over between *Alu* elements on the X and Y chromosomes. There were reports of similar unequal cross-over events that involved *Alu* elements.

Although when *Alu* is transcribed it is mainly by Pol III, there are also (about 10%) Pol II transcripts that appear among the hnRNA in the nucleus. The latter are commonly eliminated during the mRNA processing. Methylation may play a major role in the regulation of transcription. The dinucleotide CpG is a site for methylation and methylation may interfere with transcription. Here there is a difference between young *Alu* and older *Alu*. While the young *Alu* contain 9% CpG, much less CpG are in older elements. The assumption is that there is a "decay" from CpG to TpG. Notably, the overall frequency of CpG in the human genome is only 1%. Moreover, most of the young *Alu* of the *Ya5 Alu* sub-family are unmethylated, while in general the *Alu* of human tissue is fully methylated. This is true also in the germ-line cells of human females. The sperm cells and the oocytes are generally hypomethylated but there is a pattern of this methylation – it is not random. It was found that *in vitro* the *Alu* transcription is reduced by methylation but it is not clear how DNA methylation regulates the *in vivo* transcription of *Alu*. We should note that in Class II T.E. it was clearly demonstrated that methylation reduces transposition (see below). We shall also recall that recent investigation of chromatin has shown that the modification of certain amino acids has a major role in the control of gene expression by chromatin (see Appendix II). The latter may well play a role in the level of *Alu* transcription. Whatever the regulation, the *in vivo* transcription of Alu is very low. Is it because of a weak promoter or because of methylation? Only further research will tell.

The existence of an *Alu* transcript is only the first requirement for the transposition of *Alu*. The reverse-transcription is another requirement and these elements do not code for an RT protein. We can only guess that the RT is contributed by either other retrotransposons in the primate genome or from another unknown source. Obviously, while not being very frequent, transposition of *Alu* does occur and even into exons or introns of essential genes. Several cases of such transpositions of human *Alu* elements have been recorded in which such insertion-mutations have caused human diseases. These were reviewed by Deininger and Batzer (1999). In one such case an *Alu* element was inserted in an intron of the NR1 gene and caused neurofibromatosis.

We mentioned above, while dealing with LTR retroelements of plants, that with several stresses, such as *in vitro* culture of plant tissue, increased transposition of T.E. The rate of *Alu* transposition is so low that it cannot be recorded experimentally. However, the rate of *Alu* transcription can be traced. Thus, it was found that several stress conditions, such as elevated temperature and adenovirus infection, increased *Alu* transcription.

Alu elements, other *SINE*s as well as *LINE*s, were very useful in analyzing phylogenetic relationships. A very specific question was of interest to Nikaido et al. (1999): which of the land-mammals is the closest relative of whales? They thus used *SINE*s and *LINE*s to trace the phylogenetic relationships of major mammalian groups. They scored 20 loci in the nuclear genome and asked whether or not this locus is occupied by a non-LTR retrotransposon (i.e. a *SINE* or a *LINE*). The score could be "present", designated by 1 or "absent", designated by 0. One should recall that such sites are only filled, but not emptied, because of the replicating manner of transposition in *SINE*s and *LINE*s. The presence of an element could be determined positively, but in some cases it was difficult to analyze the specific site (locus). In

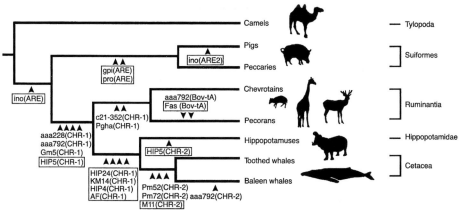

Figure 41. Phylogenetic relationships among the major cetartiodactyl sub-groups. All insertion sites of SINEs and LINEs characterized until 1999 are mapped on the phylogeny inferred from these data. Boxed loci indicate those loci discussed at length in Nikaido et al. (1999) and the specific retroposon unit inserted at each locus is given in parentheses. (From Nikaido et al., 1999).

such cases the score was 0. Obviously, this was not a very precise analysis but the editors of the *Proceedings of the National Academy of Science (USA)*, liked this presentation. So we have an answer. Out of the ancestors of the analyzed animals, the first to branch out were the camels; (from the ancestors of pigs, ruminants, hippos and whales), the pigs then branched off from the ancestors of ruminants hippos and whales. There was then a split into two main branches. One branch led to ruminants and another branch split again into hippos and whales. Hippos were therefore considered the closest relatives of whales (Figure 41). Do such phylogenetic trees have the strength to support the heavy weight of hippos and whales on their branches?

Alu sub-families were also helpful in tracing human origin and determining the age of the *Alu* elements. Kapitonov and Jurka (1996) analyzed the sequence of all major *Alu* sub-families as well as of the Free Left Monomer (FLM) of *Alu*. The FLM is considered a proto-*Alu* element that originated from a DNA complementary to a processed 141-bp internal deletion. This happened sometime at the dawn of the evolution of mammals. The analysis led to the construction of a phylogenetic tree. The timescale calculation was based on the assumption that the rate of nucleotide change is 0.16% per million years. This is obviously a rate that may not be accurate and if it should be amended the timescale attributed to the branching of *Alu* families, as well as the various animal orders, will also change. The estimate of Kapitonov and Jurka (1996) was that the primates branched off from major mammalian orders about 100 MYBP. This is also the time during which FLM branched off from all other *Alu* families. According to the *Alu* phylogeny, human ancestors branched off from lemors 50–75 MYBP. Further studies on the phylogeny of *Alu* elements in primates were performed by Zietkiewicz et al. (1998) and by Schmitz et al. (2001). Both these latter teams arrived at conclusions like those of Kapitonov and Jurka (1996).

Table 5.   List of non-LTR retrotranspons included in sequence comparisons and phylogenetic analyses.

Trans-poson[1]	Organism	Reference[2]	Accession No[3]
Amy-Bm	*Bombyx mori*	Foster et al. (unpublished)	UO7847
BS-Dm	*Drosophila melanogaster*	Udomkit et al. (1995)	555544[a]
CgT1	*Colletotrichum gloeosporioides*	He et al. (1996)	L76169
Cin4-Zm	*Zea mays*	Schwarz-Sommer et al. (1987)	Y00086
CR1-Ps	*Platemys spixii*	Kajikawa et al. (1997)	AB005891[b]
CR1-Gg	*Gallus gallus*	Burch et al. (1993)	L22152
Cre1	*Crithidia fasciculate*	Gabriel et al. (1990)	M33009
Cre2	*Cr. fasciculate*	Teng et al. (1995)	558380[a]
Doc-Dm	*D. melanogaster*	O'Hare et al. (1991)	X17551
F-Dm	*D. melanogaster*	Di Nocera and Casari (1987)	M17214
Hyp1-Cte	*Chironomus tentans*	Blinov et al. (1997)	L79944
Hyp2-Cth	*Ch. thummi*	Blinov et al. (1993)	S31175[a]
1-Dm	*D. melanogaster*	Fawcett et al. (1986)	M14954
1-Dr	*Drosophila teissieri*	Abad et al. (1989)	B36186[a]
Ingi-Tb	*Trypanosoma brucei*	Murphy et al. (1987)	S28721[a]
Jockey-Dm	*D. melanogaster*	Priimagi et al. (1988)	P21328[c]
Juan-Aa	*Aedes aegypti*	Mouches et al. (1992)	M95171
Juan-Cp	*Culex pipens*	Agarwal et al. (1993)	B56679[a]
Lian-Aa1	*Ae.aegypti*	Tu et al. (1998)	U87543
LINE1-Bg	*Biomphalaria glabrata*	Knight et al. (1992)	X60372
LINE-Hs	*Homo sapiens*	Hattori et al. (1986)	PO8547[c]
LINE1Mm	*Mus musculus*	Martin (1995)	U15647
Q-Ag	*Anopheles gambiae*	Besansky et al. (1994)	U03849
R1-Bm	*B. mori*	Xiong and Eickbush (1988b)	M19755
R1-Dm	*D. melanogaster*	Jakubczak et al. (1990)	X51968
R2-Bm	*B. mori*	Burke et al. (1987)	M16558
R2-DM	*D. melanogaster*	Jakubczak et al. (1990)	X51967
RT-Ce	*Caenorhabditis elegans*	Wilson et al. (1994)	U46668
RT1-Ag	*An. gambiae*	Besansky et al. (1992)	M93690
Sart1-Bm	*B. mori*	Takahashi et al. (1997)	D85594
SLACS	*T. brucei*	Aksoy et al. (1990)	X17078
T1-Ag	*An. gambiae*	Besansky (1990)	M93689
Tal1-1-At	*Arabidopsis thaliana*	Wright et al. (1996)	L47193
Tad1-Nc	*Neurospora crassa*	Cambareri et al. (1994)	L25662
Tart-Dm	*D. melanogaster*	Levis et al. (1993)	U02279
Tras1-Bm	*B. mori*	Okazaki et al. (1995)	D38414
Trim-Dmi	*Drosophila miranda*	Steinemann and Steinemann (1991)	X59239
Tx1-Xl	*Xenopus laevis*	Garrett et al. (1989)	M26915

[1]Arranged alphabetically; [2]For references, see Tu et al. (1998) and in the list of references in this book; [3]The accession numbers are from GenBank/EMPL, unless marked: [a]PIR accession number, [b]DDBJ accession number, [c]Swiss-Prot accession number.
(From Tu et al., 1998).

Thus, for example Zietkiewicz et al. estimated that the early dimeric *Alu* elements, formed by combining Free Left – with Free-Right monomers and linked by 19 nucleotides, were established 60 MYBP. This is the end of the Cretaceous era and it is close to the period of the origin of primates – at least according to some evolutionists.

Novel types of *SINE* elements are still being revealed in animals. N. Okada (of the Tokyo Institute of Technology) and associates (Ogiwara et al., 2002) recently reported on a new superfamily termed *V-SINEs* that have a unique and conserved central region. They estimated that *V-SINEs* were formed 55 MYBP.

A list of non-LTR retrotransposons sequences (of various animals, mostly invertebrates, but one vertebrate as well as a plant and a fungal organism) that served in phylogenetic analyses, are provided in Table 5.

5.3. SINE Elements in Plants

Yoshioka et al. (1993) found short, repetitive DNA sequences in the tobacco genome. The sequencing of these repetitions revealed four similar elements. The length of these elements was about 200 nucleotides. They were flanked by direct-repeat sequences of 7 to 15 nucleotides and all had sequences that were very similar to tRNA^{Lys} (of rabbit!). We indicated above that *SINE* elements from several animals also have sequences that are similar to tRNA. From the common characteristics of these tobacco elements the authors categorized these elements as *SINEs* and respectively termed them *TS* elements (for tobacco *SINE* elements). It was estimated that there are about 5×10^4 copies of *TS* elements in the haploid tobacco genome.

Are *TS* still mobile? There is no definite answer to this question. What was found was that *in vitro*, in a HeLa cell extract, clones of *TS* were transcribed to transcripts of 350 (!) nucleotide RNA. This transcription was probably activated by Pol III. Under the same conditions no such transcripts were formed with total tobacco DNA as template.

Plant species of the Cruciferae family have their own *SINE* elements. A detailed evolutionary study of these *S1* elements was performed by Lenoir et al. (1997).

First, a remark: the term *S1* for these *SINEs* (given by the team of Jean-Marc Deragon who started this study) is not a successful one: it is too similar to the designation of the genes for self-incompatibility of the Cruciferae family. The *S1* elements, all derived from tRNA sequences, were searched in 38 species of Cruciferae. They were found in 35 species and were missing in 3 of them. The experimental plant *Arabidopsis thaliana* does not contain *S1*. The *S1* was also missing from five species of other plant families: *Nicotiana tabacum* (Solanaceae), *Solanum tuberosum* (Solanaceae), *Cucumis sativus* (Cucurbitaceae), *Helianthus annuus* (Compositae) and *Zea mays* (Gramineae). The authors suggested that there were, in evolution, bursts of *S1* amplifications that occurred in recent periods and that these amplifications were highly dynamic, possibly contributing to genome rearrangements and to reproductive isolation.

Thus, in plant *SINEs* we do have evolutionary information. However, the molecular mechanisms of *SINE* transposition in plants (see: Lenoir et al., 2001) as well as in animals, still require investigation.

5.4. LINE Elements in Invertebrates

We shall start our summary-review of non-LTR retrotransposons in invertebrates with the protozoan genus *Trypanosoma*. Members of this genus are pathogenic parasites on mammals. It is an open question how trypanosomes made their living before the emergence of mammals. However, this is a very general question for parasitic organisms and outside the scope of our book. We do meet specific organisms that purposely cultivate other organisms, of different phylae, in order to feed themselves. The "cultivated" organisms are commonly of an earlier phylogeny; the ants of the *Ata* genus cultivate fungi and man cultivates plants and domestic animals.

Two teams studied the parasitic *Trypanosoma brucei*: Kimmel et al. (1987) of the International Laboratory for Research on Animal Diseases in Nairobi, Kenya and Murphy et al. (1987) of the Free University of Brussels, Belgium. Both teams were actually engaged in a very similar study on RIMEs (ribosomal mobile elements) that were previously revealed as one of the groups of reiterated DNA sequences in trypanosomes. RIMEs are dimers (of 512 bp), which were found inserted into a ribosomal RNA gene. Both teams reported the same phenomenon: a long element of 5.2×10^3 bp that is flanked by two halves of the RIME element. The Nairobi team gave the *T. brucei* element the name *Ingi*. The Brussels team termed their *T. brucei* elements as *TRS* (for trypanosome repeated sequences). *T. brucei* causes the deadly human "sleeping sickness" but can also affect cattle. Incidentally, although "brainless" the trypanosomes accumulated sufficient wisdom in their DNA to render them not only efficient parasites that increase their distribution via a fly (Tse-Tse), but they are also equipped with a clever mechanism to avoid the defenses of the immune system of their mammalian hosts. The long versions of the two elements, *Ingi* and *TRS*, from Nairobi and Brussels, respectively, were fully sequenced. Several overlapping ORFs were detected in these elements. They resemble the ORFs found in the LTR retroelements mentioned above. As in mammalian *LINE*s, these elements have poly (A) tails. In short, the *Ingi* as well as the *TRS* have the features of *LINE*s. They are transcribed to mRNAs with poly (A) tails. Moreover, the level of transcript differs: it was found to be more abundant in bloodstream forms than in cultured procyclics. It is thus plausible that these two elements are mobile, meaning that they are capable of retrotransposition. This activity can have an impact on the genome organization of *T. brucei*. Moreover, genetic changes in the genome of trypanosomes may have a selective advantage because it may lead to antigenic variation as a defence against the immune system of their hosts. Even if retrotransposition of these *T. brucei* elements still takes place, the estimated number of copies of *Ingi* elements, per trypanosome genome, is limited to only a few hundreds.

A Spanish team (Martin et al., 1995) characterized the cDNA that was putatively transcribed from a non-LTR retrotransposon of another trypanosome: *T. cruzi*. The cDNA was termed *L1Tc* and was fully sequenced. The sequencing indicated that *L1Tc* was derived from an element that is similar, but not identical to *Ingi* of *T. brucei*. The *L1Tc* elements were abundantly expressed and most copies had a length of 5 kbp. They were found to have 3 putative ORFs, in different frames. ORF2 probably encodes an RT. ORF3 encodes *gag*-like proteins that contain a cytosine-motif of the "zinc finger" family that was previously found in other non-LTR retroelements. The ORF1 was a surprise. It encoded a polypeptide that had a significant homology to the human AP endonuclease protein. In man, AP belongs to a family of repair enzymes that are implicated in the first step of repair of cleaved DNA at specific sites. The role of the ORF1 product in *L1Tc* is not known. What clearly emerges is that *L1Tc* of *T. cruzi* belongs to the group of non-LTR retrotransposons, even though its mobility was not proven.

Only recently were non-LTR retroelements revealed in nematodes. Youngman et al. (1996) detected an element in *Caenorhabditis elegans* that they termed *Rte-1*. It was apparently difficult to find a proper "drawer" for *Rte-1*. It had a long ORF that coded for an RT as well as some other features of retrotransposons. It also had two terminal direct repeats of 200 bp each. Thus, it could be an LTR retrotransposon. Further analysis of the sequence of *Rte-1* put it in the "drawer" of non-LTR retrotransposons. First, the direct repeats of *Rte-1* did not have the features of the regular LTRs. The RT of *Rte-1* was more similar to the RT of non-LTR retrotransposons than to LTR retrotransposons.

Later studies of Malik and Eickbush (1998) found RTE (i.e. the family to which *Rte-1* belongs) to be a "class" of non-LTR retrotransposons that exists not only in nematodes. RTE-like elements are widely distributed and also exist in arthropods and vertebrates. Also, *Rte-1* elements code for an apurinic-apyrimidinic (AP) endonuclease motif. Such a motif was noted above in *L1Tc* of *T. cruzi*.

When we were dealing, above, with the LTR retrotransposons of *Drosophila*, we mentioned *hybrid dysgenesis* – a phenomenon of sterility that was causally related to the activation of transposable elements. It happens when the sperm of a male, with "sleeping" T.E., is transferred to a female of another line (M) that never contained any T.E. Hybrid dysgenesis was also revealed in *Drosophila melanogaster* strains that carried non-LTR retrotransposons (Fawcett et al., 1986). These authors isolated and sequenced a T.E. in *D. melanogaster* that is termed *I Factor*. *I Factors* are flanked by direct repeats (of the host), they lack LTRs and their 3' terminal is A-rich. These factors have two long ORFs, meaning they have the consensus features of a mammalian *LINE* element.

Two additional non-LTR retroelements that have similarities to the *I Factor* were then reported to exist in *D. melanogaster*: the *F-element* and the *G-element* (Di Nocera and Casari, 1987; Di Nocera, 1988). These elements turned out to have a similarity to the *I Factors* of Fawcett et al. (1986). However, *Drosophila* has more *LINE*-like non-LTR retrotransposons in its arsenal. An example of such an additional element emerged from the studies on bristles and hair mutants in *Drosophila* (see: Campuzano et al., 1986). Several recessive and dominant mutations affect these fly organs that are inervated sensory elements. Many of these mutations are involved with the *achaene-scute* gene complex, located at the tip of chromosome X. The T.E. *gypsy* and *copia* (LTR retrotransposons) that were discussed above, may insert into this gene complex and change the pattern of the sensory elements. A team from Madrid, Spain, studied the insertions in detail and detected two novel insertions in this gene complex. Their sizes were of 4.5 and 2.6 kbp, respectively. Campuzano et al. (1986) names them *Sancho* 1 and *Sancho* 2, respectively, "after the faithful companion of Don Quijote (de Cervantes, 1605)". Why the authors chose this spelling rather than Don Quixote and why *Sancho* rather than the faithful horse, *Rosinante*, was not explained by the authors. We shall probably never know because the *Sancho* elements went into oblivion (while Don Quixote is probably immortal).

In a series of publications, the same Russian team that we met while dealing with *gypsy* studied another non-LTR retrotransposon type in *Drosophila*: *Jockey* (Mizrokhi et al., 1985; Priimagi et al., 1988). In the past this team showed evidence that their *mdg4* (*gypsy*) element has a strong "homing" tendency, meaning, that this element has a strong tendency to reinsert into previous insertion sites. *Gypsy* is an LTR retrotransposon of *Drosophila* and the Russian investigators found that another element can be "riding" on the *gypsy* and then transpose together with *gypsy*. The "hitchhiker" element was called *Jockey*. Detailed analysis of *Jockey* (e.g. by full sequencing) revealed a *LINE*-like, non-LTR retrotransposon of about 5 kbp, that has at its 3' end an oligodeoxynucleotide (dA) sequence. The latter sequence is preceded by two ORFs. One encodes a nucleic acid binding protein and one encodes an RT. Not all *Jockey* elements were full-length, many were truncated. Elements lacking the ORF for RT were frequently found.

We have most probably not covered all the *LINE*-like elements of *Drosophila*. Before leaving *Drosophila* we shall only mention another type of non-LTR retroelements. These are elements that were found in the non-coding spaces of the repeated gene for the 28S ribosomal RNA. This gene is repeated many times and two related elements, *R1* and *R2*, may insert into the intergenic spaces. The insertions reduce the transcription (Jakubczak et al., 1990). The elements seem to code for the common proteins of non-LTR retrotransposons. These elements were found to be closely related to the *R1Bm* elements found previously by the same T.H. Eickbush team (Xiong and Eickbush, 1988) in the silk moth *Bombyx mori*. A subsequent survey of these *R1* and *R2* elements by the Eickbush team (Jakubczak et al., 1991) revealed that insects from 9 orders harbored these elements that are integrated in their specific locations in the 28S RNA genes: 43 of 47 studied species of insects had these elements. The range of hosts spanned from parasitic wasps to beetles. The main difference between *R1* and *R2* elements is in their respective "favorable" site of integration. While *Drosophila* and silk moth harbor both elements, some other insects harbor either *R1* or *R2* elements. Are these elements also transferred horizontally? If transmission is primarily vertical, they should be a "treasure trove" for those interested in insect evolution. A detailed study of the *R2* of *B. mori* (Luan et al. 1993) actually led to the proposed mode of transposition in non-LTR retrotransposons that we saw in Figure 32.

The silk moth non-LTR retrotransposons *R1Bm* and *R2Bm* deserve special attention. While other elements of this large group, such as the human *L1* elements, insert into a wide variety of DNA targets, the silk moth elements are target-specific. Two relatively recent studies (Feng et al., 1998; Yang et al., 1999) addressed the question of the basis for this specific integration into the 28 rRNA genes. As for the insertion specificity of *R1Bm*, Feng et al. (1998) showed that this element encodes an enzyme that has characteristics similar to an apurinic-apyrimidinic endonuclease. This enzyme, when tested *in vitro*, cleaved at the site of 28S rRNA, where *R1Bm* integrates. As for *R2Bm* (Yang et al., 1999) the motif for site-specific integration is probably located downstream of the location of the code for RT.

Retrotransposable elements were discovered in mosquitoes only after they were revealed in flies. Besansky (1990) reported a non-LTR retrotransposon type in the mosquito *Anopheles gambiae*. This is the main malaria vector in Africa. The element was represented about 100 times in the mosquito genome and had the standard features of non-LTR retrotransposons in *Drosophila* and silk moth. There was one feature worth mentioning: there was no direct repeat of host sequences flanking the element. While this is unusual for retrotransposons, it is not the only reported case in which such direct repeats were missing. The lack of these direct repeats can have several explanations. One possible explanation is that at the site of integration the chromosomal break is not staggered but blunt. Also, no transcript of the *A. gambiae* was found in the adult flies, from where the RNA was taken for Northern hybridization.

While intensive work on non-LTR retrotransposons of mosquitoes started only in 1990, many studies on transposable elements in mosquito were published during the years 1990 to 1998. These were briefly reviewed by Tu et al. (1998). This intensive activity probably reflects the medical importance of the investigated mosquito species: *Aedes aegypti* is the vector of Yellow Fever and *Anopheles gambiae* is the vector of malaria; these are among the most deadly human diseases. Indeed, the intensive studies yielded T.E. One example comes from *Aedes aegypti*. In this mosquito, Mouches et al. (1992) found *Juan-A*, a putative non-LTR retrotransposon that is 4.7 kb long, and has the "standard" features of this kind of retrotransposon with the "regular" ORF encoding two polypeptides for a *Cys*-rich nucleic acid binding motif and for an RT. *Juan-A* has an adenosine-rich 3' end: it is similar to the *Drosophila Jockey*, that was mentioned above.

A detailed investigation was conducted more recently into another family of non-LTR retrotransposons of the Yellow Fever mosquito *Aedes aegypti*: the *Lian* elements (Tu et al., 1998). These elements have many copies in each mosquito genome. About 1,400 copies of *Lian-Aa1* were estimated to exist per haploid *A. aegypti* genome. With respect to several features *Lian-Aa1* was regarded as a non-LTR retrotransposon, but unique features were also revealed. The central region of *Lian-Aa1* has a long ORF. The latter is flanked by 545 bp, at the end of the 5' untranslated region and 255 bp at the 3' end of the untranslated region. The ORF codes for three putative domains, but the *gag* domain, typical of other non-LTR retrotransposons, is missing. The N-terminal of an encoded polypeptide had similarity to an endonuclease domain. An RT was revealed in the middle of the ORF and its coding sequence was similar to RT of other studied non-LTR retrotransposons. Downstream of the RT there was also a putative code for an RNaseH domain. Several *Lian*-type elements were identified in *A. aegypti*. These were, at least partially, sequenced and found to represent versions of *Lian-Aa1* that were truncated at their 5' end.

5.5. A LINE Element in an Amphibian Organism

The non-LTR retroelements of amphibians received only very little attention, although it can be assumed that these elements also exist in this phylum. An exception is *Xenopus laevis* that was studied by D. Carroll and associates (see: Christensen et al., 2000). These investigators revealed two families of *LINE*-like elements in *X. laevis*: *Tx1L* and *Tx1D*. The *Tx1L* is a site-specific element. It encodes an apparent endonuclease (EN) that is related in its sequence to the apurinic-apyrimidinic endonucleases, that were mentioned above. This EN probably introduces a single break in a specific target-DNA site and initiates there the target-primed reverse transcription (TPRT) that is probably typical to non-LTR retrotransposons. *In vitro* expression of this EN sequence indeed showed the assumed cleavage capability.

5.6. LINE Elements in Fishes

By 1998 it became evident that retrotransposons, including non-LTR retrotransposons, probably exist in all the species of metazoa. Actually, up to now there has been no report that has stated that such T.E. are absent from any such species. This makes fishes a source of an "endless" number of non-LTR retrotransposons: there are about 25,000 (or more) species of fishes and each of these is a potential source of one or more types of non-LTR retrotransposons. We shall thus not attempt to catalogue all these elements. Those who are interested in following the search for non-LTR retrotransposons in fishes will find ample references in Volff et al. (2000). Actually, the team of this latter publication, M. Schartl and associates, of the University of Würzburg, arrived at the non-LTR retrotransposons of fishes via their genetic study of a tumor: *Tu* (see: Froschauer et al., 2001). The *Tu* locus is involved with this tumor and it can be suppressed by a tumor-modifier locus. Volff et al. (1999) focused on the *Xmrk* oncogene/proto-oncogene region in chromosome Y of the platyfish genus *Xiphophorus*. The melanoma is formed "spontaneously" in certain hybrids in this genus. We already mentioned above the work of the researchers from Würzburg in connection with their study on LTR retroelements. When these researchers sequenced a region of the Y chromosome of *X. maculatus* that was apparently involved with tumor formation, they came upon three "new" RT-carrying retrotransposons. They coined them: *Rex1*, *Rex2* and *Rex3* (for retroelement of *Xiphophorus*). Of these, *Rex2* was related to a previously detected non-LTR retroelement, *Maui* of the pufferfish (Fugu), *Fugu rubripes* (Poulter et al., 1998). *Rex1* was not assigned to any group of retroelements. The investigators focused on the *Rex3*. The *Rex3* had several standard features of non-LTR retrotransposons. It encoded sequences for an apparent endonuclease and an RT. Many of the sequences that hybridized with the *Rex3* probe were truncated elements that did not contain the 5' end of the full element. Instead of the poly (A) tail in many types of such elements, *Rex3* 3' tails consisted of tandem repeats of GATG. The common target-site duplication was also found in *Rex3*. The element was estimated to exist in about one thousand copies in each fish genome. When other fish species were analyzed for the presence of *Rex3*-like elements, it was found in many teleost species. Transcripts of *Rex3* were revealed. Thus, the authors concluded that *Rex3* is an active and autonomous non-LTR retrotransposon.

The same team of investigators from the University of Würzburg, Volff, Körting and Schartl (Volff et al., 2000) went on to investigate another *Rex* element from *X. maculatus*, the *Rex1*. This element is related to the *Babar* retrotransposon, a truncated retrotransposon of the fish *Battrachocottus baikalensis*. In the *Rex1/Babar* elements (as in other non-LTR retroelements) the 5' is frequently truncated, probably due to incomplete reverse transcription. An intensive evolutionary study was performed, mainly on the basis of RT sequences. Thus, ancient lineages of *Rex1* were found. Three of these were revealed in the "experimental" fish *Fugu rubripes*. Interestingly, *Rex1* sequences were not found in some fishes such as trout, pike, carp and zebrafish that diverted from *Xiphophorus* presumably 100–120 MYBP. In any case, the *Rex1* sequencing was helpful in better understanding fish evolution.

While the Würzburg team reached the *Rex* retroelements via tumor research, another team, from the University of Dalhousie in Nova Scotia, Canada, reached the *CiLINE2* element through their interest in aquaculture (fishery). Thus, Oliveira et al.

(1999) studied a *LINE2*-type (see Smit, 1996) retroelement of the cichlied fish (a tilapia type) *Oreochromis niloticus*. By sequencing the isolated clone, the authors identified in their clone sequences of the *LINE2* elements that were previously found in man and in *Drosophila melanogaster I Factor*. Following the screening of fish DNA sequences, the *CiLINE2* element was revealed in many tilapia fishes. The authors estimated that about 5,500 *CiLINE2* copies exist in each haploid genome of *O. niloticus*. This is a higher number than estimated for *Rex3* (about one thousand) and much higher than the number of *Rex1* copies in *Xiphophorus maculatus* (about 10 to 200). Interestingly, more than one hundred years earlier (as detailed above in the Historical Background) F. Miescher used fishes to identify what turned-out to be DNA – but then DNA became useful to reveal the phylogeny of fishes.

Not all *LINE*-type elements of fishes are high-copy-number elements. An example of a *LINE* element with a low copy number (10 to 20 elements per genome) is *Swimmer 1* (Duvernell and Turner, 1998) that was found in two teleost fishes.

5.7. LINE Elements in Mammals

We have covered above the "early" studies on non-LTR retroelements of mammals. Now we shall look at these *LINE*-like elements again and update the main information since about 1988.

5.7.1. The LINE Elements of Primates
Since about 15 years ago, monkeys and apes were gradually removed from serving as models for human biomedical research. Therefore, in our coverage of primate *LINE*s we shall actually deal only with humans.

We shall start with the human elements that are also termed *L1Hs*. One of the *L1Hs* is involved in a possible impact of the transposition of a non-LTR retrotransposon on the trend of history. It is the case of Czarevich Alexis. Czar Nicolas II, the last Russian ruler of the Romanov dynasty (1894–1917), married a Princess of Hesse. The German princess became Czarina Alexandra Feodorova and brought with her the gene for haemophilia (probably Haemophilia A, carried on one of her X chromosomes). Her son, the Czarevich Alexis (born in 1904) was afflicted with haemophilia. This caused major turmoil in the ruling capability of Nicolas II and Alexandra and possibly contributed to the termination of the Romanov dynasty.

Haemophilia A is an X-linked disorder of blood coagulation caused by the deficiency of Factor VIII. The gene encoding Factor VIII has 26(!) axons (separated by introns). Kazazian et al. (1988) of the John Hopkins University, in Maryland, screened 240 unrelated males inflicted with haemophilia A and studied two of them in detail. In both these males the investigators found that exon 14 was not normal. The exons were interrupted by an *L1* element. The lengths of the *L1*s in the two patients were 3.8 or 2.3 kbp, respectively. In both cases the inserts were flanked with a target-site duplication of 13 (or more) nucleotides. No insertions in the exon 14 were revealed in the parents of the patients. This strongly suggested that the *L1* was transposed *de novo* into the Factor VIII gene.

Was the transposition of an *L1* element into an exon of the gene for Factor VIII the cause of haemophilia in Czarevich Alexis? If it was, in which generation of Czarina Alexandra did it happen? These are questions for which there are no sure answers but the possibility that the transposition of a mobile element can affect human history cannot be denied. With respect to our concern in this book, the study of Kazazian et al. (1988) clearly indicated that *L1* elements can cause insertional mutations and these can cause human maladies.

The defective Factor VIII of human patients was then instrumental in identifying for the first time, intact and active transposable elements in man, by the team of Kazazian (Dombroski et al., 1991). It was known, before the study of this team from the John Hopkins School of Medicine, that the diversion from consensus sequence is not the same in randomly analyzed *L1* elements and in *L1* elements that are active in transposition. Hence, the regular diversion is about 5%, while the cDNAs of *L1* from a tumor cell line divert only less than 1% from the consensus sequence. This already indicated that an element that is still capable of transposition should have a sequence that is very close to the consensus sequence. The help came from patient JH-27, who had an *L1* insert in his gene for Factor VIII. The researchers traced the element from where the *L1* hopped into the Factor VIII site and found it on chromosome 22. It was a full-size element and was termed *L12A*. The 3' part of *L12A* was identical to the new insertion in patient JH-27. The question could then be asked whether the RT encoded by *L12A* really is active in reverse transcription (Mathias et al., 1991). For that the ORF2 (that includes the code for the RT) of *L12A* was engineered into a plasmid that included (upstream) a selectable marker and ORF1 from the *Ty1* element of yeast. The question could then be asked if the expression of the fusion protein would be functional in the yeast cells, resulting in the production of VLP (retrovirus like particles). The answer was affirmative. When, as control, a modified ORF2 from *L12A* was used that had an inactivating mutation in its code for RT, no VLP were detected. The same team of the John Hopkins Medical School (Wood-Samuels et al., 1989) reported another *L1* insertion into the human Factor VIII. The investigation went "backwards". A patient (JH-25) had haemophilia and an *L1* element was found in an intron upstream of exon 11. Significantly, while the insertion was also detected in his maternal predecessors, the latter did not have haemophilia.

L1 elements are involved in several other human maladies (see Ostertag and Kazazian, 2001 for extensive review). Miki et al. (1992) in a collaborative study of scientists from Tokyo, Japan and the John Hopkins Oncology Center, Maryland, studied the APC gene. This is a gene that suppresses polyposis coli, the development of numerous polyps in the colon, some of which develop into colorectal tumors. When the APC gene from patients with colorated cancers as well as from healthy humans was investigated, an insertion was revealed in the APC gene of the patients. Although the insert was short (about 750 bp), several regions of this insert were almost identical with the consensus sequence of *L1*. The insertion was into an exon of the APC gene and it was flanked by target-site duplications. An earlier case of an apparent *L1* element was reported by D. Givol and associates (Katzir et al., 1985). These investigators revealed an insert into the oncogene *c-myc* of the veneral tumor of dogs. Several features of this integrated element indicated that it could be categorized as a "*Kpn*I" element, meaning a *LINE1* element. This was a truncated element of 1.8 kbp. No search for a full-length element in the dog was apparently reported.

The team of H. Kazazian and associates, from John Hopkins, revealed another "new" *L1* element. The latter came to light in a family (JH-1001) with Becker/Duchenne muscular dystrophy. Affected patients develop muscle weakness at about the age of 10 years and become wheelchair-bound at about 20 years. These investigators (Holmes et al., 1994) found that the dystrophy gene of an affected individual of the JH-1001 family had an insertion with the characteristics of a rearranged *L1* element. This led the investigators to a "new", apparently still transposing, *L1* element on human chromosome 1q. This element, *LRE2*, exists in most humans that were analyzed but does not exist in other primates such as chimpanzee and gorilla. It seems that *LRE2* appeared 1–2 million YBP. It produces a cDNA that has two ORFs as well as other consensus features of a *LINE1*. It has a short truncation at its 5' end of 21 bp, but this does not seem to affect its transcription. Whether *LRE2* is autonomous in its transposition or its mobility is assisted by host factors, is not yet known. We again face the irony of nature: a defect in human mobility (Duchenne muscular dystrophy) led to the better understanding of the mobility of DNA.

Insertions of *LINE*s (as well as *Alu* elements) into the exons and introns of human genes will most probably be detected in the future when diseases will be correlated to specific nucleotide sequences of the human genome. However, there is another, general means by which mutations in the human genome may cause specific diseases, and this is by unequal recombination. Some of these may lead to deletions in essential genes. Such unequal recombination could result from pairing between *LINE* elements that flank an exon. Such a case was reported by Burwinkel and Kilimann (1998). These investigators studied the human gene that encodes the β subunit of phosphorylase kinase (*PHKB*). Deficiency of *PHKB* gives rise to a glycogene storage disease. They found that by the unequal homologous recombination, a deletion of 7,574 bp occurred that included exon 8 of the gene for *PHKB*.

Another genetic disease is attributed to a *LINE1* element-induced molecular rearrangement from studies by J. Zhou and associates (see: Segal et al., 1999). The *Alport syndrome* (*AS*) is a predominantly X-linked disorder, also termed *hereditary nephritis*. It is marked by hematuria and progressive renal failure in affected human males. It is often accompanied by hearing loss and ocular abnormality. Female carriers, who are heterozygous for the affected X chromosome, usually exhibit a milder phenotype. In addition, a small subset of affected individuals manifest an associated syndrome: *diffuse leiomyomatosis* (*DL*) that is expressed by smooth-muscle tumors in the esophagus, upper airways and female genitals (because *DL* affects both males and females). Most *AS* cases are attributed to the gene COL4A5 that is arranged head-to-head with COL4A6. Patients of *AS/DL* frequently have deletions in these genes, and this is probably causing defected collagen. Unequal homologous recombinations that resulted from fusion between *L1* elements that are located in this locus caused deletions of 13.4 and ~40 kpb, respectively, in two analyzed cases.

5.7.2. The LINE Elements in Rodents

There is probably nothing unique with respect to *LINE*s of rodents, in comparison to other mammals. Rodents are serving frequently in biomedical research and therefore have gained attention of those who have studied *LINE*s. Full sequencing of a rodent (rat) *LINE* was already achieved in an early phase of *LINE* studies (D'Ambrosio et al., 1986). From these and other studies it emerged that the full-length *LINE*s of rodents have the same general structure as human *LINE*s. The specific *LINE3* studied by these authors is probably represented 40,000 times in the rat genome and is transcribed by Pol II. The transcripts of *LINE3* were found in the nuclei of rat livers.

The testis is an organ in which both somatic and germ-cells are located. Branciforte and Martin (1994) used the testis of mice to search for *LINE* activity. They found that the expression of *L1*, manifested by both *L1* RNA and *L1* proteins, is strongest in the testis of postnatal, 14-day-old mice. The proteins encoded by ORF1 of the *LINE1* of mice could be detected in spermatocysts that were in the leptotene and zytgotene phases of meiosis. In the 14-day postnatal mice these meiotic cells comprise a relatively high proportion of the cells in the seminiferous epithelium of the testis: about 14%. In mature male mice this proportion is reduced to 2%. This means, probably, that there is a "window" of active *LINE* transposition in male mice that is not only restricted spatially to the germ-line, but also temporarily to a very young age.

Later, detailed analyses of mice *L1* elements revealed additional information (Goodier et al., 2001). One finding concerns the 5' untranslated region (UTR) of mice *L1*. It has a tandem repeat of about 200 bp. The repeats were termed "*monomers*". It was found, by linking the *monomers* to reporter genes, that they have a promoter activity; additional *monomers* enhanced the transcription of reporter genes. Actually, several *LINE1* lines of mice differ in the details of structures of their 5' UTR. It appeared that the number of *LINE1* sub-families in mice is in a process of rapid amplification (DeBerardinis et al., 1998). One such *LINE1* line was termed G$_F$. The latter line differed not only in its 5' UTR but also in its ORF1. The authors concluded that G$_F$ is a recently evolved *LINE1*. This brings the sub-families of *LINE*s in mice to three: T$_F$, A and G$_F$, with a total of about 3,000 full-length (and potentially active in transposition) *L1* elements. This renders the mice *L1* elements in a great excess over the number of *L1*s estimated to be active in the human genome.

Casavant et al. (1996) were engaged in solving an enigma. It appeared that only a very small fraction of the *LINE*s in mammals are still actively transposing. Is there an evolutionary process going on in *LINE*s? This may mean that at any stage of the evolution of a species there is only one *master LINE* that propagates and produces transposition-active *LINE*s. These investigators chose the deer mouse *Peromyscus* for their study. They found two *LINE1* lineages in deer mice; these lineages split before the *Peromyscus* species radiated from other rodents. Hence, while in humans there is probably only one *master L1* line, there are two *master L1* lines in *Peromyscus*.

5.7.3. L1 Elements Transposition in Cultured Cells
After the Boeke/Kazazian team isolated human *L1* elements that transposed into specific genes, they devised an elegant (though rather elaborate) procedure to follow retrotransposition in cultured human cells. For that they attached a reporter cassette to an *L1* element (*L1.2*). This reporter should enable to select transposed *L1.2* elements because the plasmid was constructed in a way that only after retrotransposition (removal of an intron) will the reporter be expressed. The elaborated plasmid was then introduced (and expressed) into cultured human cells as well as into a mouse fibroblast cell line. With this system Moran et al. (1996) could show that *L1.2* and *LRE2* are capable of retrotransposition. They could also obtain an answer to the question of which ORF encoded genes are essential for this

retrotransposition. It was found, for example, that certain point mutations in the sequences that code for proteins drastically reduced retrotransposition.

The same approach was utilized by the team of Kazazian and associates (Kimberland et al., 1999) with two additional *L1* elements: *L1β-thal* and *L1RP*. These were full-length elements (not truncated at their 5' end) that were previously found as integrated into the genes for β-globin (β-thal) and retinitis pigmentosa-2 (RP2), respectively. Both elements maintained a high capability for retrotransposition in cultured human cells (HeLa cells).

5.7.4. The Impact of LINEs on the Mammalian Genome

The ability of *LINE*s to insert into DNA sequences, to cause unequal homologous recombination and some other interferences with the faithful replication and expression of the genome, were mentioned for specific cases above. Kazazian (2000) brought the capability of *L1* to shape mammalian genome into genetic perspective. In this summary, several phenomena of *L1*-induced changes in the genome were noted. Examples of two of them have been detailed above: insertions of *L1* into genes and mispairing leading to unequal crossing-over between *L1* elements, that cause deletions. Another phenomenon is the transduction of 3' flanking sequences that have been carried along with *L1* elements. The poly (A) signal of *L1* RNA is weak. This may lead to a failure of *L1* RNA-cleavage at this poly (A) site. Cleavage could, consequently, occur after the next (downstream) poly (A) signal. The mammalian genome can also be shaped by a reintegration of a reverse-transcribed mRNA that is not derived from a *LINE* element. Such cases were described in detail by Esnault et al. (2000). Pickeral et al. (2000) provided evidence that suggested that about 15% of the full-length *L1* elements bear evidence of flanking DNA segment transduction. As they estimated 600,000 *L1* elements to exist in the human genome, about 1% of this genome is DNA transduced by *L1*. This is about the same fraction as occupied by exons. *LINE* elements can mobilize transcribed DNA that is not associated with a *LINE*. This happens when the enzyme machinery of an active *LINE* acts on the "alien" mRNA and causes it to "retrotranspose" into the genome. This will generate *pseudogenes*. These "pseudogenes" obviously lack introns as well as promoters. They do contain a poly (A) end and have a target-site duplication. Interestingly, this "retrotransposition" of pseudogenes can be activated by *LINE* elements only if they have no deleterious mutations in their ORFs; furthermore other retroelements such as retroviral-like elements, are not able to cause this retrotransposition of transcripts.

Finally, the *LINE* elements are probably responsible for the transposition of *Alu* elements. This is assumed to occur by *trans* activity supplied by the proteins encoded by *LINE*s, although it should be noted that up to present there is no experimental proof that *Alu* are mobilized by *LINE*s. As *Alu* elements occupy a very high proportion of the human genome (10 to 12%), their role in shaping the genome by the mechanisms mentioned above, for *LINE*s, such as insertional mutations, mispairing and unequal crossing over, should be taken into consideration.

5.7.5. Concluding Remarks on Mammalian LINEs and a Note on Fungal LINEs

In order to obtain the overall picture on what is known on mammalian *LINE*s we shall describe the "consensus" of these elements, as represented by the human *L1* element. It is a DNA sequence of 6 kbp. It has a 5' untranslated region (5' UTR) that

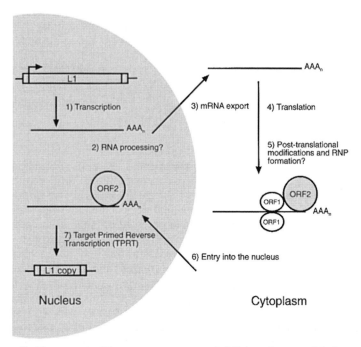

Figure 42. The steps in L1 retrotransposition. A full-length active L1 element is transcribed from its internal promoter (bent arrow) to produce a bicistronic mRNA. It is currently unknown if the RNA undergoes processing or how the RNA is exported from the nucleus. Once in the cytoplasm, the ORF1 and ORF2 proteins are translated and specifically function on the RNA that transcribed them (cis preference). At least one L1 RNA molecule, one ORF2 molecule, and one or more ORF1 molecules may assemble into a ribonucleoprotein (RNP) complex that is an intermediate in retrotransposition. Both the ORF2 protein and associated L1 RNA must gain access to the nucleus, where the L1 RNA is reverse transcribed and integrated into a new genomic location by a process called target primed reverse transcription (TPRT). Many L1 elements undergo 5' truncation or 5' inversion and truncation during the TPRT process, resulting in an inactive DNA copy of the original element. The TPRT process creates 7–20-bp target site duplications that flank the L1 element. (From Ostertag and Kazazian, 2001).

contains an internal (probably weak) promoter. There are two ORFs separated by a short intergenic region, a 3' untranslated region (3' UTR), a poly (A) signal (AATAAA) and a poly (A) tail (A(n)). The *L1* elements are often flanked by 7–20 bp target-site duplications (TSD). The ORF2 encodes RT and EN proteins as well as a conserved cystein-rich motif (C). *L1*-like elements that are shorter than 6 kbp are usually truncated at the 5' end. These constitute the majority of the *LINE*-like elements.

The transcription of the *L1* from an internal primer takes place in the nucleus. The transcript then moves out of the nucleus and the translation of the ORF regions takes place in the cytoplasm, where translation of proteins that are required for transposition (e.g. RT, EN and C) takes place. A complex that includes the transcript and transposition-related proteins is then formed. This ribonucleoprotein

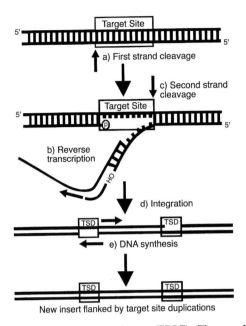

Figure 43. Target primed reverse transcription (TPRT). The non-LTR retrotransposons are thought to integrate by TPRT. This TPRT model is based upon the mechanism worked out in vivo for the R2 retrotransposon. (a) During TPRT, the retrotransposon's endonuclease cleaves one strand of genomic DNA at its target site, producing a 3'-hydroxyl (OH) at the nick. (b) The retrotransposon RNA inserts at the nick and the retrotransposon's reverse transcriptase uses free 3'-OH to prime reverse transcription. Reverse transcription proceeds, producing a cDNA of the retrotransposon RNA. (c) The endonuclease cleaves the second DNA strand of the target site to produce a staggered break. (d) The cDNA inserts into the break by an unknown mechanism. (e) Removal of RNA and completion of DNA synthesis produces a complete insertion flanked by target site duplications (TSDs). (From Ostertag and Kazazian, 2001).

complex returns to the nucleus. There it undergoes a Target-Primed-Reverse-Transcription (*TPRT*, Figure 32) and the element is integrated in the host DNA. In most cases, the cut in the host DNA is staggered, thus direct repeats of the host DNA flank the inserted element. This summary of the transposition of non-LTR retroelements is schematically shown in Figures 42 and 43. Several components of this transposition are not yet fully understood. For example, how the nucleoprotein complex returns to the nucleus is still enigmatic. Also, the reason why most transposed elements are truncated at their 5' end is not clear.

Another note concerning the non-LTR retrotransposons is that our coverage of this element has not been in phylogenetic order. We started with the discovery of these elements in mammals and then provided examples from various invertebrates and vertebrates, reaching, finally, these elements in mammals. On the way we skipped several major groups of organisms in which these elements have been reported. As already noted, non-LTR retrotransposons probably exist in all eukaryotes. We shall finally mention one additional group of eukaryotes: fungal organisms. While detailed studies on LTR retrotransposons have been performed in yeasts since 1979 (see above), the non-LTR retrotransposons were revealed in a hyphal fungus (*Neurospora crassa*) only 10 years later. The study of *LINE*-like elements in *N. crassa* (see: Cambareri et al., 1994) was initiated by J.A. Kinsey and associates of the University of Kansas. They termed these elements *Tad*. The *Tad1* was studied in detail and has a length of about 7 kbp with the common ORFs typical of the *LINE*-like elements. The transposition activity of *Tad1* could be clearly shown due to the phenomenon of *heterocaryon formation* in hyphal fungi. Briefly, when hyphae of two genetically different fungal lines meet each other (e.g. in the border between the two respective colonies) the inter-colonial hyphae may fuse. The nuclei in ascomycete fungi inhabit the hyphae without cell-walls between these nuclei. Therefore, by hyphal fusion, the two populations of nuclei meet and a heterocaryon is formed. Kinsey and associates utilized this heterocaryon formation. They included in one heterocaryon genetically marked nuclei that were devoid of *Tad1* as well as marked nuclei that did have *Tad1*. It was found that *Tad1* was transposed from one nucleus to another. The transcribed *Tad1* was most probably moving out of the donor nucleus (see Figure 42) and complexed in the cytoplasm of the heterocaryon hyphae, with the required ORF proteins. The complexed mRNA then entered the host nucleus and integrated by TPRT into the host DNA. In fungal organisms such as *Neurospora*, it is a routine procedure to "label" nuclei with genetic (selectable) markers. Then, when spores (each with a single nucleus) are plated on selective media, the genotypes of the respective nuclei can be identified and analyzed for any additional DNA sequences, such as inserted *LINE*s.

The same *Neurospora* fungus that was instrumental in the hands of Beadle and Tatum (in 1940) in initiating the one-gene-one-enzyme concept (and providing a Noble Prize) was also useful, after 54 years, in providing proof for the mobility of a *LINE* element! The source of *Neurospora* is spoiled bread. In biology, an unpleasant source can yield beautiful results.

5.8. LINE-like Elements in Plants

The discovery of *Cin*, a long, non-LTR retrotransposon in maize, by Schwarz-Sommer et al. (1987) of Cologne, was quite a surprise. Before 1987, no such elements were revealed in maize, although maize was already known then as the host of other Class I and Class II T.E. Actually, the *LINE*s were first thought to be confined to primates, then to mammals, thereafter to vertebrates and only finally they were also found in invertebrates. Because *LINE*s do not produce enveloped, retrovirus-like infectious-particles, their horizontal movement is not likely. This suggests that this maize *LINE* element was carried along during evolution since about the formation of the early eukaryotes, into an advanced monocot plant. These authors found several *Cin*s. One of them was found inserted in the *A1* gene, but they recovered several families of *Cin*, and even several *Cin-4* elements. The *Cin-4* element family ranged in length from about 1 kb to about 6.5 kb and most had a length of about 4.5 kb. As in animal *LINE* families, the individual elements were similar at their 3' end but were truncated at the 5' end. They were generally very similar to animal *LINE*s and we shall therefore not describe the *Cin-4* elements in detail.

Six years elapsed until another *LINE*-like element was reported in another plant,

Lilium speciosum (Leeton and Smyth, 1993). This element, termed *del2*, was found in many copies in the lily genome: approximately 250,000! The structure of *del2* was similar to animal *LINE*s and its ORFs coded for the same proteins as in these elements. The code for RT was similar to the code for *Cin-4* RT. The *del2* elements occupy about 4% of the large *L. speciosum* genome. There are other *del* elements in other species of *Lilium*, but with much less copies per genome.

Soon after the discoveries of *LINE*-like elements in the above-mentioned two monocot plants (maize and lily), such elements were revealed in numerous dicot plants. However, we shall first go back on the phylogenetic ladder to algae.

Higashiyama et al. (1997) of the Hiroshima University, Japan were analyzing the small chromosome I of the unicellular algal species *Chlorella vulgaris*. They revealed there the *Zepp* elements. In this case it was of greater interest where *Zepp*s were found rather than the characteristics of *Zepp*s. The *Zepp* element was discovered in the subtelomeric region of chromosome I of *Chlorella vulgaris*.

We shall devote a special, short section, on *telomeres* because T.E. are involved in these structures. Here we shall only note that telomeres are vital structures at the chromosome terminals. They protect chromosomes from DNA degradation and avoid incomplete replication of DNA. At least in animals there is a correlation between the shortening of the telomeres and the cessation of cell division. In many plants there are telomeric repeats that consist of repetition of the seven bases 5'-TTTAGGG. In chromosome I of *C. vulgaris* the telomeres consist of about 500 bp, meaning about 70 repetitions of the seven bases motif. The *Zepp* elements were revealed on one arm of chromosome I, proximal the telomeric structure. There were several *Zepp* elements. In many cases the *Zepp* element inserted into a previous *Zepp* element (nested insertion). This means that the *Zepp* itself served as a preferred target for *Zepp*-integration. The direction of the *Zepp*s was, with the poly (A), towards the centromere, except for one element, which had its poly (A) tail towards the telomere. A similar cluster of *Zepp* elements was revealed in an additional end of a *C. vulgaris* chromosome. The *Zepp* elements were clearly of the *LINE*-type, although most of them were truncated at the 5' end. Interestingly, mostly full-size transcripts of *Zepp* appeared after the alga was exposed to stress: a heat-shock.

Finally, it should be noted that while the *Zepp* elements were found in this specific location, most *Zepp* elements of *C. vulgaris* reside at non-telomeric locations and only in two, out of the 32 algal chromosomes, were the *Zepp* elements associated with the telomeres. The full-length *Zepp* was found to be 8,943 bp, and about 140 copies reside in the genome of *C. vulgaris* (Noutoshi et al., 1998).

We now return to angiosperm plants. After *LINE*-like elements were reported in monocot plants, the first such elements were reported in a dicot plant: sugar beet (*Beta vulgaris*) by Schmidt et al. (1995) of the John Innes Centre, Norwich, UK. These investigators termed the *LINE*-like elements of sugar beet *BNR*s. The genome of *B. vulgaris* consists of 750 Mb and it was previously estimated that 60% of the genome are repeated sequences. The name *BNR* stands for beet non-viral retrotransposon. Several *BNR*s were found and these had similarities with the previously revealed maize element *Cin-4* and the *del2* element of lily.

One element, *BNR1* was studied in more detail. It was found in clusters and highly amplified in the genome of sugar beet. Fluorescent, *in situ* hybridization indicated that *BNR1* exist in all chromosomal arms but only of one chromosome pair. It was also found that *BNR1* is excluded from the nucleolar organizing region.

There are sometimes considerable gaps in time between submission and publication of articles. The article of Schmidt et al. (1995) was received in January 1995 and accepted for publication in *Chromosome Research* in February 1995. This quick acceptance is in line with the name of the publisher of this journal: Rapid Communications of Oxford Ltd. A less "rapid" acceptance was the fate of the article of Wright et al. (1996). This article was with the editors of *Genetics* for four months and then four additional months with the printers. Thus, Wright et al. could rightfully claim that theirs was the first report on a *LINE* element in a dicot. Wright et al. (1996) focused on the genome of *Arabidopsis thaliana* and stressed the small size of its genome, about 100 Mb (10^8 bp) in comparison to some monocot genomes that contain 10^{11} bp. The authors end their introduction with an interesting statement: ".. the small size of the *A. thaliana* genome is (at least in part) the consequence of the failure of retrotransposons to successfully amplify and colonize its genome".

Nevertheless, T.E. do exist in this species and one *LINE*-like element was located near the ABI 3 gene. It was termed *Ta11-1*; it should have had a good scent because it was isolated from a cosmid (a λ-phage vector that can contain about 35–47 kbp of alien DNA) that has the name *4711*! Its length was 6.2 kb and it encoded the proteins of animal *LINE*s in two overlapping ORFs. Once isolated the *Ta11-1* could serve to look for similar additional *LINE* elements. Such elements were indeed found in the *A. thaliana* genome. One of these, *Ta7*, was found in the mitochondrial genome. The copy number of *Ta* elements was low and they are not apparently active in transposition.

The *Ta11-1* was analyzed in detail. Its total length was found to be 6,077 bp and it was flanked by two host duplications of 15 bp each. Two ORFs were revealed that were separated by the *LINE*-typical, −1 frame shift. ORF1 encoded a $CX_2 CX_4 HX_4C$ motif that is characteristic for *gag* proteins. ORF2 encoded an RT. In general, *Ta11-1* had similarities to animal non-LTR retrotransposons. Further search in the *A. thaliana* genome revealed additional *LINE*-like elements. By the decoded amino acid sequence of the RT, these elements were closely related, but somewhat different from the RT of monocot *LINE*s and animal *LINE*s. The uniqueness of the *LINE*-like elements of *A. thaliana* was their low copy number: none of the 17 different *Ta* elements appeared more than 7 times in the genome. The authors suggested that during the evolution of *A. thaliana* there was a reduction of the copy number of *LINE*s and this could be an explanation for the small size of the *Arabidopsis* genome.

When, by 1999, Noma et al. (1999) and Schmidt (1999) reviewed the *LINE*s of plants, several additional plant species were found to harbor these elements. True, the reports first included mainly crop plants, such as rice, maize, wheat, barley and ryegrass (of the Gramineae family) as well as turnip, spinach, tobacco, potato and eggplant (of dicot families). But *LINE*s obviously are not restricted to crop plants. *LINE*s were also revealed in a conifer, the pine.

When a considerable part (17 Mb out of about 130 Mb) of the nucleotide sequence of *A. thaliana* became available, a further search for T.E. in this species could be performed (Le et al., 2000). The results did not cause a major change in the picture that had emerged previously, from the study of Wright et al. (1996).

Additional reviews dealt in detail with *SINE*s and *LINE*s of plants (Kumar and Bennetzen, 1999; Bennetzen, 2000) and a detailed survey of T.E. in rice was provided by Turcotte et al. (2001). The sequencing of the *Arabidopsis* and rice genomes is now complete. The picture of the T.E. populations in these plants will therefore soon become rather clear. We can guess that in the future *LINE*-like elements will be found in almost all investigated plant species. We may also understand in the future the control of the copy number of T.E. Did plants that have a small genome developed systems to reduce *LINE* element transposition, or even to eliminate the existing *LINE* elements?

CHAPTER 5

TELOMERES AND TRANSPOSABLE ELEMENTS

1. CHARACTERISTICS OF TELOMERES

Eukaryotic organisms developed means to protect the DNA of chromosomes from terminal degradation. These termini are occupied by *telomeres*. Up to the late eighties of the last century, telomeres were defined more by their *role* than by their structure. There was a good reason for such a definition: very little was known of the molecular details of telomeres. What was known was that they are at the tips of all chromosomes and they must have a role and function in maintaining the integrity of the chromosomal termini. They should avoid the stickiness that was observed by Barbara McClintock, between broken ends of chromosomes (see Historical Background). Also, conventional DNA polymerases cannot complete the synthesis of both strands of a blunt-ended DNA template: one of the DNA strands will not replicate to its end, resulting in a single stranded "tail". In addition, a double stranded DNA end that is not protected is exposed to exonuclear degradation by the cellular (actually nuclear) enzymes. In his excellent *Genetic Manual*, G.P. Redei (1998) furnished a description of telomeres that started with the brief definition: "Telomeres are special terminal structural elements of eukaryotic chromosomes rich in T and G bases". Redei already noted that there are exceptions to the "regular" telomeres, in *Drosophila*, where transposable elements constitute the building stones of telomeres. Therefore, although telomeres are outside the scope of this book, these entities cannot be ignored by us. For an update on this subject, the comprehensive review of Shore (2001) is recommended. Since intensive research on telomeres and especially on the chromatin structure and function at the telomere regions is in progress, we shall only provide a partial picture of these entities.

In most analyzed organisms chromosome DNA ends with multiple tandem repeats of a few bases. In mammals it is 5'-TTAGGG-3' while in yeast it is TGGG. A list of telomeric-DNA sequences, in several groups of organisms, is provided in Table 6. Usually there are several G bases at the 3' end of the repeats. The repeats have different total lengths. This length is only 300 bp in yeast, 10 kbp in humans and 50 kbp in mice. These numbers are means; the lengths are variable in different cells of the same organisms. This means that the telomeres undergo cycles of (programmed) elongation and shortening. The above-mentioned numbers refer to the germ-line cells of animals. From the early work on yeast (*S. cerevisiae*) it emerged that the G-rich single-strand overhang of DNA is required as substrate for the enzyme that extends the telomere: the *telomerase*. The molecular structure of telomerases was uncovered by two pioneering studies of T.R. Cech and associates (Lingner et al., 1997; Nakamura et al., 1997). The telomere repeats are generated by this telomerase that is a nucleoprotein holoenzyme that acts as a kind of reverse

Table 6. Sequences of telomeric DNAs. The sequence of the strand running 5' to 3' from the end of the molecule forward to its center is presented first.

Organism	Sequence
Protozoa	
Tetrahymena	C_4A_2/T_2G_4
Paramecium	$C_3{}^C_AA_2/T_2{}^G_TG_3$
Oxytricha	C_4A_4/T_4G_4
Plasmodium	$C_3T^A_GA_2/T_2{}^T_CAG_3$
Trypanosoma	C_3TA_2/T_2AG_3
Giardia	C_3TA/TAG_3
Slime molds	
Physarum	C_3TA_2/T_2AG_3
Didymium	C_3TA_2/T_2AG_3
Dictyostelium	$C_{1-8}T/AG_{1-8}$
Fungi	
Saccharomyces	$C_{2-3}ACA_{1-6}/T_{1-6}GTG_{2-3}$
Kluyveromyces	$ACAC_2ACATAC_2TA_2TCA_3TC_2GA/TCG_2AT_3GAT_2AG_2TATGTG_2TGT$
Candida	$ACAC_2A_2GA_2GT_2AGACATC_2GT/ACG_2ATGTCTA_2CT_2CT_2G_2TGT$
Schizosaccharomyces	$C_{1-6}G_{0-1}T_{0-1}GTA_{1-2}/T_{1-2}ACA_{0-1}C_{0-1}G_{1-6}$
Neurospora	C_3TA_2/T_2AG_3
Podospora	C_3TA_2/T_2AG_3
Cryptococcus	$A_2C_{3-5}T/AG_{3-5}T_2$
Cladosporium	C_3TA_2/T_2AG_3
Invertebrates	
Caenorhabditis	GC_2TA_2/T_2AG_2C
Ascaris	GC_2TA_2/T_2AG_2C
Parascaris	$TGCA_2/T_2GCA$
Bombyx, other insects	C_2TA_2/T_2AG_2
Vertebrates	C_3TA_2/T_2AG_3
Plants	
Chlamydomonas	C_3TA_4/T_4AG_3
Chlorella	C_3TA_3/T_3AG_3
Arabidopsis	C_3TA_3/T_3AG_3
Tomato	$C_3A^T_AT_2/A_2{}^A_TTG_3$

(From Zakian, 1995).

transcriptase and contains its own internal RNA primer. The telomerase of yeast was found to contain several protein components that are encoded by the respective genes. Its activity ensues with the termination of the conventional DNA replication, during the late S-phase of the cell-division cycle. Proteins of the SIR family (especially the *Sir2* protein that is a NAD-dependent deacetylase) are probably involved in the programming of the telomerase activity but details are still under investigation. It appeared, though, that telomeres, by themselves, may not hold the only key for the prevention of end-degradation of the chromosomal DNA. Specific proteins (capping proteins) associated with the chromosome termini may play an essential role in this protection, possibly also serving to link these termini to the nuclear envelope (in fungal organisms the nuclear envelope does not disintegrate during the M-phase of cell division). Such a "capping" protein, *Ku*, was also found at the termini of mice chromosomes and it probably has a role in preventing telomere/telomere fusion. Eukaryotes have an elaborate "check-point" system for double-strand breaks (DSB) in their DNA. This system polices the DNA for DSB and stops replication when it encounters a DSB. It appears that such "check-point" systems are also involved in telomere elongation.

Why should we deal with telomeres in a book on transposable elements? There are two good reasons for that (here the author feels uneasy: some scholars claim that whenever more than one reason is suggested, all the reasons are dubious). First, the reverse transcriptases of telomerases that were analyzed have phylogenetic relations with the RT of non-LTR retrotransposons. The second reason will be presented below: the telomeres in *Drosophila* are composed of transposable elements rather than of tandem repeats of oligonucleotides.

2. THE TELOMERES OF *DROSOPHILA*

The telomeres of *Drosophila* are unique. We do not know yet if they exist only in *Drosophila* or if similar telomeres will be found in other insects or in other eukaryotes. Whatever, the telomeres of *Drosophila* are useful for scholars who have a *teleological* approach to evolution. Their claim is that transposable elements, that commonly have negative effects by insertion into vital genes, preventing their expression, should also have a beneficial role (a "purpose"). The utilization of transposable elements for the construction of telomeres in *Drosophila* is a case-example for the claim of these scholars. A full coverage of the telomeres of *Drosophila* was recently presented by M.-L. Pardue of the MIT, Cambridge, MA (Pardue and DeBaryshe, 2002).

There is irony in the fact that early studies on telomeres focused on *Drosophila* telomeres, while later these telomeres turned out to be an exception: they differ substantially from all other known telomeres. Historically, after the pioneering studies by H.J. Muller of *Drosophila* telomeres, the molecular approach to telomere structure was applied to *Tetrahymena*, a ciliated protozoon. This organism has a macronucleus that disintegrates into minichromosomes; the latter end with tandem repeats of C_4A_2/T_2G_4. Further molecular studies revealed similar structures in all analyzed organisms until the study turned back to *Drosophila* telomeres. These flies do not have the standard repeats in their telomeres. Instead, the *Drosophila* chromosome ends are maintained by repeated transposition of two special non-LTR retrotransposons: *HeT-A* and *TART* (for a brief review of the early studies on *Drosophila* telomeres see Levis et al., 1993). While these two elements could be grouped with non-LTR retrotransposons, they differ substantially from each other and are rather unique among this group of T.E. There are also a few similarities between *HeT-A* and *TART*. Both have poly (A) tails and proximal to their tails there is a 3' untranslated region that is quite extensive. As is typical for populations of *LINEs*, many of the *TART* and *HeT-A* elements are truncated at their 5' end while only a few retain their full length. *TARTs* have several features that make them rather similar to non-LTR retrotransposons. They have a pair of non-terminal repeats and two ORFs. *TARTs* encode a reverse transcriptase (RT) and a *gag* protein. The *HeT-A* elements have only one ORF. The promoter of *HeT-A* is unique and appears to be an evolutionary intermediate between the promoters of non-LTR retrotransposons and those of LTR retrotransposons. How this evolved is mysterious because these two T.E. diverted very early in the evolution of eukaryotes. *HeT-A* elements do not encode an RT and actually lack the *pol* code. It is assumed that both *TART* and *HeT-A* integrate by target-primed reverse transcription (TPRT). By its RT the *TART* element is related to the *Jockey* non-LTR retrotransposons of arthropods. As *HeT-A* has no RT it is not clear where to put it on the phylogenetic tree of T.E., but in some features it is close to certain non-LTR retrotransposons of insects, possibly also to the *Jockey* elements. Obviously, the most typical feature of *TART* and *HeT-A* is the specificity with which they transpose to chromosome ends. It is not clear what leads these elements to this specific location. It is probably not the DNA sequence at the chromosomal termini because it was observed that these elements can "heal" broken ends of chromosomes. There are sites into which these elements will not integrate when the chromosomes are intact.

It is noteworthy that neither *TART* nor *Het-A* exist in the euchromatic regions of *Drosophila* chromosomes. It would be interesting to learn how these elements behave when transferred (by genetic transformation) to other insects that have the conventional composition of telomeres. In any case, the molecular basis for the confinement of *TART* and *HeT-A* to heterochromatic regions is not yet clear. Could it be that they *produce* the heterochromatin wherever they insert?

The exact sequence of these elements is presently based on only a very few full-length elements. Thus, two *HeT-A* elements that have been fully sequenced differ in a region that encodes 55 amino acids. The *gag* protein among *HeT-A* elements can differ substantially (by about 20%). The *gag* proteins encoded in the single ORF of *TART* also differs among individual elements of the same fly.

Interestingly, *TART* produces both antisense and sense transcripts and the former is much more abundant than the latter. *HeT-A* does not produce antisense transcripts.

As noted above, the genus *Drosophila* is the only genus known to have retrotransposon-telomeres instead of telomerase-telomeres. We already described cases of non-LTR retroelements that have associations with telomeres in other organisms. Such are the Y' elements of yeast and two elements of the silk worm. The latter insert specifically into the TTAGG telomerase-repeat-arrays of this insect. We also mentioned above the *Zepp* element of *Chlorella* (a unicellular green alga) that may occupy a telomeric position on one of the chromosomes.

Finally, for the telomeres of *Drosophila*, the present author allows himself a speculation. Is there an advantage for the *TART* and *Het-A* elements to become telomere components? Telomeres that are derived from telomerase activity are vulnerable to the usual risk of mutation. As there are only a few genes in each genome that code for telomerase components, mutations in these genes could impair the formation of telomeres. Not so in telomeres built of *TART* and *Het-A*; there is an "endless" number of these elements and mutations will not eradicate them. However, here comes the obvious question: if these elements are such a good solution for the "immortality" of telomeres, why were they found only in *Drosophila*? True, there are at least 15 million species of insects, and analyzing even a fraction of them for the existence of T.E. in their telomeres is far beyond a reasonable approach. Possibly *Drosophila* is not the only genus with T.E. in the telomeres.

CHAPTER 6

CLASS II TRANSPOSABLE ELEMENTS
IN EUKARYOTES

1. GENERAL CHARACTERISTICS

The (re)discovery of *Minos*, in Heraklion, Crete, was heralded in 1994. This was not the mythological king of Crete and son of Zeus and Europa, who's wife conceived the monster *Minotaurus*, after mating with a bull. The (re)discovered *Minos* (Franz et al., 1994) is a Class II T.E. that we shall mention below when dealing with T.E. of insects. I do not know why investigators are inspired by Greek mythology when they give names to T.E. We saw this inspiration above, when Russian scientists gave mythological names to Class I T.E. of insects.

As the title implies, this part will deal with the large group of T.E. termed Class II elements. As narrated in the Historical Background, the awareness of T.E. started with this group of elements through the studies of Barbara McClintock in the early nineteen forties.

The Class II T.E. are a large group of quite different elements that are represented in almost all eukaryotic organisms, with very few exceptions, such as vertebrate animals. It is legitimate to deal with them under one heading because they do have common denominators. First, they differ from the bacterial Insertion Sequences, by definition they are eukaryotic elements. They also differ from the retrotransposons, because Class II elements transpose without an RNA intermediate. Finally, they all have (usually short) inverted-repeat terminal sequences.

We shall start with three families (or superfamilies) of T.E. that were revealed in plants: the *Ac/Ds* family, the *En/Spm* family and the *Mu* (Mutator) family. All three were discovered by genetic studies in maize. We shall then handle Class II T.E. in other plants (angiosperms) that are related to the three aforementioned families. We shall then proceed to Class II T.E. in fungal organisms and in animals.

While we call the elements T.E. it is ironic that we have evidence for only a small minority that they are really mobile; the vast majority are "fossils". They came under the name T.E. because they have sequences that are similar or identical to mobile elements (for example in their terminal inverted-repeats). This is so for a large and versatile group of elements termed MINEs, "miniature" elements that have inverted repeats but their total length is about 300 bp (or less) and they all lack ORFs.

There are some issues related to T.E. that require special attention. The first is *chromosome breakage*. This issue actually prompted Barbara McClintock to delve into the subject of Controlling Elements (i.e. T.E.). Another issue is mutagenesis and gene tagging. Mutagenesis and gene tagging by T.E. became major tools for molecular genetics and breeding (see Appendix III). The structure and function of chromatin also requires attention. This is an emerging subject and it is therefore handled in Appendix II. The chemical modifications that can occur in the histones probably have a profound impact on the expression of genes. Most molecular-genetic studies on T.E. did not take the chromatin issue into consideration. This should not surprise us because the most interesting information on chromatin modifications and their impact on gene expression has accumulated only in the last five years. Indeed, very recent results suggest T.E. methylation can be separated from the modifications of chromatin, although both may require a common factor (Lisch et al., 2002). An additional issue that has emerged in recent years is the specific destruction of certain RNA sequences by a mechanism termed *RNA silencing* or *RNAi*. This issue probably has a considerable impact on transposition, as *RNAi* could destroy the mRNA that encode transposase.

The three issues of mutagenesis and gene tagging, chromatin modifications and *RNAi* are relevant to many types of T.E. A thorough discussion of these issues is beyond the scope of this book. Therefore these three issues will be represented in the Appendices.

2. THE CLASS II T.E. FIRST DISCOVERED IN MAIZE

We now come back to the end of our Historical Background, namely to the discovery of transposition. We already stressed the decisive role of Barbara McClintock in this discovery. Nina Fedoroff is probably the best source for the scientific achievements of McClintock. Before launching her own outstanding research on T.E., Fedoroff devoted a lot of efforts to become familiar with all the details of McClintock's work, including the study of McClintock's unpublished records. From my own experience it is a formidable task. Thus, for those interested in the details of McClintock's work it is recommended to first read carefully the reviews of Fedoroff (1983, 1989a, 1998) and only thereafter to delve into the many annual reports and other publications of McClintock herself. Fedoroff gave us the *time* of the discovery of transposition by McClintock: *the late spring of 1948*. For about 10 years McClintock was the only investigator who dealt with the *Ac/Ds* T.E. (termed *Controlling Elements* by her). Then another type of T.E. was revealed in maize. Peterson (1953) revealed the *En* elements and McClintock (1954) reported on the *Spm* element. The studies of Peterson and McClintock were performed independently, but both revealed the same kind of T.E. Only about 15 years later was a third type of T.E. revealed in maize: the *Mutator (Mu)* T.E. There are reports on numerous additional T.E. families that were either clearly grouped as Class II T.E. or were not fully defined. Such families were mentioned by Fedoroff (1983) and listed in the review by Feschotte et al. (2002) and will not be detailed in this book.

We shall summarize the above three types of T.E. of maize sequentially. There are several good reviews that cover these T.E. and those interested in more details should consult these reviews as well as the references mentioned in them (Fincham and Sastry, 1974; Campbell, 1980; Federoff, 1983; Freeling, 1984; Peterson, 1987; Wessler, 1988; Gierl et al., 1989; Fedoroff, 1989a; Weil and Wessler, 1990; Kunze, 1996; Fedoroff, 2000, 2002; Feschotte et al., 2002). I do not recommend the original publications of McClintock, starting from 1942, for the ordinary reader. These publications should be read only by those who are willing to take an educational voyage in genetics and are ready for hard reading.

2.1. The Ac/Ds Controlling Elements of Maize

After about 12 years of studies at the Cold Spring Harbor Laboratories of the Carnegie Institution of Washington, B. McClintock characterized the *Ac/Ds* system of T.E. She achieved this by genetic and cytogenetic methodologies, which we shall not detail here. Summaries of these studies were presented by McClintock (1951, 1956, 1961a). For those willing to read the original publications of McClintock, I recommend for them to have the review of Peterson (1987) open while reading McClintock's reports, because Peterson provided ample illustrations as well as full lists of *gene symbols* and of *terms*. Peterson also listed the 10 T.E. that were known by 1987.

2.1.1. Sexual Reproduction in Maize

The *Ac/Ds* T.E. were studied in maize. Therefore, a few features of the sexual reproduction of this plant will be helpful for the uninformed. The cereal species *Zea mays* (maize, in the Midwest of the USA called corn, an abbreviation of Indian Corn) is a monoecious plant. It has staminate (male) flowers on its terminal inflorescence (tassel) and its pistillate (female) flowers are on the "ears". In maize, as in many other organisms, the male organs are less esoteric than the female organs, therefore a maize plant can easily be emasculated by removing the terminal inflorescence. The ears can be covered before the female flowers are sexually mature. After maturation, the long stigmata of the pistillate flowers ("silk") that protrude out of the ear, can be pollinated and the ear can be covered again to avoid uncontrolled pollination. An embryo sac develops in each pistil; this sac contains haploid nuclei. Fusion of one of the pollen-tube nuclei with the egg-cell nucleus results in the diploid zygote. In parallel, another nucleus of the pollen tube fuses with two synergid nuclei of the embryo sac and from this fusion the triploid endosperm is formed. The grain (kernel) of maize is, in botanical terms, a fruit rather than a seed. Its outer layer, the *pericarp*, is a maternal tissue and not a result of fertilization. If the pericarp is not darkly pigmented, the color of the outer layer of the endosperm, the *aleurone*, can be observed through the covering pericarp. Characters such as the starch composition and the aleurone color of the endosperm are thus the results of the fertilization and they can be scored already a few weeks after cross-pollination. Each ear can provide hundreds of grains within a short period and as there are usually two ears on each maize plant; two types of cross-pollinations can be performed on each maize plant. All this means that aleurone phenotypes that are not too rare (10^{-3} or higher) can be scored on a single plant. More details on the reproduction of maize can be found in Fedoroff (1983).

2.1.2. The Main Findings of McClintock on the Ac/Ds System

The main findings of McClintock can be summarized as follows:
- The *Ac/Ds* T.E. are composed of two types of "loci": the *dissociation* (*Ds*) and the *activator* (*Ac*). These were also termed *receptor* and *regulator*, respectively.
- The *Ac* is *autonomous*, meaning it can change its location (by transposition) in the genome of maize, even in the absence of *Ds*.
- The *Ds* is not autonomous in its mobility: it can transpose only when *Ac* is also present in the genome. Thus, in the absence of *Ac*, the *Ds* is stable in its location (and does not cause dissociation).

- Both *Ac* and *Ds* can affect the expression of genes; the genes that were the main subjects of the studies were those affecting pigmentation and starch characters of the grains (kernels) endosperm.
- In the presence of *Ac*, *Ds* has two possible effects. It can cause chromosome breakage (dissociation) at the location of *Ds* and it can cause instability in gene expression.
- Not all *Ds* elements are capable of causing chromosome breaks. *Ac* by itself never causes chromosome breaks.
- Because the endosperm is triploid there are four possible doses of *Ac* in the endosperm: none, one, two and three. The ability of *Ac* to mobilize *Ds* (and by that to cause changes in gene-expression, such as from colorless to pigmented aleurone cells) *decreases* from one dose of *Ac* to three doses of *Ac*. This seemed paradoxal but it was amply documented by McClintock. Three doses of *Ac* rather than a single dose of *Ac* cause less frequent transpositions and also delays the transpositions of itself and of *Ds*. On the aleurone layer that means small spots (late transposition) and less spots (less transpositions).
- The *Ds* elements are different from each other. One clear difference is that some can cause chromosomal breaks while others cannot. While all the *Ac* elements have the same basic effects (transposition of themselves, causing gene instability and activating the mobility of *Ds* elements), there are certain variabilities (epigenetic) between *Ac* elements.

When insertional mutagenesis is concerned, in addition to the two regular alleles (e.g. a dominant "wild-type" and a recessive allele), there is an insertional mutagenized allele. For example, for the gene *Bz* that is involved in aleurone pigmentation, there is the colored *Bz* allele and a recessive colorless allele: *bz*; in addition there is the *bz-m* allele that has an insertional mutation. Upon transposition (of *Ac* or *Ds*) away from the *bz-m* locus, the phenotype will revert to the colored wild-type. Similarly, the *wx-m* endosperm (containing only amylopectin but not amylose) and the *sh-m* endosperm (that lacks sucrose synthase and is shrunken), can revert to the respective wild-types when the T.E. is transposed out of these loci.

The notion that the non-autonomous *Ds* is a defective derivative of the autonomous *Ac*, has already been expressed by McClintock but the molecular "proof" of this had to wait many years – until molecular genetics became the tool for investigating T.E.

McClintock was also on the right track when she distinguished between the genetic information that is stored in a given locus and specific elements that have the capability to *control* or to *regulate* the expression of this information. This concept was beautifully documented in bacteria by Jacob and Monod, although they were not aware of the previous investigations of McClintock when they studied bacteria. McClintock, Jacob and Monod probably exchanged information at the Cold Spring Harbor Symposium in 1961 that they all attended.

The phenomenon of chromosome breakage and the breakage-fusion-bridge (BFB) cycle, which included the formation of dicentric chromatids, started to draw McClintock's attention several years before she studied the controlling elements. However, when engaged with the *Ac/Ds* elements she found that the break occurred at the *Ds* locus and that it occurred only when an *Ac* was in the genome and that not all *Ds* elements cause this breakage.

The amount of effort invested and the quantity of detailed cytogenetic data that

McClintock accumulated were enormous. Here it is appropriate to look again at the *motto* of this book, which indicated that only keen and accurate observation combined with analysis of the mind will yield new and meaningful knowledge. It should be noted that much of the data of McClintock was not published. We owe some of the unpublished information to Fedoroff (1983). An example is the determination of 13(!) locations to which *Ds* moved, from the original location (proximal to the *Wx* locus) in chromosome 9 to various sites distal to *Wx* and even distal to the *C* locus (close to the terminal of the short arm of chromosome 9). McClintock made use of her knowledge that *Ac* will prompt *Ds* to change location, but when *Ac* is removed from the genome (by appropriate cross-pollinations), the *Ds* becomes stabilized in its location.

The cytogenetic studies of McClintock also showed that *Ds* can be involved in various chromosome rearrangements. These were revealed especially in the short arm of chromosome 9, where there are several loci that affect grain characters and pigmentation (*Wx, Bz, Sh, C*) as well as a potential mobile (and breakage causing) *Ds*. In this arm the intrachromosomal transposition could be followed precisely by the use of appropriate genetic lines, as indicated above. *Ds* transposition was, in several cases, associated with chromosomal rearrangements.

Studies by McClintock and others, but performed before the molecular aspects of maize T.E. became evident, all indicated that *Ds* mutations can affect a given locus in several ways. For example, the *Ds* mutation can lower the level of transcription; it can shift the timing of the expression and alter the tissue specificity. However, the explanations for these changes in expression levels, timing of the expression and the change in tissue specificities had to wait until the molecular tools became available.

2.1.3. The Mutator Function of Ac

The *Ac* element has a potentially dual effect. As indicated above, it triggers the mobility of *Ds* and the chromosome-breakage by *Ds*; but it plays also a role that is not related to a *Ds* element residing in the same genome: it can cause unstable mutations. What *Ac* cannot do is to break chromosomes. The unstable mutations caused by *Ac* are exemplified by the loci for *Bz* and *Sh*: in the *bz-m4* allele *Ac* causes sectors of *Bz* (pigmented) aleurones; in the *sh-m* allele *Ac* causes non-shrunken parts in the otherwise shrunken endosperm.

2.1.4. Mechanism of Transposition

We should remember that during most of McClintock's intensive studies, namely until the early sixties of the last century, nothing was known on the mobility of T.E. elements that reside in organisms other than maize. The first reports on Insertion Sequences in bacteria appeared during 1966–69 (see above). It took an additional ten years or more until efficient molecular methodologies unveiled the process of transposition of bacterial Insertion Sequences, and the understanding of the transposition of retroelements was delayed even longer. Thus, the transposition of *Ac/Ds* was based entirely on genetic and cytogenetic studies until the early eighties of the last century. However, elaborate genetic studies could furnish a surprising amount of information.

To give an example we shall leave the *Ac/Ds* system and shift our attention to the *Mp* element (that actually belongs to the same family of T.E). The *Mp* is an autonomous element, as is the *Ac* element. Moreover, in *Mp* there is the same

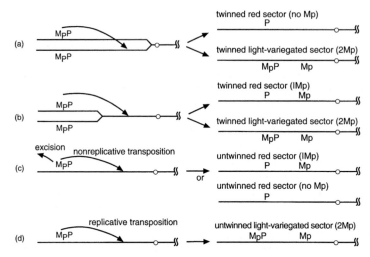

Figure 44. A diagrammatic representation of various types of intrachromosomal transpositions. The transpositions diagrammed in (a) and (b) occur during or after replication of the Mp element at the P locus and result in its removal from the P locus on one sister chromatid and its insertion either in a replicated site on the other sister chromatid (a) or in an unreplicated site on the same chromosome (b). The chromosomes received by the daughter cells derived from the cell in which each transposition has occurred are indicated on the right, as are the expected phenotypes of kernels on ears of plants in which such transpositions have occurred at premeiotic mitoses. Transpositions of the type diagrammed in (a) and (b) have been documented. Other possible types of transpositions, as well as simple excision of the element are diagrammed in (c) and (d), and their genotypic and phenotypic consequences are indicated at the right. (From Fedoroff, 1983).

dosage effect as in *Ac*: more *Mp* elements in the genome reduce the effect of *Mp* and also delay it. *Mp* can be tightly linked to the *P* locus. The latter locus affects the pigmentation of the grain's pericarp (a maternal component of the maize kernel, see Brink and Nilan, 1952). Thus, P-RR-*Mp* is an unstable locus; it will cause variegation of pericarp color. Genetic studies established that when *Mp* "leaves" the *P* locus, the pericarp cells become red but while simultaneously leaving one locus, *Mp* will settle in another locus. This and additional genetic studies clearly indicated that *Mp* and *Ac* transpose in a "cut and paste" mechanism. That the "cut" is usually not "clean" but rather leaves footprints behind, was revealed only much later when the excision-site could be sequenced. When an *Mp* element is tightly linked to *P* and suppresses the pericarp pigmentation, there are several types of possible intrachromosomal transpositions with different phenotypic results. The *Mp* can move away from *P* after DNA replication and insert in either a replicated site or in a site that has not yet replicated. The *Mp* can also transpose after replication. The various possibilities are schematically shown in Figure 44. Experimental evidence indicated that the great majority of transpositions of *Mp* are as in *a* and *b* of Figure 44 (see Greenblatt, 1974, for details and references). There is a complicating factor: as in *Ac*, two doses of *Mp* will reduce the inhibition of pigmentation and cause a light-variegation of the pericarp. The transposition is commonly to a nearby site on the same chromosome. Therefore, the *Mp* that has been excised from a known locus can easily be found in an adjacent location.

Returning now to *Ac*, it became clear from genetic studies that one *Ac* can "activate" two *Ds* elements at different locations in the maize genome. Moreover, the activation appeared simultaneously. It was thus assumed that *Ac* produces a *factor* that can move in the nucleus and cause this activation. Obviously, the genetic studies did not characterize this factor, although one suggestion was that the factor is an enzyme. There was one observation that concerns *Ac* effect that was rather obscure in McClintock's investigations. She found that *Ac* can undergo "heritable" changes so that the same *Ac* could have a stronger or weaker impact on *Ds*-induced chromosome breakage (accentric-dicentric formation) and on the timing of *Ac* activated mutation. A name was given to these changes: *states of Ac*. A "weak" state of *Ac* was similar in its effect to more than one *Ac* in the genome. Again, the meaning of these changes of *state* in *Ac* was not clear before the molecular era of the *Ac/Ds* system.

We have already noted the similarity between *Ac* and *Mp*. Thus, investigators such as Brink and others that studied *Mp* intensively actually revealed important information that is also true for *Ac*. We shall not detail these studies with *Mp* but the *Mp* also has "state" characters. The notion that when *Mp* is inserted at different exact sites within a locus it has different effects on the respective gene, is probably equally applicable for the *Ac/Ds* system.

2.1.5. Ac can Undergo Reversible Inactivation and "Mutate" into a Ds Element

Even though, based only on the genetic and cytogenetic evidence of McClintock, it was strongly suggested that *Ac* can change from an active element into an inactive one. This conclusion was derived from the *wx-m7* allele that has *Ac* at the *Wx* locus and the *a-m3* allele, that has *Ds* at the *A* locus. Neither mutable allele showed somatic mutation in most of the kernel progeny. Return of the *Ac* to the active phase did occur in a few cells, as inferred from simultaneous destabilization of both alleles in the same cell lineage. It was also observed that the inactive *Ac* did not contribute to (the negative) dosage effect and that the *wx* mutation could mutate somatically if an active *Ac* was introduced together with the inactive *Ac* at the locus. The genetic studies of McClintock revealed three cases by which it became rather clear that *Ds* elements were derived from previously fully functional *Ac* elements.

2.2. Molecular-Genetics of Ac/Ds Elements: Pioneering Studies

The "swallow" that brought the "spring" of the molecular era to plant T.E., came from the Brookhaven National Laboratory in Upton, N.Y. There is only a short flight from Upton to Cold Spring Harbor, where B. McClintock was working for many years. This "swallow" was a publication by Benjamin Burr and Frances Burr (1980). This short publication was the written version of a lecture at the Cold Spring Harbor Symposium on Quantitative Biology. The material on which the Burrs worked was the *Sh* locus. Notably, the *sh* allele causes shrunken endosperms and it was discovered in maize in an Indian Reservation in Nebraska, from where Rollins Adam Emerson and four of his assistants arrived when they established the famous maize genetics laboratory at Cornell. The *sh* mutation was an old discovery and the mutation was well characterized biochemically and genetically. In brief, in the wild-type *Sh* there is an abundant production of an enzyme (sucrose synthase) and the kernels have a normal shape. In *sh* there is a deficiency in sucrose synthase and the grains are shrunken. This also happens when *Ds* is in the *Sh* locus (e.g. in *Ds sh-m6233A2*). The Burrs obtained maize lines from B. McClintock that were either mutations resulting from *Ds* insertions or revertants of mutations as well as wild-type (*Sh*). From the cDNA coding for sucrose synthase a probe was produced and this probe served to analyze endonuclease-digested DNA that was run on Southern blots. The DNA from the different maize lines showed different patterns of hybridization. The conclusion was that *Ds* "intervention" caused the change in pattern. Further studies of the Burrs (Burr and Burr, 1981, 1982) detected modifications at the nucleic acid level, caused by *Ds*. Since no exact sequencing was performed the amount of molecular information was limited. However, detailed restriction maps and heteroduplex analyses between genomic DNAs and the mRNA for sucrose synthase not only revealed changes in the DNA at the *Sh* locus, but also differences between the different mutated lines.

The maize mutants at the *Sh* and *Bz* loci, caused by *Ds*, also reached the University of Cologne. There Döring et al. (1981) performed experiments that were similar to those of the Burrs. The former investigators also had evidence, from restriction-endonucleases digestions and Southern blot hybridization, that the insertion of *Ds* caused changes in the *Sh* and *Bz* loci, respectively, but these changes were not analyzed at the nucleotide-sequence level. In a note added to the proof of Döring et al. (1981), it was reported that N. Fedoroff had also obtained similar results by Southern blot hybridizations.

While the years 1980 and 1981 brought the first reports on the molecular meaning of the *Ac/Ds* elements, the real "revolution" came during 1982–1984. We should recall that while in the early nineteen-eighties DNA sequencing was already an established procedure, it was still cumbersome; sequencing was "hand made" and required skilled experimentalists. Sequencing was done with short DNA clones. Therefore, the respective clones first had to become available. This was performed with different maize lines that included *Ds* and *Ac* insertions that could be isolated and identified with the respective genetic loci.

For example, a clone that encoded the sucrose synthase gene and should also contain the inserted element (as indicated by genetic tests) could be identified by molecular means and propagated in a respective plasmid. This was performed by Geiser et al. (1982) in the Institute for Genetics of P. Starlinger at the University of Cologne. In a subsequent study by the same team (Courage-Tebbe et al., 1983), but now in collaboration with N. Fedoroff of the Carnegie Institution in Baltimore, the work was carried on to a further level. It was found, still mainly by Southern blot hybridization, that the *sh-m 5993* allele had a *Ds* insertion and that in revertants the *Ds* is excised from this locus. However, still no exact sequencing was reported.

Another locus, the *Wx*, was then investigated by N. Fedoroff and her associates (Shure et al., 1983). The DNA fragment that encoded the mRNA for the synthesis of the starch was isolated and cloned. One clone which, by previous genetic studies, should have a *Ds* insertion, was found to have an additional DNA fragment of about 2.4 kb at the *Wx* locus; this insertion was found near the 3' end of the mRNA coding sequence. In a subsequent study by Fedoroff et al. (1983), clones with additional *Wx* insertional mutations (with *Ds* and *Ac*) were isolated. In this latter study, partial sequencing and heteroduplex analyses were employed. Now it became clear that the *Ac* is a larger element than *Ds* but that the two elements had clear similarities: they shared the same sequences at their terminals. One of the *Ds* elements was actually very similar to the Ac (in *wx-m9*) only that the *Ds* was shorter than the *Ac* by 0.2 kbp. Moreover, it became evident that the sequences of the termini of *Ac* and *Ds* are quite frequent in the maize genome but the full-length *Ac* sequence is rare.

2.3. Further Molecular Studies on Ac/Ds Elements

The year 1984 was a very good year for the molecular biology of the *Ac/Ds* system. The first DNA sequence of a *Ds* element appeared in a January issue of *Nature* (Döring et al., 1984). The article came from the Institute for Genetics of Starlinger in Cologne. These investigators sequenced 4.2 kbp of an insertion into the *Sh* gene of maize (*Sh-m5933*). The *Ds* in *sh-m5933* was revealed as a rather complex element. The *sh-m5933* contains an insertion of 30 kbp that is flanked on one side by a double *Ds* and on the other side by a deleted double *Ds*. Then, in March 1984, the publication of Sutton et al. (1984) appeared in *Science*. This publication came from the laboratory of W.J. Peacock in Australia. It was a collaborative publication with Drew Schwartz, a veteran maize geneticist from Bloomington, Indiana, who had ample experience with the *Adhl* gene. The Australian study consequently dealt with the *Adh1* locus and promoted the understanding of the *Ds* insertion one step further. It was found that the insertion of *Ds* into *Adhl* locus was of 405 bp. This 405 bp of *Ds* had inverted repeats of 11 bp at both ends and there was a 8 bp duplication of the host DNA at both sides of the insertion. Upon excision of *Ds* from *Adh1*, the footprints of these 8 bp were left behind.

Back in Germany the investigation of *Ac* was continued by P. Starlinger and associates (Behrens et al., 1984). Two maize lines, with *Ac* insertions in the *Wx* locus, were investigated (*wx-m7* and *wx-m9*). Detailed restriction maps and heteroduplex analysis of hybridization between the DNA of wild-type *Wx* and *wx-m7* clearly established that *Ac* is 4.3 kbp long and that the *Ac* of *wx-m7* is inserted 2.5 kbp upstream of the insertion site of *Ac* in *wx-m9*.

The full sequencing of 4,812 bp that included the *Ac* from *Ac9* (*Wx*) was performed in a collaborative effort of the laboratory of J. Messing (then at the University of Minnesota) and N. Fedoroff (Pohlman et al., 1984). The exact length of *Ac* was found to be 4,563 bp. This *Ac* had (imperfect) terminal inverted-repeats of 11 bp. The insertion site was CATGGAGA. In revertants from which *Ac* was excised there were the footprints of a tandem of these 8 bp of the insertion site, at both sides of the *Ac* element. This study also indicated that the *Ac* contains two ORFs. Both ORFs seemed to be functional in the expression of proteins because they were both bounded by sequences that are typical of functional plant genes. The authors assumed that one of these ORFs encodes a transposase. When loci with *Ds* insertions were compared to *Ac*, none of the former had the full sequence of the *Ac*: fragments of different lengths were missing in the *Ds* elements and in no case was an intact ORF1 found in a *Ds* element.

On the other hand, the sequence of *Ds9* had similarity to the sequence of *Ac9* (*wx-m9*), suggesting that *Ds9* was derived from *Ac9*. The common feature of all the sequenced *Ac* and *Ds* elements was the 11 bp inverse-repeats at the terminals of these elements (Figure 45). Since the same *Ac* can mobilize all the *Ds* elements, the authors concluded that these inverted-repeat domains are the minimal sequences required for transposition. The overall picture that was derived from this study, was

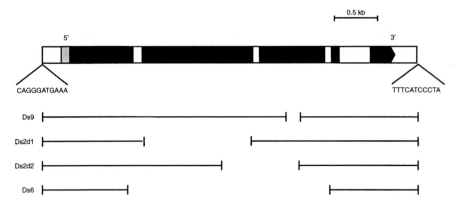

Figure 45. Structure of the Ac element and several Ds elements.
The Ac element is 4.6 kb in length and has 11-bp imperfect terminal IRs. The 5' and 3'
ends of the Ac transcription unit are indicated, and the left and right ends of the
element are correspondingly designated the 5' and 3' ends. The stripped box represents
the region of the element within which transcription start sites map. Filled boxes
correspond to exons, and the open boxes between them represent introns. The
interrupted lines below the Ac element represent the structure of the several Ds
elements. The line corresponds to the part of the Ac sequence, and gap represents the Ac
sequence missing from the Ds element. (From Fedoroff, 1989c).

that there is similarity between *Ac* and bacterial Insertion Sequences (e.g. Tn3).
While the Pohlman et al. (1984) study concerned the *Ac* in *wx-w9*, the laboratory of
Starlinger (Müller-Neumann et al., 1984) focused on another *Ac* in *wx7* (mentioned
above). The latter publication was submitted a few months after the former and
actually reported on the same findings. Müller-Neumann et al. indicated that the
DNA sequence of their *Ac* differed in certain important bp from the previously
published *Ac* in *wx-m9*, but when they checked the two sequences again they found
that the two sequences were identical. Since there are phenotypical differences
between *wx-m7* and *wx-m9*, Müller-Neumann et al. (1984) concluded that the
difference in the phenotypes of the two alleles may be attributed to the difference of
insertion sites of the *Ac* into these loci.

McClintock studied several *Sh* mutations in which *Ds* was inserted. We already
mentioned the *sh-m5933* allele that was found by Döring et al. (1984) to have a
"double" insertion of *Ds* in this allele. Now the Starlinger team, in collaboration
with Fedoroff (Weck et al., 1984), focused on another *Ds*, in *sh-m6233*. It became
evident that in both *sh-m5933* and *sh-m6233* the insertion is into an intron of the *Sh*
gene but the insertions are into different introns. The insertions, in both cases, are of
a double *Ds* but the total length of the insertion is also different: about 30 kbp in
the case of *sh-m5933* but only about 4 kbp in *sh-m6233*. More precisely, *sh-m6233*
is an insert into the first intron of the *Sh* gene while the *sh-m5933* is inserted about
2.5 kbp downstream of the *sh-m6233* insertion. The *Ds* "complex" in the two alleles
seemed to be similar: two *Ds* elements inserted into an intron of the *Sh* gene. The
Ds elements all had an 11 bp inverted repeat at the terminals and the *Ds* was
bordered by an 8 bp duplication of the "host" DNA. The "unit" *Ds* in both cases
was about 2,000 bp. In *sh-m6233* the elements were arranged in a way that one *Ds*

was inserted in the middle of the other, but in reverse direction. This caused a situation that two adjacent 11 bp terminal repeats were not inversed but in the same direction. If the transposase complex identifies and binds to these 11 bp, the incorrect binding could lead to events that could cause a dicentric situation and the latter could result in chromosome breakage.

The *Ds* from a completely other gene complemented the above-mentioned results. A large team of investigators from Cologne, Berkeley and Bergamo (Döring et al., 1984) investigated the *Adh1-2F11* mutant of maize. This is a *Ds* mutant that does not produce *ADH1*. Due to a selection method devised originally by Drew Schwartz and associates, *Adh1* mutants can be selected at the pollen grain stage. Wild-type pollen will be killed by allyl alcohol while pollen having the mutation (and which do not produce *Adh*) is resistant. Millions of pollen grains can be screened and resistant (mutant) pollen can be used in cross pollination. In the presence of *Ac* the *Adh1-2F11* allele is unstable: about 10^{-4} to 10^{-3} of the pollen will revert to wild-type. The authors tested about 4 million pollen grains and found that 1,436 grains reverted in the presence of *Ac* in the genome. No reversions were observed in most mutant plants (but $\sim 3.10^{-5}$ reversion was recorded in some maize plants). The *Adh1* gene is on chromosome 1 and the length of the *Ds* insertion was 1.4 kbp. Unlike in the *Sh* insertions mentioned above, the insertion in *Adh1-2F11* was into a transcribed region of the gene and the mRNA of the mutated gene included the inserted *Ds*.

The Starlinger team (Courage et al., 1984) then published a review of the findings concerning *Ac* and *Ds* in three different genes: *Sh*, *Wx* and *Adh1* (a publication of 12 (!) authors in alphabetical order). These authors first characterized the *Ac* and indicated that all those analyzed are almost identical: their length is ~4.5 kbp and they all have an 11 bp inverted repeat that is identical in all *Ac* elements but may vary for the outermost nucleotide of these elements. The *Ac* have two ORFs in one direction and a third, ORF3, in another direction. An 8 bp duplication of the recipient DNA is formed on both sides of the insertion. The DNA, inside of the inverse-repeats, contain segments of about 140 bp that can be arranged in a mismatched inverted repeat structure: a "stem" of about 140 bp with the large "loop" that contains the bulk of the *Ac*. The *Ds*, on the other hand, are of various kinds (see Figure 45) and all can be regarded as *Ac* elements from which various central parts are eliminated.

Kunze (1996) updated the structure of *Ac* (for more detailed information see Kunze et al., 1987; Kunze and Starlinger, 1989; Fusswinkel et al., 1991; Kunze et al., 1993). He noted that this autonomous element has a simple structure spanning 4,565 bp, with 11 bp terminal inverted repeats (IRs), whose outermost nucleotides are not complementary (they are C and A, rather than C and G). The G + C contents of the different regions of *Ac* vary considerably. About 240 nucleotides at the 5' end and at the 3' end they are relatively poor in G + C, containing only 45% and 40%, respectively, of these nucleotides. The 5' end contains 26 CpG and one GpC and the 3' end contains 24 CpG and one GpC. In contrast, the long, untranslated leader is 68% G + C, without a bias towards CpG. The coding region is again poor in G + C (only 38%) with very few CpG. As we shall see the CpG dinucleotides, that are subjects of possible methylation may serve in the regulation of the activity of *Ac*. The *Ac* is schematically shown in Figure 46.

In summary, the molecular study of the *Ac/Ds* system that started in 1980 and culminated during 1984, revolutionized our understanding of this system; shifting it

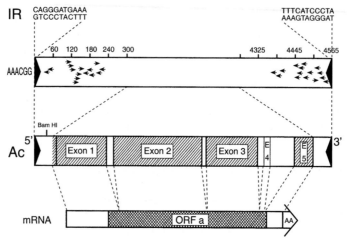

Figure 46. The Ac element is 4,565 bp in length and has 11-bp imperfect terminal inverted repeats whose sequences are shown in the upper line (IR). The distribution of AAACGGs and closely related motifs, the Tpase-binding sites, within the sub-terminal Ac regions is indicated by arrows underneath (AAACGG). The left Ac end containing the unique BamH1 site at position 181 is designated the 5'-end. Transcription of Ac is initiated at multiple sites between positions 280 and 380. The 3.5 kb mRNA consists of 5 exons. The 5'- and 3'-untranslated regions are indicated as stippled boxes; ORFa is the 2,421 nucleotide-long Tpase open reading frame. (From Kunze, 1996).

from a genetic level (that was quite abstract to many biologists) to the level of nucleotide sequences that are the "reality" to molecular biologists.

2.4. Transposition of Ac/Ds Elements

2.4.1. Sites of Insertion
Although in certain cases a domain for preferred insertion target has been reported, there is generally no specific insertion target for *Ac/Ds* elements, such as was detected for some bacterial Insertion Sequences (Tn elements). Thus, insertion can be into the following types of regions, relative to the plant gene.
− Insertion into a site that is not a gene and also not known as a region for controlling the activity of the gene.
− Insertion into the 5' controlling domains of a gene such as the promoter or the leader of a gene.
− Insertion into the 3' controlling domain (termination of transcription) of a gene.
− Insertions into the gene itself that can be divided into insertions into coding regions (exons) or introns.

Clearly, when the insertion is far away from any gene that is either vital for the plant or that is manifested in function, form or pigmentation, it will not be detected by genetic studies. Thus, it is no wonder that in the past there were no reports on such insertions. With the progress of genome sequencing in some plants as *Arabidopsis* and rice, it is expected that when *Ac/Ds* are introduced into these plants by genetic transformation, such insertions will be found and located.

With respect to the locations of insertions there are two considerations: (1) how will the insertion affect the function of the gene; (2) will excision of the inserted element bring the gene-activity back to the wild-type situation, or will the footprints have an effect on gene activity. Weil and Wessler (1990) reviewed this subject and since this review additional studies have been performed. From these studies no general rule can be derived. There are also reported cases in which the insertions of a *Ds* element into the coding region of a gene (*Wx*) did not abolish wild-type expression. In these cases the *Ds* was spliced from the primary transcript (Wessler, 1988). Obviously, splicing of the transcript can "correct" insertions, especially when the element is inserted into an intron of the gene, but in many cases the expression of the gene was reduced or even prevented. We should note that although the number of studies on this subject was quite substantial, they were performed with only a handful of genes that are easily scored, such as *Wx*, *Sh* and *Adh1*.

One phenomenon is worth noting: when *Ds* is inserted in the gene in the antiparallel orientation, splicing of the element may take place; while when inserted in parallel orientation to the gene there is no complete splicing, and a truncated transcript is produced.

There is also no general rule for insertion and excision of elements upstream of the coding region of genes. Moreover, the same insertion may affect the expression of a gene differentially: for example, husks and seedlings may not be affected in the same way. This should not surprise us because the spatial regulation of a gene is commonly located upstream of the coding region.

2.4.2. Transposition and Excision Requirements
This subject was reviewed by Kunze (1996). Here we shall only present the main information. The most obvious information is that the correct (or almost correct) sequences of both IRs are required for transposition. The outermost base may be mismatched without abolishing transposition. In certain cases specific other mismatches are also tolerated. Also, about 240 subterminal bp (noted above) are essential for normal transposition. Elimination of parts of these subterminal regions have a progressive negative effect on transposition. The role of chromatin in the transposition of *Ac/Ds* elements has not been explored in detail, but it appears that transposition can occur from a variety of chromosomal locations.

The complexity of chromatin and especially the role of histone modifications in gene regulation is a subject that was investigated in depth only recently (see Appendix II of this book). Therefore, we will have to wait until we shall be able to understand fully the role of chromatin in transposition.

On the "negative" side it can be concluded that the direct (usually 8 bp) "host" DNA repeats flanking an *Ac/Ds* element are not essential for excision. Actually, there are exceptions to the 8 bp direct repeats. The footprints may vary in the number of nucleotides. We should also recall that if the element is excised from a coding region, footprints that are a multiple of 3, may have only a low effect on expression. Each group of 3 nucleotides may add one amino acid to the encoded protein and such an addition may not be detrimental to the functionality of the protein. A case in which only 4 bp of a footprint was retained after an *Ac* was presumably excised from a coding region was recorded (Schiefelbein et al., 1988). The additional amino acid destabilized the encoded enzymes.

2.4.3. Features of Ac/Ds Transposition
The following are the general features of *Ac/Ds* transposition.
- *Ac/Ds* transpose by non-replicative (cut and paste) mechanisms.
- The elements transpose primarily during or shortly after DNA replication.
- The elements transpose from only one of the two daughter chromosomes.
- The elements reinsert into either replicated or unreplicated regions of the chromosome.
- The *Ac/Ds* have a clear preference for short-range transpositions.

Thus, it is quite common that the transposition of an element is confined to several sites in the same gene. However, while this short-range is common, it is not obligatory: long-range transpositions were revealed. Much of the data on distance of transposition came from experiments in which the elements were transferred, by genetic transformation, into other plants such as tomato, tobacco and *Arabidopsis*. Such transfers have an advantage because the host plants do not contain *Ac/Ds* elements, thus the newly introduced elements can be traced. We shall return to T.E. in transgenic plants in a separate section.

2.4.4. The Transposase of Ac
The primary transcript of *Ac* spans almost the whole length of the element; only about 300 bp and 260 bp of the 5' and 3' terminals, respectively, are not transcribed (see Kunze et al., 1993). The transcript is then processed to a 3.5 kbp mRNA that has a poly (A) tail. The promoter lacks CAAT and TATA; it is considered a weak promoter that acts constitutively. The first 600–700 bp are not translated and are considered to be an (unusually) long leader. The coding region is a 2,421 bp ORF that encodes the *Ac* transposase (Tpase). The Tpase has an apparent MW of 112 kD. It is a rare protein even in cells (e.g. endosperm cells) that contain an active *Ac*: about 1,000 molecules per cell. There is an additional protein, in small and variable amounts, that is encoded by *Ac*. This protein has an apparent MW of 70 kD. The Tpase can aggregate into 2 μm rods that are probably not active in transposition. The 807 amino acids of the Tpase were divided into domains with probable tasks in the transposition. The N-terminal 102 residues are not only dispensable but probably reduce the transposition. The increase of Tpase concentration (above a certain minimum) decreases transposition (we noted above that three *Ac* elements in the genome may inhibit transposition, when compared to only one *Ac*).

By *in vitro* experiments it was concluded that Tpase binds to repeats of the AAACGGG motif in the target DNA. At least six repeats of this motif are required for sustained binding (Becker and Kunze, 1997). It appears that several Tpases have to bind to each end of *Ac* during transposition and form a transposition complex.

Methylation strongly affects the binding of Tpase to the AAACGG motif. When C is methylated in both strands there is no binding. But there is a reduction of binding even when the C is methylated only in one strand, hence:

AAA C GGG
| | | | | | |
T T T G CCC   is bound to Tpase

While   AAA ᵐCGGG   is poorly bound to Tpase
| | | | | | |
T T T   GCCC

This led to a hypothesis that suggests that before replication of the chromosome there is full methylation; with replication there is an asymmetric demethylation so that there is a difference between the 5' end and the 3' end of the element. This difference enables the Tpase to recognize the two ends.

The awareness that methylation may have a decisive role in controlling the activity of *Ac* was already reported many years ago (Schwartz and Dennis, 1986). These authors observed that an *Ac* element, derived from the *wx-m9Ac* mutant, can be either inactive or active. The active *Ac* is methylated only at one end of the element while the inactive one is methylated at both ends. Very similar observations were made by Chomet et al. (1987), who submitted their manuscript five months after the former authors. The data of the latter authors were on the *Ac* of *wx-m7* (rather than on *wx-m9*). There are indications that the active Tpase is an oligomeric protein. Several studies were conducted on this subject (Scofield et al., 1993; Heinlein et al., 1994; Boehm et al., 1995; Becker and Kunze, 1997; Wang and Kunze, 1998), but the exact chemistry of *Ac/Ds* transposition, in similar details as was revealed in bacterial Insertion Sequences, still awaits further studies. From a study by Wang and Kunze (1998) it appears that hypermethylation of the 5' end of an element does not abolish Tpase binding, while high methylation of *Ac* correlates with the inactivity of the *Ac* promoter, reducing Tpase production.

There is a "logistic" problem that has to be solved by the maize cell. The transposition of *Ac* and *Ds* takes place in the nucleus, during the S phase of the cell cycle, while the production of Tpase, which is required for the transposition, takes place outside the nucleus, in the cytoplasm (probably during the interphase period). Therefore, the Tpase has to find its way from the cytoplasm, across the nuclear membrane, into the nucleus. Indeed, the *Ac* Tpase is equipped with nuclear-localization signals (NLS) that are instrumental for the trafficking of Tpase into the nucleus (see Boehm et al., 1995).

2.4.5. Is there an Ac Transposome?

In our deliberations on bacterial Insertion Sequences we detailed the synaptic complexes in which Tns proteins, DNA and possibly "host" factors, interact in the process of transposition. Are there analogous processes and complexes in the transposition of the *Ac/Ds* elements? Surely, there are differences. For example bacterial Insertion Sequences usually produce more than one Tns, while only one Tpase was revealed in the *Ac/Ds* system. What is the role of the 70 MW protein that is also encoded by the *Ac* transcript?

Essers et al. (2000) have started to approach this question. Their results indicated that at the C terminal of Tpase of *Ac* there is a dimerization domain. Oligomerization of Tpase is probably essential for transposition but may also lead to aggregates that are not active. The structure of a hypothetical transposome in the *Ac/Ds* is not yet known, but there are indications that Tpase binds to DNA sites that are upstream of the transcription start of *Ac*. Such binding could regulate transcription.

However, this "autoregulation" is not sufficient to explain cases of reduced transposition, while the level of transcript is high. There is a riddle: why is there a "window" of Tpase levels, between a very low level and a too high level, in which transposition is optimal. This riddle has been approached in some investigations (e.g. Heinlein, 1996) but a satisfactory answer awaits further studies.

2.4.6. Transposition Frequency
As elements may insert into vital genes and thus partially or completely eliminate their expression, there is a selective advantage to a low rate of transposition. Indeed, the frequency, as evaluated by the number of transpositions per cell in a given division cycle, is low. In the germinal lines of cells it is commonly in the range of 0.01% but it may be higher in some cases and in some variants of the *Ac/Ds*. For example, the frequency of *Ds2* excision from the *bz2::Ds2* allele was found to range from 0.2 to 1% (Levy and Walbot, 1990). Several hypotheses were suggested for the low rate of transposition, but the bottom line is that this is not known. What is quite sure is that under normal conditions, there is no lack of transcript of *Ac*. We noted above that additional *Ac* elements increase the transcript but reduce transposition, causing later transposition and a reduced frequency. In the aleurone pigmentation, the excision of an element from a mutated Bz allele will then be manifested in smaller and fewer pigmented spots. While this negative *Ac* dosage effect on excision was reported in maize, the situation is very different when the *Ac/Ds* system is moved into other plants such as tobacco and where no such negative dosage effect was revealed. This clearly indicates that host factors are involved.

2.4.7. A Special Note on the P Gene of Maize
This gene was probably the subject of (unpublished) studies of Gregor Mendel, as can be derived from a letter of Mendel to Nägeli in 1867 (mentioned by Rhoades, 1984). In a way, this *P* gene issue accompanied the history of genetics (or at least of maize genetics) for over 130 years! Some of this history was illustrated exceptionally well in the review of Rhoades (1984). The *P* gene issue re-emerged in the early years after the term genetics was coined (see Historical Background, above) with the detailed publication of Emerson (1914), who described the variegation of the maize-grain pericarp due to the unstable P^{vv} gene.
      Many years later Brink and Nilan (1952) found that a transposable element, then termed *Mp*, situated near the P^{vv} allele, inhibited the functionality of the alleles. The *Mp* is actually an *Ac* element. We shall "jump" about 40 more years and review briefly the studies of Thomas Peterson and associates with the *P/Ac* interactions. The product of *P* is essential for the red pigment in the pericarp of the maize kernel and of the cob. The P^{vv} causes variegated pericarp. It contains an inserted *Ac* in the second intron of the coding region of *P*. Excision of this *Ac* restores pigmentation. The P^{vv} mutated to P^{ovov} and caused orange variegation. P^{ovov} permits substantial cob and pericarp pigmentation. As P^{vv}, P^{ovov} also contains (the same) *Ac* but in a different location: 161 bp away from the original insertion site. This indicated that the same insertion but in a slightly different site can change the expression of an unstable mutation (Peterson, 1990). If P^{vv} and P^{ovov} are included in the same genome, homologous recombination may occur and a null allele may be produced by deletion in the *P* gene (Athma and Peterson, 1991). The *Ac* can move to various additional locations in the *P* gene and consequently cause different unstable pigmentations in the pericarp and the cob. The *Ac* can also insert into an intron of the *P* gene. The short-range mobilization can occur in both directions (towards the 5' end or the 3' end of the gene) and reinsertion can be in the reverse direction (Athma et al., 1992). Although movements of *Ac* to various sites in the *P* genes were detected, there is a hot spot for *Ac* insertion and an *Ac* may reinsert at a site from which it was previously excised (Grotewold et al., 1991). We noted above that

certain *Ds* elements can cause chromosome breakage, while *Ac* does not break chromosomes. This is true for "simple" *Ac*. When a full-length *Ac* and a fractured *Ac* are located near each other (e.g. in an intron of the *P1* gene), this combination can cause chromosomal rearrangements, including breakage and rejoining of chromosomes (Zhang and Peterson, 1999). The rate of homologous recombination, which may lead to deficiencies, is obviously dependent on the exact locations of the two *Ac* elements that are involved in this recombination (Xiao et al., 2000).

Finally, when *Ac* is even only slightly mutilated, it loses its own mobility but maintains the capability to cause *Ds* to "jump". This was shown by Xiao and Peterson (2002) with a mutation of *P1^{vv}*. The derived mutant *P1*-vv-5145 has an *Ac* insert in the same location as *P1^{vv}* but the Ac in *P1*-vv-5145 is missing the last nucleotide of the 3' end (A) and also a C of the flanking host sequence. This slight deletion of 2 nucleotides prevented further mobility of the *Ac* from *P1*-vv-5145, while the latter allele could cause transposition of a *Ds* element. I apologize for the inconsistency of the nomenclature: the maize geneticists used different designations during the years: first *P*, then *P1* and also *p1*; also, the same allele was termed *P^{vv}* and *P1-vv* or *P1*-vv.

2.5. The Ac/Ds Elements in Transgenic Plants

When one wishes to trace a T.E. that was excised from a known locus but has landed in an unknown location, it is best to use a plant that normally does not harbor such T.E. Also, if *Ac/Ds* should serve for gene tagging outside the maize system, the *Ac* and *Ds* elements should be transferred to transgenic plants. Therefore, for these reasons, as well as out of scientific curiosity, the fate of these T.E. was explored in genetically transformed plants.

2.5.1. Transgenic Tobacco and Potato

Baker et al. (1986) transformed the maize element *Ac* into transgenic tobacco. This was a pioneering collaboration between J. Schell of the MPI near Cologne and N. Fedoroff of the Carnegie Institution in Baltimore. In this study, performed before the genetic transformation was streamlined, only transgenic cell cultures and not functional transgenic tobacco plants were obtained. Even from this experimental system it became evident that the *Ac* of maize can be transferred to tobacco cells and can then be transposed from the site of original integration into new sites. This means that *Ac* is also functional in a dicot cell. In the very same month (July 1986) in which the former study was published, Fedoroff and two French experts on *Agrobacterium rhizogenes* transformation (J. Temple and M.A. Van Sluys) submitted an extension of the previous study (Van Sluys et al., 1987). The latter study was with carrot and *Arabidopsis thaliana*. The genetic transformation procedures, which included the use of *Agrobacterium tumefaciens* and *A. rhizogenes*, were still in their infancy in 1986 (see Breiman and Galun, 1997, for details on the development of genetic transformation) but the results were clear: also in carrot and *Arabidopsis* the alien *Ac* is capable of transposition.

To understand the rapid progress of research on *Ac/Ds* elements in transgenic plants one has to be aware of the advances in genetic transformation of plants that took place from about 1988, and the vast improvement in molecular-genetic methodologies that paralled these advances. One of the first steps was to show that when *Ac* is engineered into an appropriate plasmid that contains the promoter of a baculovirus, insect cells can be infected. These insect cells will then produce the protein that is encoded by the *Ac* element of maize (Hauser et al., 1988). It was soon also reported (Knapp et al., 1988; Coupland et al., 1988) that it is possible to select for transgenic potato and tobacco plants that harbor *Ac*. For that, the *Ac* was engineered into a transformation vector in a way that it was inserted into a gene for antibiotic-resistance. Transgenic plants were then obtained. As long as the inserted *Ac* was maintained in its original site there was no resistance, but upon excision from the gene for resistance, the cells became resistant and could be selected. Similar results to those obtained with transgenic tobacco were also obtained with transgenic *Nicotiana plumbaginifolia* that contained an *Ac* element (Marion-Poll et al., 1993).

A further refinement of the system was reported by Jones et al. (1989). These authors utilized the sensitivity of the cotyledons in tobacco seedlings to streptomycin. In the presence of streptomycin these cotyledons are chlorotic ("white"). When they harbor a gene for streptomycin-resistance (SPT) they maintain the green pigmentation. When the authors introduced the SPT gene that had an *Ac* insertion into tobacco, they expected that the cotyledons would still be sensitive to streptomycin – showing white cotyledons when germinated in a medium with streptomycin. However, when the *Ac* was excised, the SPT became active and green spots appeared in the cotyledons of seedlings grown in streptomycin. The expectations of the authors were fulfilled and green-spotted cotyledons were observed. There was an interesting aspect in these experiments: increasing the dosage of *Ac* did not reduce the number of spots; hence the negative *Ac* dosage effect that is the norm in maize was not observed in tobacco. This observation was also verified in a later study of Keller et al. (1993).

The *Ac* dosage effect was further studied in transgenic plants. For example, Scofield et al. (1992) introduced a *Ds* element into the SPT gene and obtained transgenic tobacco plants that contained this construct. In parallel they engineered plasmids that contained the coding region of Tpase behind different promoters (e.g. CaMV 35S promoter or the *ocs* promoter), and also obtained the respective transgenic tobacco plants. When the two types of plants were crossed to obtain in the same plants the *Ds*/SPT and the Tpase code with different promoters, it was expected that the *Ds* would be excised from the SPT, causing streptomycin-resistant spots on the cotyledons of seedlings. It was found that the CaMV 35S promoter caused the largest spots and also the highest rate of Tpase transcript. Smaller spots and less transcript was recorded by the use of the weaker promoters (*ocs* and *nos*). The wild-type *Ac* caused the transposition of *Ds* to happen early but not frequently (e.g. a few large spots), while the *nos* promoter caused more frequent and later transposition (many small spots). This and a latter study in this line (Scofield et al., 1993) added suggestions for the basis of the *Ac* dosage effect, but no unequivocal solution.

The β-glucuronidase (GUS) reporter gene can be expressed in transgenic plants. When an *Ac* element is inserted into the leader of GUS, there should be no such expression, unless the *Ac* is transposed away from the GUS. This experimental system was used by Finnegan et al. (1989). These authors found the same phenomenon in transgenic tobacco as the one reported by Jones et al. (1989). In the case of the GUS reporter, the enzymatic activity could be revealed after appropriate staining. In these GUS experiments it was also shown that *Ac* could mobilize a *Ds* element in the transgenic tobacco. In a further study of Jones et al. (1990), with the SPT gene having an *Ac* insertion, these authors found that as in maize the *Ac* preferentially transposed to linked sites. With a smile we note that the authors also underwent transitions; two of the authors moved to other locations and the company where the research was conducted changed ownership, from Advanced Genetic Sciences to DNA Plant Technology (meanwhile the latter company was dissolved).

While the early studies on the fate of *Ac/Ds* in transgenic plants were confined to tobacco, *N. plumbaginifolia* and potato, which can be transformed with relative ease, with the advancement of transformation procedures additional transgenic plants were used such as tomato, *Arabidopsis*, rice, petunia, barley and *Lotus*.

2.5.2. Transgenic Arabidopsis thaliana

By the end of the nineteen eighties *A. thaliana* was already an established model plant for molecular biology and developmental studies. Thus, Schmidt and Willmitzer (1989) introduced *Ac* into transgenic *A. thaliana* plants and found a low rate of germinal-excision frequency. A further study with transgenic *A. thaliana* was conducted by Swinburne et al. (1992). The main purpose of these authors was to attempt a more massive transposition, induced by *Ac*, to render *A. thaliana* an efficient subject for gene tagging. They used *Ds* in a gene for streptomycin-resistance that should be excised by alien Tpase and result in "islands" of streptomycin-resistant cells. The Tpase code was driven by different promoters as in Scofield et al. (1992), mentioned above. It was found that the system indeed works in *A. thaliana*: the CaMV 35S caused the production of the Tpase transcript and the transposition of *Ds* resulted in streptomycin-resistant spots. There was a correlation between transcript abundance and *Ds* excision frequency. The prospect of gene tagging in *A. thaliana* by *Ac* was also investigated by Dean et al. (1992). These authors searched the germinal and somatic activities of *Ac* in *A. thaliana* transformants. They also used the rates of change from streptomycin sensitivity to streptomycin-resistance (by the excision of the *Ac* from a transgenic streptomycin-resistance gene) as a measure of *Ac* transposition. There was quite an extensive variability among the *A. thaliana* transformants but the general trend was that with the increase of *Ac* dosage there was an increase in somatic and germinal excision. The excisions were followed to the fourth generation and did not diminish. Individual plants that were genotypically identical seemed to vary in *Ac* activity.

More recently, Xiao and Peterson (2000) devised a system to increase the incidence of homologous, intrachromosomal recombination in *A. thaliana* by transposition. For that they engineered a special vector in which two partially deleted GUS genes were separated by an insert that contained a *Ds* element. The *Ds* element could be activated by the introduction, by sexual cross pollination, of an *Ac* element into the same genome. The engineered construct (*Gu-Ds-US*) was transferred to *A. thaliana* by genetic transformation. This transformation became rather efficient when the *in planta* procedure was followed (see Galun and Breiman, 1997; Galun and Galun, 2001). The incidence of homologous recombination could be scored by GUS staining. It increased about 1,000-fold by the supply of Tpase that was imposed by the addition of an alien *Ac* element.

2.5.3. Transgenic Tomato

In several studies the *Ac/Ds* elements were transferred into transgenic tomato plants.

Yoder et al. (1988) and Yoder (1990) introduced an *Ac* element into tomato by *Agrobacterium*-mediated genetic transformation and followed the presence of *Ac* for several sexual generations. While in most tomato lines the number of *Ac* per genome remained low and stable, in one line the number of *Ac* elements increased from the original one copy to more than 15 copies in two sexual generations. However, when a *Ds* element was introduced into transgenic tomato plants (without the introduction of an *Ac* element), the *Ds* element was not mobile.

Rommens et al. (1991) of the Free University of Amsterdam launched a more complicated analysis. They inserted a "double" *Ds* into transgenic tomato. This "double" element had one *Ds* inserted into the other (from the maize allele *sh-m5933* that was mentioned above). In maize this double *Ds*, in the presence of *Ac* can cause several rearrangements and even break chromosomes. When only the double *Ds* was inserted into the transgenic tomato, this element maintained its position and no rearrangements were detected. However, when *Ac* was added, changes did occur. Notably, the double *Ds* was not excised from its location; nor was a chromosomal break reported (the article did not deal with cytological observation; it was based primarily on restriction endonuclease digestions and Southern blot hybridization). The dominant phenomenon in plants with double *Ds* and *Ac* was rearrangement of the DNA in the region of the double *Ds*, leading to disruption of the original double *Ds* sequence. Because the changes in the region of the double *Ds* were not followed by detailed DNA sequencing, the information on the changes is limited.

A team of investigators (Carroll et al., 1995) from the John Innes Centre (Norwich, UK) that previously showed the fate of *Ac/Ds* in tobacco, used similar approaches to follow these elements in transgenic tomato. Basically, the approach was to integrate a plasmid with *Ds* into transgenic tomato and then to cross such tomato plants with a tomato that harbors an *Ac* element that will mobilize the *Ds*. Germinal transposition of *Ds* was observed. Investigators from The Netherlands (Stuurman et al., 1998) introduced an agrobacterial T-DNA fragment into transgenic tomato. This fragment contained a *Ds* element as well as the selectable markers, *Hpt* and *Bar*, for hygromycin and phosphinotricin resistances, respectively. *Cre-lox* site-specific recombination sequences were also added to promote site-specific recombination. After insertion into the genome of a transgenic tomato (in chromosome 6), the transgenic tomato was crossed to another transgenic plant that harbored an *Ac* element. The fate of the *Ds* was then followed in the sexual progeny. Briefly, the *Ds* apparently did not transpose randomly: some genes, even when linked to the site of integration may be difficult to tag.

More recently a team that included investigators from the Czech Republic and the John Innes Centre (e.g. J.D.G. Jones and associates) launched an extensive research study on the transposition of *Ds* elements in transgenic tomato (Briza et al., 2002). The main questions that were investigated were whether the transposition is restricted to "hot spots" in the genome and if the transposition is primarily or exclusively to nearby locations. This study may be summarized by the answer that there is no unequivocal preference for transposition of *Ds* from a donor site to one specific chromosome, although the transposition is also not completely random. This "sounds" like a typical biological answer. So it is!

2.5.4. Insertion of Ac/Ds into Rice and Other Cereals
The main purpose of inserting the *Ac* and *Ds* elements into rice (e.g. Shimamoto et al., 1993; Nakagawa et al., 2000) was to develop functional genomics in these

plants. The main rationale for these studies was to show that these elements can serve in gene tagging. Now that the genomes of *Japonica* rice and *Indica* rice are fully sequenced, any isolated DNA sequence can be located on the rice genome with precision: the exact chromosomal location can be determined.

Takumi (1996) investigated the *Ac/Ds* system in wheat. He saved himself the effort of obtaining stable transgenic wheat plants, which is a difficult task. Thus, cell cultures of *Triticum aestivum* and *T. monococcum* were transformed. To achieve transposition of the *Ds* in wheat, Takumi co-introduced a truncated *Ac* element. This element had no inverted repeats and was therefore not mobile. It did produce Tpase, thus it could mobilize the *Ds* elements.

The *Ac/Ds* system was also introduced into barley. McElroy et al. (1997) also did not deal with functional plants: they introduced the *Ac* and *Ds* elements into callus cultures. The strategy was similar to the one mentioned above. To evaluate the transposition of *Ds* it was engineered into a GUS gene so that this gene can be expressed only when *Ds* is excised. The excision of *Ds* was induced by cointegration of an *Ac* element. Indeed, spots of GUS staining were revealed, indicating that *Ac/Ds* elements are operational in barley. Functional transgenic plants of barley that contained an *Ac* element were obtained by Scholtz et al. (2001) from the University of Hamburg, where the "art" of regeneration of transgenic barley plants is well established. These authors reported that the *Ac* was indeed introduced into the barley genome and that it was mobile in the barley genome. Hence, the *Ac* can be useful in the future for gene tagging in this plant, which has one of the largest genomes among all organisms.

2.5.5. *The Introduction of Ac/Ds into Other Transgenic Plants*

The appropriate plasmids that contain *Ac* and *Ds* elements with different reporter genes and selectable genes are now available. Therefore, it is a simple task to obtain transgenic plants that contain these elements. The only prerequisite is that genetic transformation procedures should be available. In some plants transformation is possible – but difficult. This is the case in some legumes such as *Lotus japonicus*, *Glycine max* (soybean) and *Vicia faba*, where in spite of the difficulty transgenic plants with *Ac* were obtained (Thykjaer et al., 1995; Zhou and Atherly, 1990; Bottinger et al., 2001). Genetic transformation is relatively simple in *Petunia*. Thus, protoplasts of *Petunia* served to investigate specific aspects of *Ds* transposition, such as which components of the *Ac* coding region are most efficient in driving *Ds* excision (e.g. Becker et al., 1992), the role of aggregates of Tpase in transposition/mobilization of *Ds* (e.g. Heinlein et al., 1994) and the role of specific DNA methylation in transposition (Ros and Kunze, 2001). The introduction of the *Ac/Ds* elements into transgenic plants was also performed in numerous other plants. The claimed purpose of such work was commonly to develop a gene tagging system. These additional reports will not be listed by us but for one publication that is in a way exceptional: the introduction of *Ac* into a tree: *Populus* (Fladung and Ahuja, 1997). A brief review on Gene Tagging is presented in Appendix III.

3. THE *SPM/EN* ELEMENTS

We shall start our deliberation on the *Spm/En* T.E. with a short review on the discovery and nomenclature of these elements. Although who first discovered a biological phenomenon does not contribute to the increase of scientific knowledge, this issue is of major concern to the scientists involved. It is therefore fair to furnish, briefly, the available information.

The genetic and cytogenetic studies on *Spm/En*, mainly by B. McClintock and P. Peterson, were very intensive from about 1951. These studies yielded an enormous amount of data and provided convincing evidence that indeed these are mobile controlling elements. While McClintock did not bother to compose a comprehensive review of all her studies on T.E., Peterson did. His extensive review (Peterson, 1987) provides information on his own studies of about 25 years, as well as on others. This review is recommended for those interested in the details; it is amply illustrated but not easy reading for the novice. Interestingly, as already indicated above, the studies of both investigators were probably not known to the scientists who, during 1966 to 1969 revealed the Insertion Sequences of bacteria that are likewise mobile controlling elements. Much later, two German investigators who were involved in the discovery of bacterial Insertion Sequences, P. Starlinger and H. Saedler, moved from bacteria to maize, and started molecular investigations on the *Spm* and *En* sequences, respectively. They and others then led the way to our present understanding of the *Spm/En* system. Here we shall provide references to the detailed genetic studies of B. McClintock and P. Peterson, but will not detail them in this book. We shall start the more detailed description of the *Spm/En* elements from the point of the beginning of the molecular studies, when the genetic results became more meaningful.

3.1. Discovery and Nomenclature

The *En/Spm* elements were revealed independently by B. McClintock and P. Peterson. Some years before 1953, Peterson revealed the *En* element while he was a graduate student of the University of Illinois. He then worked at the University of California in Riverside and finally settled at Iowa State University in Ames. The first report on a transposable element appeared in Peterson (1953). Reports on *Spm* by McClintock started to appear in the *Annual Reports of the Carnegie Institution of Washington Year Books*, at about the same time Peterson started his *En* studies. Information on mutability of the $a_1{}^{m-1}$ locus was presented in a Cold Spring Harbor Symposium (McClintock, 1951). At that time B. McClintock already sensed that this locus does not represent a usual gene. A typical statement was phrased by her: "It is not reasonable to regard such changes in expression and action as being produced by changes in a single gene". Additional information on $a_1{}^{m-1}$ was provided by McClintock (1953b). In a subsequent report (McClintock, 1954), the element that controls the $a_1{}^{m-1}$ locus was termed as *Spm*, for Suppressor-mutator. The $a_1{}^{m-1}$ allele is manifested by spots of dark anthocyanin pigmentation on a pale background. Such "spots" appear on the aleurone but also on somatic tissue (e.g. anthers). A full discussion of the *Spm* system (as well as the *Ac/Ds* system) was provided in McClintock (1956). In the latter (rather extensive) publication the two-component entity of *Spm* was detailed in which the *Spm* proper, that is autonomous and mobile, is the controller, while a derivative of this element is controlled by the autonomous *Spm*. The controlled element is exerting its direct effect on the mutable locus ($a_1{}^{m-1}$). The prophetic suggestion of McClintock (1956), written much before the classical work in bacteria by Jacob and Monod (1959), is amazing: "It might be considered that a controlling element represents some kind of extrachromosomal substance that can attach itself or impress its influence in some manner at various positions in the chromosome complement, and so affects the action of the genic substances at these positions". This is only one of the several "prophetic" statements in this publication. The source of the maize line in which the $a_1{}^{m-1}$ appeared was not detailed, but it was in a "Cold Spring Harbor Culture".

The *En* (enhancer) element was revealed in an abstract of Peterson (1953). *En* was found to affect the *Pg* (pale green) locus, causing the pg^m (dark stripes on pale-green leaf blades) phenotype. Already in this short publication the results of extensive genetic studies were summarized and it was concluded that the system has *two components*: the mobile *En* that activates another element, and the other element, *I* (inhibitor), that is located at the mutable locus. The maize line in which the two-component *En* was revealed originated from the atomic-bomb test-field on the atoll Bikini. Hence, the same atoll furnished the two-component bathing suit as well as the two-component *En/I* system, but whether or not *En/I* is causally related to the atomic blast is enigmatic.

A more detailed report on the pale-green mutations and *En* was presented by Peterson (1960) and in subsequent publications (Peterson, 1961, 1965, 1970) and was reviewed by Peterson (1987). In the first few years Peterson investigated the impact of *En/I* on pg^m but then he focused on the mutable a₁ gene. The "transition" from *Pg* to *A1* rendered the genetic study much more efficient. With *A1* the *kernels* could already be scored a few weeks after test-pollination. Moreover, to score 500 progeny only two ears are required. On the other hand, for scoring *Pg/pg* the seeds, after test-pollination, have to be harvested, then replanted and 500 plants have to be grown in the field. In an early study, Peterson (1965) found that the effects of *Spm* and *En* on mutable loci are identical; thus the two elements should be either the same or very similar. From the publications of Peterson and McClintock it became clear that these two controlling elements were discovered independently and during about the same period. The two names *Spm* and *En*, were maintained by McClintock and Peterson, respectively for many years. Peterson and his German collaborators (e.g. Saedler and associates) kept the name *En*, while McClintock as well as some other investigators (e.g. N. Fedoroff and associates) used the *Spm* term.

It has to be pointed out that at least until 1960 the claims of McClintock and Peterson that genetic elements move away from one locus to another site, were met with scepticism. Geneticists (though not all of them) argued that while they see evidence that an element (*Spm* or *En*) "disappeared" from a locus, there was not sufficient evidence that showed to which new locus the elements were mobilized. Because transposition is relatively rare and the maize genome is very extensive, it was very difficult to locate a transposed element. The maxim of "looking for a needle in a haystack" comes to mind. Hence, both investigators had a hard time defending their claims in genetic meetings.

3.2. Basic Characteristics of the Spm Family

First, for brevity, because *Spm* and *En* are virtually the same element or at least both belong to the same family of transposable elements – we shall therefore use in the following only the *Spm* name. We should recall that *Spm* has non-autonomous versions and we shall term them *dSpm*. The non-autonomous versions of *En* were termed by P. Peterson as *I*. The *dSpm* and *I* were probably derived from *Spm* and

En, respectively. All the *Spm* elements are structurally conserved while the *dSpm* elements are not conserved – there are different kinds of *dSpm* elements (and different kinds of *I* elements).

The genetic studies of Peterson and McClintock, during more than 15 years, clearly established that the *Spm* system consists of two components: the autonomous *Spm* and the inhibitor/mutator element *dSpm*. They also recognized that there is one kind of *Spm* but several different *dSpm* elements. The *Spm* as well as the *dSpm* elements can insert at (or adjacent to) a gene-locus. When this happens *dSpm* will stay there unless an *Spm* is in the genome; while an inserted *Spm* can excise from its location and be transposed to another location. Many genetic studies were conducted with the *A1* gene. The *A1* gene therefore serves to illustrate the interaction of *Spm* with a structural gene. When the latter gene is active a red pigment is produced. In maize kernels *A1* is manifested in a red aleurone. The regular *a₁* (recessive) allele causes colorless kernels. It was found, by genetic studies, that when *dSpm* is inserted in the *A1* locus (the resulting allele is termed a_1^{m-1}) the pigmentation is intermediate: pale pigmentation. When, in addition to the *dSpm* at the *A1* locus, there is an *Spm* element in the maize genome, there is a new phenotype: the aleurone is generally colorless but there are deeply pigmented sectors (spots) on the aleurone. The conclusion was derived that without the presence of *Spm* the *dSpm* is stable in the gene and it reduces the *A1* product but does not prevent it. When *Spm* is present it has two effects: first *Spm* causes the *dSpm* to suppress completely the activity of *A1* and then it also occasionally causes the excision of *dSpm*, away from the *A1* locus. When the excision happens, then in the cell where the excision occurred, and in all the descendants of this cell, the *A1* resumes its activity. This is the *mutator* effect. When the mutator effect is happening early in the endosperm development, the spots are large; if the excision is late, the spots are small. When excision is frequent, the spots are numerous.

Obviously *Spm* and *dSpm* can affect numerous genes and the preference of their integration into specific sites in the maize genome is unknown. One can record the impact of *Spm/dSpm* integrations only in a limited number of cases. When the integration causes lethality (as by inhibiting the synthesis of a metabolite that is indispensable and vital for the cells) – it will not be revealed. Also, reduction of the expression of many genes will not be clearly visible. Thus, we are left with genes that have an impact on non-vital activities such as pigmentation and starch quality, that are easily detectable. In spite of these limitations, geneticists detected numerous *Spm* elements in the maize genome; Fincham and Sastry (1974) listed 13 *Spm* in maize alleles. Clearly, such limitations in the detection of *Spm* and *dSpm* do not exist in the molecular era.

One approach to detect such elements is by PCR methods. Briefly, there are two main ways. In one, the 5' and 3' terminals of the *Spm* serve as primers to obtain DNA fragments that represent *Spm* and *dSpm*s. In the other way, one of the primers will be from a *Spm* (or *dSpm*) terminal but the other will be from the DNA sequence of a specific gene. The PCR fragments are cloned and then sequenced.

Moreover, in genomes with known DNA sequence, such as rice (Yu et al., 2002; Goff et al., 2002) the search for *Spm* and *dSpm* can be performed by computer analysis. Indeed, when the *indica* rice genome was searched for *Spm* sequences, three such sequences were found, but none of them was an intact *Spm* (Yu et al., 2002). These considerations will become clear after dealing with the molecular structure of *Spm*.

Turning back to the genetic studies of McClintock and Peterson, we already indicated that the non-autonomous elements (*dSpm*) come in different versions. The manifestations of these differences become clear when we observe the resulting phenotypes. In other words, focusing on the a_1^{m-1} locus, a kernel may have different patterns of dark spots on a white background. For example, there may be many small spots or a few large spots. Fedoroff (1983), who reviewed the *Spm* elements, termed these different patterns as "alleles" of the A_1^{m-1}, meaning that the locus may have different *dSpm*s inserted in it. The *Spm* that induces the mutation (excision) is structurally (i.e. in its nucleotide sequence) always the same, but it also may have different states. We shall understand the state after dealing with the molecular aspects of *Spm*. What could not be determined by the genetic studies is whether or not the different "alleles" (i.e. the different phenotypes) resulted from different structures of the inserted *dSpm* or that the trans-acting *Spm* also caused a short-range transposition of the *dSpm*. This short-range transposition could retain the *dSpm* in the same locus, but a short distance away from the original insertion site. The new site could affect the frequency of excision of the *dSpm* and its exact location within the gene and therefore cause a change in the pattern of pigmentation. An explanation for the differences in "allele" expressions had to wait until the molecular methods were applied to *Spm*, but genetic studies already showed that each "allele" had its unique pattern and unique change during development.

It was also observed that a "mutation" can change from "unstable" to "stable". In the stable mutation in the *A1* locus the aleurone will have a uniform pigmentation. This may be colorless, pale or almost fully pigmented. When the trans-*Spm* is removed from the genome the "mutation" becomes stable. There are other means by which an unstable mutation may become stabilized. In such cases the *dSpm* is no longer responsive to the impact of *Spm*.

As for the autonomous *Spm*, there is an apparent paradox: all the *Spm* elements are structurally similar or even identical, but they could be classified according to their interaction with the *dSpm* elements:
– the *Spm-s* is the standard autonomous element;
– the *Spm-w* is a weaker element (causing fewer and smaller spots).

The *Spm-s* is dominant to *Spm-w* and there are actually several versions of *Spm-w*. One of these is a truncated *Spm-s*. Another complication was revealed: there is also a *modifier* element that interacts with *Spm-w* and *Spm-s*. There may also be *mediators*. One such *Mediator* was characterized by Muszynski et al. (1993) and found to mediate between an *En* element and an inserted element (*I*). A special name was assigned to it: Irma, for *I*nhibitor that *R*equires *M*ediator *A*lso. Additional interactions will not be detailed here but one should take into consideration that the states of *Spm* are far from simple. For example, there are *Spm* elements that undergo cyclic changes (*Spm-c*). The cycles are with respect to the two activities of *Spm*: in causing the suppression of expression of the structural gene (*A1*) by the *dSpm* and in causing the transposition of the *dSpm* away from the structural gene. The *cycles* are reversible; causing active and non-active states of the *Spm-c* during plant development. During the active phase of the cycle the element was designated *Spm^a* and during the inactive phase – *Spm^i*. To further indicate the complexity we shall note that if there is more than one *Spm-c* in the maize genome, each of these may have its own cycle of activity. Also, the inactive phase of a *Spm-c* can last more than one generation but then the *Spm^i* becomes *Spm^a* again. There are two additional types of *Spm*: cryptic *Spm* and inhibited *Spm* (see Fedoroff, 1989a). We shall deal briefly with these *Spm* elements when dealing with methylation. Some *Spm* elements are strongly affected by the tissue in which they reside. One such case was described by Peterson (Peterson, 1987) in which the *En* (Peterson's designation of *Spm*) had a very different activity in the crown of the kernel than in its sides. This *Spm* was termed by Peterson as *En-crown*. There are other *localizations* of *Spm* activity but we shall avoid the details.

A final example of peculiar interactions between *Spm* and *dSpm* reminds us of *Alice in Wonderland*. More specifically, of the Cheshire cat in this story: first there is a grinning Cheshire cat on the tree but then the cat disappears gradually and finally only the grin of the cat remains. McClintock did not call it the Cheshire Cat phenomenon – she used the term *Presetting*. In combinations where an *Spm* affects the *dSpm* there are cases in which even after the *Spm* disappears from the genome (e.g. by segregation or excision) the impact of the disappeared *Spm* still stays with the *dSpm*. This simplistic description is far from providing the actual conditions under which presetting can happen, but it does provide the main message that it is possible for a *Spm* to set the stage for an influence, but then the influence continues even without the presence of this *Spm*. Notably, not only *Spm-s* can cause presetting; *Spm-w* and *Spm-c* are equally effective in this phenomenon.

Other features of the *Spm* system were also investigated during the *genetic era* of *Spm* studies; such as transposition. Since they became much more understandable by molecular investigations we shall now turn to the *molecular era* of *Spm* research.

3.3. Adding Molecular Methods to Spm Research

For the determination of the structure of a genetic entity, at the nucleotide-sequence level, it is first necessary to isolate and clone this entity. To perform this with the *Spm* elements (i.e. *Spm*, *dSpm*, *En* and *I*), clones, which contained these elements as insertions, were isolated. Even before full sequencing the presence of the elements could be verified: when an insert exists in a structural gene such as *Wx* or *A1*, the total length of the DNA is increased, respectively. This could be evaluated by methods such as restriction endonuclease mapping and heteroduplex hybridization between a clone of DNA that included an *Spm* element and a clone of the same gene that did not include the *Spm* (or its derivative). Once clones were verified to include *Spm* (or derived elements), these could be fully sequenced.

DNA fragments with inserted *Spm* (including *En*) elements and their derivatives were indeed cloned (e.g. Schwarz-Sommer et al., 1984; Fedoroff et al., 1984) and partially sequenced (e.g. Banks et al., 1985; Pereira et al., 1985; Schwarz-Sommer et al., 1985) and finally the full lengths of *En* (Pereira et al., 1986) as well as of *Spm* and of several *dSpm* elements (Masson et al., 1987) were sequenced.

The results of these sequencings (see reviewers of Fedoroff, 1989a; and Gierl, 1996) are summarized in Figure 47. These sequences brought at once the understanding of *Spm* system to a much higher level than could be achieved by using only genetic investigations.

3.3.1. The Overall Structure of the Autonomous Spm

The *Spm* element emerged as a DNA fragment of 8,287 base pairs. The sequence is virtually the same in *Spm* and *En* (Figure 47a). The *Spm* has terminal inverted

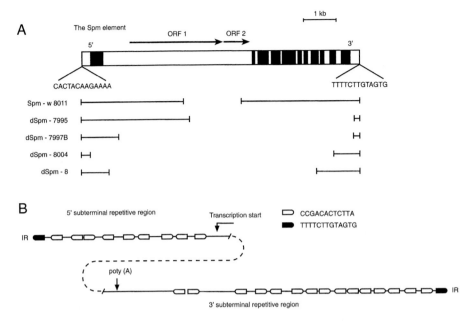

Figure 47. The Spm element. A. Structure of the Spm element. The Spm element is represented by the open box. The 5' and 3' ends of the transcription unit are indicated, and the left and right ends of the element are correspondingly designated the 5' and 3' ends. The filled areas between the ends of the transcription unit correspond to exons, and the empty areas correspond to introns. The two large ORFs within the element's major intron are represented by arrows over the element. Below the Spm diagram are interrupted lines representing the various dSpm elements whose designations are shown at the far left. The interruption in each line represents the internal sequence deleted from each dSpm element. Structure of Spm ends. The repetitive structure of the ends of the Spm element is represented diagrammatically; top line, 5' end; bottom line, 3' end. The transcription start and poly(A) addition sites are indicated. Filled arrows represent the 13-bp inverted terminal IRs; open arrows represent the location and orientation of sequences and marked homology to the 12-bp consensus sequence shown in the diagram. The regions comprising the closely spaced repeats represented by open arrows are designated the sub terminal repetitive regions. (From Fedoroff, 1989c).

repeats (*TIRs*) of 13 bp (CACTACAAGAAAA). The *TIRs* are very conserved, not only in the *Spm* system; they are identical to *TIRs* of other System II T.E., such as the family of the *Tam* elements of snapdragons (see below). Inside the 5' terminal and the 3' terminal there are sequences of about 180 and 300 bp, respectively, that are termed subterminal repetitive regions.

3.3.2. The Subterminal Repetitive Regions

These are schematically shown in Figure 47b. The 5' subterminal region is shorter than the 3' sub-terminal region but both are composed of repetitive elements. These elements have a 12 bp consensus sequence of CCGACACTCTTA that is 75 to 90% conserved. These are in either direct or inverse orientation. Two of the more terminally located repeated elements are oriented in a back-to-back orientation, which Gierl, Saedler and associates regard as essential in the process of transposition (see below and reviews of Gierl et al., 1989; Fedoroff, 1995; Gierl, 1996). The overall nucleotide sequences of the two subterminal regions is such that they can be drawn as an interrupted stem-loop structure (Schwarz-Sommer et al., 1984). Again, such a configuration may be essential for the process of transposition (Fedoroff, 1995; Gierl, 1996).

3.3.3. Exons, Introns and Translation

As shown in Figure 47a, the *Spm* has many introns and exons. One intron is especially long. Two "major" proteins are encoded in *Spm*: the TNPA of 67 kD and the TNPD of 131 kD. After transcription the respective mRNAs are differentially spliced from the same pre-mRNA. Both proteins are not produced in abundance but much more TNPA is produced than TNPD. This TNPA/TNPD ratio is maintained in maize, but when *Spm* is transferred to some dicots, such as *Nicotiana tabacum* and *Solanum tuberosum*, more TNPD than TNPA is produced, indicating different splicings. Figure 47a also shows that *Spm-w* is a truncated version of *Spm*(s), missing a middle region. The *dSpm* elements shown in the figure and actually all others that were sequenced, may differ in the length of the subterminal repetitive regions but all are extensively truncated in the middle region. It should be noted that *Spm* produces two additional proteins: TNPB and TNPC. The mRNA of these proteins are also spliced from the same pre-mRNA. The function of TNPB and TNPC is still engimatic.

3.3.4. Structure and Function of Spm and dSpm

The sequencing of *Spm* and *dSpm* opened the way for understanding a vast amount of genetic data. Before elaborating on the genetic/molecular relationship, the reader is reminded on the general structure of a gene as it emerged during the years 1960–1980. This is shown in a highly schematic way in Figure 48.

Obviously, while in this figure there is only one intron and two exons, in actual cases there may be many introns and many exons in a gene. Also, in this figure, only the insertion of *dSpm* was shown, while *Spm* may also insert in a gene. What should be noted is that the insertion may be upstream of the promoter, inside the promoter, in the leader, in one of the exons or in one of the introns. A scheme based on actual findings (Figure 49) is from the review of Fedoroff (1989a); not all the potentially possible insertion sites indicated in Figure 48 are represented in Figure 49.

Nevertheless, the available genetic data enabled Masson et al. (1987) and Fedoroff (1989b,c) to explain the differences between *Spm*-suppressive alleles and *Spm*-dependent alleles in a scheme (Figure 50). Some notes may render this scheme clearer. First, in the *Spm*-suppressible allele the indication "gene expressed" does not mean it is fully expressed. In the case of alleles of the *A1*, the kernel may have different levels of (uniform) pigmentation from very pale to very dark. In the presence of *Spm* the scheme indicates "gene not expressed". Obviously, this is the case of "spots", meaning that most of the aleurone is completely devoid of pigmentation, but the excision of *dSpm* will convert the respective cell (and the cells derived from it) – back to wild-type. The exact location of the insertion of *dSpm* in the *Spm*-dependent allele is not specified in the scheme of Figure 50. It can be in an

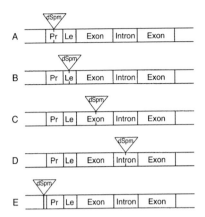

Figure 48. Schemes of insertion sites of dSpm elements in structural genes.
A. insertion into the promoter; B. insertion into the leader; C. insertion into an exon;
D. insertion into an intron; E. insertion 5' of the promoter. Pr, promoter; Le, leader.

intron, an exon or on the border between them. When the insertion of *dSpm* is into an intron, the *dSpm* could be spliced out during the processing of the pre-mRNA of the gene. This seems to happen when the direction of the inserted *dSpm* is anti-parallel to that of the gene but not if the *dSpm* is parallel to it.

Why is excision of *dSpm* in the *Spm*-suppressible allele causing gene expression in the "spots"? This would happen if the *dSpm* disappears *completely* from the gene so that normal mRNA, and consequently normal protein, will be produced. This is usually not the case: excised T.E. leave footprints behind. Such footprints could abolish normal expression of the gene. However, there is a special feature of the *Spm* system; this system usually leaves footprints of three nucleotides (or a multiplication of three nucleotides). When this happens one or more codons are added, leading to one or more additional amino acids – but no damaging frame shift. If the additional amino acid is tolerated for the functionality of the protein, then the *Spm* footprints will have no major impact on gene expression.

Obviously, not all cases of *dSpm* insertions are explained by the scheme in Figure 50. The *dSpm* can insert into an intron of the gene that is otherwise spliced out. If such an insertion happens, this may have one of several results. The splicing may remove the *dSpm* and normal expression will proceed. The insertion may also abolish the splicing and by that suppress expression. Methylation can interfere in these cases. We shall deal with methylation below.

3.3.5. Transposition

By transposition of *Spm* (and *dSpm*) elements we mean the excision of the element from the host locus and the insertion into the new locus. This process, as already repeatedly indicated above, can occur with the *Spm*, while it will happen with the *dSpm*s only in the presence of an *Spm* in the same maize genome. We should recall that transposition is a relatively rare phenomenon. Therefore, and because of other limitations, the analysis of the various steps and the investigation of the chemistry of the mechanisms of excision and reinsertion are problematic. Studying *Spm*

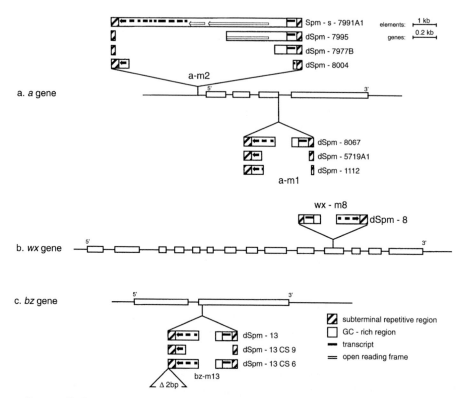

Figure 49. Spm insertions at the a, wx and bz loci. The structure of the transcription unit is shown for the a, wx, and bz genes in parts a, b and c; open blocks represent the exons, and the lines between them correspond to introns of the respective genes.
The location and structure of several Spm insertions, relative to each of the transcription units are represented above and below the corresponding gene diagram for the a-m1, a-m2, wx-m8 and bz-ml3 alleles. The top diagram in each set of insertions corresponds to the element inserted in the original mutant allele bearing the corresponding designation. (From Fedoroff, 1989c).

transposition in transgenic plants (e.g. tobacco) has advantages. The frequency may be higher and since such plants do not harbor *Spm* and *dSpm*s, their mobility can be traced with ease. In brief, the details of *Spm* transposition are not yet fully known. The understanding is much less than our acquaintance with the transposition of several bacterial Insertion Sequences.

One additional difficulty, which is rarely mentioned by those who investigated *Spm* transposition, is that we are not dealing with excision and reinsertion in "naked" DNA but rather with chromosomes in which the DNA is complexed with histones in chromatin structures. The chromatin may undergo elaborate modifications, and these modification may strongly affect the process of transposition (see: Appendix II on Chromatin Remodeling). Because of all these limitations, our description of transposition in *Spm* will not be precise. For some more details and discussions of this process, the reader is referred to the reviews by

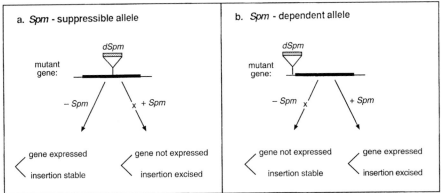

Figure 50. Effect of the Spm element on genes with dSpm insertion mutations.
Two different types of interactions between the transacting Spm element and a gene
with a dSpm insertion mutation are illustrated. a. The response of an Spm-suppressible
allele. In the absence of an Spm element, the mutant gene (filled box) is expressed at an
intermediate level and the insertion is not excised. In the presence of an Spm, the gene is
not expressed, and the insertion is excised in some cell lineages to give revertant
sectors. b. The response of an Spm-dependent allele. The mutant gene is not expressed in
the absence of an Spm and the insertion is stable. In the presence of an Spm, the mutant
gene is expressed and the insertion is excised to give revertant sectors.
(From Masson et al., 1987).

Fedoroff (1989a; 1995) and Gierl (1996) that provide ample references on this subject.

As indicated above, the terminals of *Spm* and *dSpm* have conserved sequences: the 13 bp *TIR* and the subterminal repetitive regions with their repeats of a conserved 12 bp sequence. It appears (Gierl et al., 1988; Masson et al., 1991; Trentmann et al., 1993; Schläppi et al., 1996) that the TNPA units bind to the subterminal repetitive region and the TNPA units that are bound on the 5' subterminal repetitive region, then approach the TNPA units that are bound to the 3' subterminal repetitive unit. This causes the two subterminal repetitive regions to become aligned and complexed with the TNPA units, and from a "stem" as part of the hairpin structure in which the middle region of the *Spm* or the *dSpm*, forms the "loop". It appears that TNPA has two binding domains. One for TNPA-DNA binding and another for TNPA-protein binding. In the structure of the TNPA/*Spm* complex suggested by Gierl (1996), there is a region in which there is no binding between the TNPA that is attached to the two pairs of back-to-back 12 bp motifs so that in this region the stem is open again; but the terminals of the *Spm* are again brought close to each other by complexing with TNPD and the TNPD that are complexed with one terminal has protein-protein binding with the TNPD that is complexed with the other terminal. It became evident that during insertion the "host" site is undergoing a staggered dsDNA cut and that the repair mechanism that terminates the insertion generates a 3 bp site duplication. This duplication constitutes the footprint that the element leaves behind when it is re-excised from the new locus. These footprints are characteristic for each T.E. family. One can thus ask if an *Spm* (or a *dSpm*) "visited" a DNA sequence in the past.

By the use of transgenic tobacco, potato and *Arabidopsis* (e.g. Cardon et al., 1993) the components that are essential for transposition could be identified. In such transgenic systems the production of TNPA and TNPD could be achieved by the respective cDNAs for these proteins, which were transported into the transgenic plants. The transposition of a *dSpm* could be revealed when a reporter gene was expressed upon excision of the *dSpm*. Support for molecular methods came from genetic studies (Dash and Peterson, 1994) in which several reporter alleles became expressed by excision. The essential components for the transposition were found to be the following: TNPA, TNPD and the terminal as well as subterminal regions of the *Spm* (or *dSpm*). As for the TIR, these must be intact; removal of some nucleotides may abolish the transposition, but one or two changes in the nucleotides of TIR are tolerated. The correct 12 bp motifs with their typical repeats and inverse direction are also essential for transposition. The 5' and 3' terminal regions are not symmetrical and are therefore, most probably recognized by the transposition complex, so that when in a synthetic element the 3' terminal is identical with the 5' terminal there is no transposition.

The transcription levels of *Spm* were analyzed in transgenic tobacco cells (Raina et al., 1993). These studies indicated that the constitutive promoter of *Spm*, that spans 0.2 kbp upstream of the transcription start site, is rather weak. Moreover, it is not sensitive to upstream enhancers (such as the strong 35 S CaMV promoter). The authors suggested that this insensitivity assures that the activity of *Spm* is insertion-independent; meaning that even when it were inserted downstream of a strong promoter, the activity of *Spm* would not be enhanced.

In a study by Frey et al. (1990) it was shown that both TNPA and TNPD are essential for transposition. These authors used a transgenic tobacco system. In this system the *dSpm* was inserted into the non-translated leader of the GUS gene. When such a *dSpm* is retained the GUS is not expressed but when the *dSpm* is transposed GUS activity is visible by appropriate staining. Adding to the transgenic tobacco the codes for TNPA, for TNPD or for both showed that transposition occurred only in the presence of both TNPA and TNPD.

From the suggested model of excision it emerges that the binding of TNPA and TNPD to the terminals of *Spm* causes a strong bending of the DNA of the *Spm*. We mentioned such a bending when we discussed the transposition of specific bacterial Insertion Sequences. This indicates that there is some analogy between the transposition processes of the two types of T.E. However, we should remember that the details are different. The details of dsDNA cuts and repairs in eukaryotes are now being revealed, mainly in yeast and mammals and they appear to be more elaborate in eukaryotes than in prokaryotes (see Karran, 2000 and Lichten, 2001, for reviews and literature). The cut and repair mechanism in the DNA of plant chromosomes has similarities with those of mammals but the details of this mechanism in plants are still under investigation.

3.3.6. Methylation and the Regulation of Transposition

The foregoing discussion emphasized that the *Spm* elements can undergo transitions in activity. Such transitions were already revealed by genetic studies of B. McClintock. The *Spm* could "cycle" between active and inactive states. It could also be inactive during maize development, being in a cryptic state, but then occasionally become active. Finally, *Spm* can be *inactive* inheritably, but then, by some triggering, the *inactive Spm* resumes its activity. The possibility that methylation has a decisive role in the changes of *Spm* activity emerged from findings in the *Ac/Ds* system, where methylation and demethylation was revealed by restriction endonucleases that are able to cleave only unmethylated sites. A series of studies, mainly by N. Fedoroff and her associates, established the importance of methylation in the alteration between inactive and active states of the *Spm* element. These studies were summarized in a review of Fedoroff (1995).

The subject of DNA methylation in plants was reviewed by Richards (1997). It was estimated that up to 20 to 30 percent of the cytosines are methylated in the nuclear genomes of angiosperms and most of these modifications appear in CG and CNG sequences. The methylation takes place post-replicatively by plant DNA methyltransferases. Generally, plant DNA is more methylated than animal DNA and there are many potential methylation sites in the DNA of an *Spm* element. From analyses of active and inactive *Spm* elements it was concluded that there are two regions of the *Spm* in which there was a high correlation between methylated/hypomethylated and inactive/active states. These were termed upstream control region (UCR) and downstream control region (DCR), respectively. The UCR is the region between the 5' end of the element and the transcription start site. The DCR extends downstream from the transcription start site (nucleotide 578). In the DCR there are 11 repeats of CGGGCGGGCGGCCTGC, which are subject to possible methylation.

The binding of TNPA to *Spm* was found to be reduced if in the TDNA-binding region the GC and the GNC are methylated. As this binding is essential for transposition, this methylation inhibits the transposition of *Spm* (and *dSpm*). Indeed, in inactive *Spm* the UCR was found to be highly methylated. In cryptic *Spm* the GC-rich DCR region was found to be highly methylated. When, in a population of plants, a genetic selection was performed for inactivity of *Spm*, there was an increased methylation of DCR and UCR. The UCR was methylated before the DCR. From such experiments, Fedoroff (1995) concluded that methylation of UCR is correlated with the elements inactivity. Other studies also revealed correlations between methylation during certain developmental phases of maize and the level of the element's activity.

Transcription of *Spm* probably requires an hypomethylated promoter. Thus, when UCR is methylated, there should be a lack of TNPA. But here comes a "loop-control": it was found that TNPA by itself suppressed the capability of a hypomethylated *Spm*-promoter to cause transcription. This repression most probably keeps a tight control on the *Spm* activity. TNPA was also found to demethylate an inactive (methylated) *Spm*-promoter. These apparently contradicting effects of TNPA may be explained by different binding domains of the TNPA and by the interference of TNPA binding with the binding of transcription factors.

We should recall that the "positive" role of TNPA is to bind to the subterminal repetitive region of the *Spm* (and *dSpm*) in order to initiate the transposition. We should also recall that for any transcription the promoter region of a gene should be receptive to the polymerase as well as to transcription factors. This should also be true for the promoter region of *Spm*. Thus, binding of TNPA probably interferes with binding of polymerase and transcription factors.

In short, there is a negative interaction between transcription and transposition. How exactly the balance is reached and transposition is ensured is not yet clear. It could be a matter of exact and correct level of TNPA and/or differential binding to

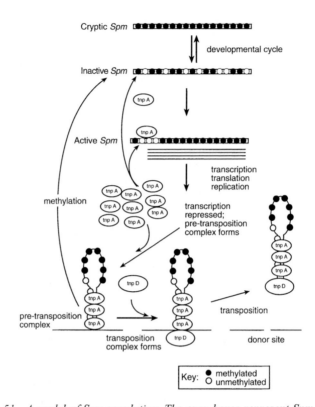

Figure 51. A model of Spm regulation. The open boxes represent Spm elements.
Open and filled circles represent unmethylated and methylated sites, respectively.
Spm elements are represent as undergoing changes in the extent of methylation during
the developmental cycle which results in reduced methylation early in development,
followed by increasing methylation during development in a tissue-specific manner.
TnpA is represented as interacting with the element's 5' end to maintain it in a
transcriptionally active hypomethylated state relative to the remainder of the element.
Accumulation of TnpA is postulated to inhibit transcription of the element and promote
formation of a pre-transposition complex in which the element ends are associated with
each other through TnpA-mediated interactions. Conversion of the pre-transposition
complex to the transpositionally competent form is postulated to occur with the
addition of TnpD to the complex, followed by transposition to a new site. An occasional
alternative fate of the pre-transposition complex is proposed to the dissociation,
releasing the element in a methylation-sensitive form. (From Fedoroff, 1995).

specific repetitive motifs (12 bp) of the subterminal region. Perhaps the binding of
TNPA to certain sites at the UCR and DCR causes demethylation of adjacent sites.
    Furthermore, TNPA binding and methylation/demethylation could cause changes
in the configuration (bending, coiling) of the DNA; and finally, one major factor has
not yet been investigated: what happens with the modification of the histones as a
result of TNPA binding and methylation/demethylation of the DNA and the UCR
and DCR? Such possibilities and others that take into account the heritability of
active and inactive states of Spm, developmental programming of Spm activity and

the existence of a population of *Spm* elements in a single plant, each with its own rate of activity – all these issues were discussed by Fedoroff (1995) but there is as yet no full understanding of these issues. In the latter review a general model for *Spm* regulation was presented (Figure 51). This model can serve as a summary of the present knowledge of *Spm/dSpm* transposition.

4. THE FAMILY OF *MuDR/Mu* TRANSPOSABLE ELEMENTS

The *MuDR/Mu* elements constitute the third major family of T.E. in maize. Maize lines that exhibited the characteristic features of *MuDR/Mu* were earlier termed *Mutator* lines. These latter lines originated from one stock. The review of Chandler and Hardeman (1992) summarized the origin of this first *Mutator* stock. It was traced to a maize line in the laboratory of Dr Alexander Brink (University of Wisconsin). This line contained an *Ac(Mp)* element inserted into the *P* locus. The latter locus codes for the phlobaphene synthesis that is required for dark pigmentation in the cob and the pericarp of the grain. A revertant of this mutant was transferred into an inbred line (W23) by repeated backcrosses until no evidence for the *Ac(Mp)* was found. Then a self-pollination was performed and one progeny line that segregated for a pale-yellow endosperm was given to Dr Donald S. Robertson (in 1961), who maintained it at Ames, Iowa, as a heterozygot, because the homozygots were not healthy plants. Here the story of *Mutator* started. Robertson termed the yellow mutant *y9*. As noted above, the mutant was maintained as a heterozygot. He used the pollen of this heterozygot to cross-pollinate homozygous marker lines. The progenies were selfed and the resulting seedlings were scored. In such segregating populations Robertson found an "unusually high number of new seedling mutants" (that were not *y9* mutants). The "unusually high number" was 50–100 times more recessive seedling mutants than in regular maize stocks. In other words, for those who are not maize geneticists or corn-breeders, this means that if commonly one among 10^6 plants is a recessive mutant, then among the tested progeny of Robertson's stock there were one or two mutants among 10^4 plants. To score 10^6 seedlings one can use "sand-benches", where the seedlings can be planted in a very high density – still, quite an extensive area of "sand-benches" is required. If you would like to score all the 10^6 mature plants, you would need already a small farm, but then you would have 2×10^6 cobs with a total of up to 5×10^8 grains for scoring. Robertson followed the tradition of maize geneticists since Rollins Adam Emerson and provided the Robertson stock to other investigators, and these used them and also found the same phenomenon of increased mutation frequency. Still, from the time Robertson received this stock until his first publication on this *Mutator* system (Robertson, 1978), there was a gap of 17 years. Well, maize genetics has its own pace; from the time B. McClintock established, verified and explained genetically the Controlling Elements of the *Ac/Ds* system – until she was awarded a Nobel Prize for it there was a gap of almost 30 years.

One more note before we enter the *Mutator* (*MuDR/Mu*) system. Out of the three main families of Class II transposable elements in maize, which were studied in great detail, three were heralded from Ames, Iowa, and two from Cold Spring Harbor, New York. This is an apparently strange statement – but it is correct. We mentioned above two such T.E., *Ac/Ds* and *Spm*, that were revealed by Barbara McClintock in Cold Spring Harbor. P. Peterson of Ames, Iowa revealed the *En/I* elements, but since these were later found to be identical or very similar to the *Spm* elements of McClintock, they surely belong to the same family. Then D.S. Robertson, also of Ames, Iowa, found the *Mutator* (*MuDR/Mu*) elements, and finally, P. Peterson and associates (Schnable and Peterson, 1988; Schnable et al., 1989) revealed the *Cy/rcy* T.E. system. The latter elements were found to be part of the *Mutator* family.

The *MuDR/Mu* system has been amply reviewed (e.g. Walbot, 1991, 1992; Chandler and Hardeman, 1992; Bennetzen et al., 1993; Bennetzen, 1996; and recently this subject was treated thoroughly by Walbot and Rudenko, 2002). Readers interested in more details on this system than provided below will find references in these reviews. The subject is complicated not only for "outsiders". Bennetzen (1996) noted in his review: "For many investigators, the most confusing aspect of the *Mutator* system has been its complicated pattern of inheritance". Well, this statement says several things. First, that there is more than one confusing aspect in *Mutator*. Then, that the inheritance is confusing even to the investigators who studied this system. Consequently the novice reader should be patient.

4.1. Early Genetic Studies

We mentioned that Robertson recognized the *Mutator* line by scoring new mutations in maize seedlings. In the *Mutator* lines these mutations occurred at high frequencies: one or two orders of magnitude higher than in regular, non-*Mutator*, lines. Based on previously studied T.E., it was inferred that such *forward mutations* could result from the insertion of a transposable element into genes or close to them. The genetic procedure to detect such mutations was simple. The *Mutator* stock was used as pollen parent to cross-pollinate non-*Mutator* lines. Controls were also performed: the *Mutator* stock was self-pollinated and also the second ear of the non-*Mutator* was self-pollinated. This was done to assure that the mutations in the F_2 of the *Mutator* × non-*Mutator* are really novel and not recessive mutations that were already existing in the parental lines. The F_1 plants that showed no novel mutations, were self-pollinated and the F_2 was scored for the *Mutator*-induced mutations. Seeds of this F_2 generation were thus planted in high density in sand-benches and scored for seedling mutations such as albino, yellow, yellow-green, pale-green, glossy and dwarf. Robertson (see Robertson, 1980, 1981, 1983, 1985; Robertson and Mascia, 1981) found that typically about 10% of the *Mutator* plants caused novel seedling mutations in the F_2. He propagated his *Mutator* stocks in a way that the mutagenic potency was maintained for many generations. We should remember that only mutations that can easily be scored in seedlings were taken into account. Mutations that were lethal or had no visible impact on seedlings could not be taken in account – although these could be the majority of mutations. Typically, most progenies of crosses between a *Mutator* plant and a non-*Mutator* plant retained the high mutation frequency: the ratios were about 90% retained to 10% lost. Being aware of the two-component *Ac/Ds* and *Spm* T.E., Robertson considered that the *Mutator* system is also composed of two components, and if mutation capability is lost, it probably indicates that the autonomous component is lost. However, the genetic results strongly suggested that in *Mutator* stocks there is more than one autonomous element, probably three; moreover, the number of autonomous elements could increase prior to meiosis.

In addition to this *forward mutation* there was another activity of the *Mutator* system: the reversion of a mutation to wild-type. These somatic reversions could be best evaluated (but not only) in the aleurone layer of the grains. Pale (or colorless) aleurones may exist in a maize line due to a mutation in a gene encoding an enzyme that is required for pigmentation. A reversion to pigmentation in one or more aleurone cells would indicate that the mutation was caused by the insertion of a transposable element, and that the excision of this element in the colored cells (or their predecessors) caused the reversion to wild-type. An experienced maize geneticist can look at colored spots of a maize grain and, in a Sherlock Holmes manner, provide a lot of information. She (or he) would be able to identify the allele involved and whether or not the gene was reverted completely or partially; she (or he) could also tell you when, during the development of the endosperm, this reversion (excision) happened.

Actually, the number of cell divisions from the first endosperm nucleus to the mature endosperm is very stable in maize: it is 18 divisions (Levy et al., 1989; Levy and Walbot, 1990). Thus, if one counts the number of pigmented aleurone cells of a spot, one can tell when, exactly, the reversion happened. In practice, the test for *Mutator* capability to cause excision is done in the following general manner. Lines that are mutated in an endosperm trait are identified. Among these, a specific line that is known to be capable of reversion is used. A plant of this latter line is crossed with a plant for which the investigator would like to evaluate the *Mutator* (reversion) capability. The cross-pollination is performed and the grains resulting from this pollination are scored for revertant spots. The data obtained included two kinds of information. First, the fraction of the grains on this ear that contained spots was determined. The number of spots in each grain as well as the size of the spots were then evaluated. For additional information, the tester-line could have, in addition to endosperm mutation, another marker that is linked to the same locus. For example, the recessive mutation that is expected to revert as a result of the excision of the T.E. element could be *bz-1* (inability to have a bronze pigmentation) and the linked gene could be *Sh1/sh1*, meaning plump/shrunken grains. The scheme of a possible cross is:

$$\frac{bz1 - m * Sh1}{bz - 1 \quad sh1} \times \frac{bz - 1 \quad sh1}{bz - 1 \quad sh1}$$

In this scheme the *Mutator* capability of the pollinator (on the right) is evaluated by scoring the grains that developed on the cross-pollinated ear (on the left). Robertson, and then other investigators, tested the reversion of several mutations in alleles such as *bz-1m* and *a1-mum3* (see Robertson and Stinard, 1987, 1989; Walbot, 1986, 1991). It was found that there was a fair (but not absolute) correlation between the two activities of *Mutators*: the ability to cause forward mutations and the ability to cause reversion to wild-type. Obviously the two types of analyses measure different processes. In the forward mutations the analysis is of many alleles, as many as can be scored by the investigator, while the reversions are commonly focused on one specific allele that is frequently an allele affecting the endosperm (aleurone). Even more importantly, the two assays measure the activity at different developmental stages. Additional differences may exist. Some of these will become evident after we learn more about the *Mutator* system. Now, just one note: in both forward mutation and reversion the same enzyme, transposase, is involved but the level of transposase required for forward mutation could be much higher than the level required for reversion.

For a number of years, Robertson and his associate were the carriers of the *Mutator* banner, but then this subject attracted several other highly talented investigators (e.g. Bennetzen, 1984; Freeling, 1984; Taylor and Walbot, 1985; Chandler and Walbot, 1986). These investigators and their associates added a wealth of information to the *Mutator* system. It became evident that there are several kinds of non-autonomous Mutator elements, but one of them, the *Mu1*, is by far the most prevalent in known insertional mutations. This already indicated that while in reversion-assays one measures mostly the excision of *Mu1*, in the forward mutation other *Mutator* elements, including the autonomous element, are also involved. More important information was revealed in the forward mutations: the *Mutator* element's transposition was not restricted to linked loci, the "jumping" capability of these elements was found to be much greater than in the elements of *Ac/Ds* and *Spm*. However, there seemed to be a preference to insert into genes or at least an avoidance of repetitive DNA sequences. The high frequency of forward mutations, transposition into non-linked loci and preference for gene-integration caused V. Walbot and others to use the *Mutator* system for gene tagging.

The insertion of a *Mutator* element (*Mu1*) into genes was actually the key to much further studies. M. Freeling and associates (Strommer et al., 1982) found that an *Adh1* mutation was caused by the insertion of such an element into the gene. Because the coding sequences of the *Adh1* were known, these could be used to probe Southern blots of *mutants* and *non-mutants* with such a sequence and then see that a specific DNA sequence was inserted in the mutant. This led ultimately to the full sequence of *Mu1*. Moreover, when the issue of inactivation/methylation came up, the investigators could test, by appropriate endonucleases (that are not cutting in sites with methylated cytosine) whether or not the inserted *Mu1* is methylated.

We mentioned above, while dealing with *Spm*, that T.E. insertion can take place in several regions of a gene: in the promoter, in exons or in introns. Each of these insertions has its consequences on the gene-expression (see Bennetzen, 1996, for detailed discussion). Generally, insertion in the promoter may change the developmental regulation of a gene; insertion in an exon will commonly abolish expression and insertion in introns may have different consequences such as interference with normal splicing. Actually, the first *Mutator* element (*Mu1*) that was located in a gene (Strommer et al., 1982; Bennetzen et al., 1984) was found to be an insert into the first intron of *Adh1*. However, observing a mutated phenotype will not provide accurate information on where in the gene the insertion took place. For example, some insertions into introns could be overlooked because they cause no or only a minor change in expression. Here is where molecular techniques are very helpful. Once we know the nucleotide sequence of a gene (or even a part of it) and the nucleotide sequence of a transposable element, we can use part of these sequences as primers, and perform a PCR to obtain an amplified DNA that includes parts of the element and parts of the gene. The amplified DNA can then be sequenced. Moreover, we shall see that while the non-autonomous *Mutator* elements differ substantially, they have very similar (almost identical, in some parts) inverted terminal repeats (TIRs). Thus, consensus sequences from such TIRs can serve as PCR primers to detect *Mutator* elements in known genes.

Once the locations of insertions and excisions could be identified and sequenced, a plethora of additional information became available. It was found that, as in other T.E. systems, insertion is accompanied with host-DNA duplications that flank the insert. On the other hand, the excision of *Mutator* elements frequently involves rearrangements of the abandoned site. We shall provide information on transposition of *Mutator* elements in a subsequent section but should note already that this transposition in the *Mutator* system differs from that of the *Ac/Ds* and *Spm* systems. In germinal tissues the *Mutator* elements do not transpose by the cut and paste manner but by replicative transposition, meaning that one copy of the element stays behind. It was thus observed that the *Mu* element excised from germinal tissues in much lower frequencies than the transposition into new sites during gametophytical development.

4.2. The Structure of Mutator Elements

Following the detection of the *Mu1* element, as noted above, several additional elements were revealed in subsequent years (see reviews of Walbot, 1991; Chandler and Hardeman, 1992; Bennetzen et al., 1993). All these (*Mu2, Mu3, Mu4, Mu5, Mu7, Mu8*) were non-autonomous elements. Their length was found to be 1 to 2.2 kbp. They all had TIRs of about 200 bp and commonly created a 9 bp duplication of the target DNA flanking the inserted element. Although all these non-autonomous elements have similar or identical TIRs, their internal regions are frequently very different. This is rather different from the situation of the previously handled maize T.E., where the internal regions are commonly merely truncated versions of the autonomous element (Figure 45 for *Ds* elements and Figure 47 for *dSpm* elements). The internal regions of the non-autonomous *Mu* elements are usually not related to each other, although some of them (e.g. *Mu1* and *Mu2*) share parts of these internal regions.

Only in 1991 was the sequence of the autonomous *Mu* uncovered (Chomet et al., 1991; Hershberger et al., 1991; Qin et al., 1991). First it was given different names (*MuA2, MuR1, Mu9*) by different authors, but then, in honor of D. Robertson, it was termed *MuDR*. The *MuDR* spans 4,942 bp, it has the usual TIRs and its internal region has similarity to one of the non-autonomous *Mu* elements (*Mu5*). Inside the TIRs there may be internal repeats of a few hundred base pairs. The nomenclature was sometimes confusing. Thus *Mu2* was previously termed *Mu1.7* and a *rcy: Mu7* element (inserted in the *Bz-1* gene) was detected that was regulated by the *Cy* autonomous element of P. Petersen (Schnable et al., 1989), but it had sequences homologous to *Mu7*. A detailed description of these elements and their sources, updated to 1992, is provided by Chandler and Hardeman (1992). The possible evolution of various non-autonomous *Mu* elements from a *MuDR*-like progenitor was presented by Bennetzen (1996). This was mainly based on the sequences of the TIRs.

A scheme of the structure of *MuDR* is provided in Figure 52. This 4,942 bp element encodes two major transcripts, one of about 1.0 kb and the other of about 2.8 kb. These transcripts are convergently transcribed from promoters in the right and left TIRs, respectively (see Benito and Walbot, 1994, for details). The 1.0 kb transcript is encoded by a gene that contains three introns, the last of these is rarely spliced. The 2.8 kb transcript also has introns and its fully spliced mRNA codes for a protein of 823 amino acids. The translation products of the partially spliced transcript are shorter. This apparent paradox is explained by stop-codons in the introns that cause shortened translation products, as shown in Figure 52. Between the open reading frames of the 1.0 kb and 2.8 kb sequences (*mudrB* and *mudrA*, respectively) there is an *intergenic region* that is usually not transcribed but in some

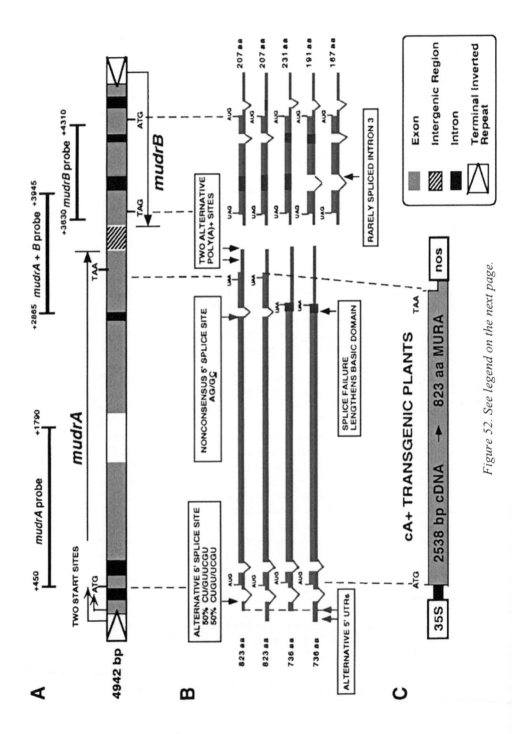

Figure 52. See legend on the next page.

cases it is transcribed, meaning that the transcription continues into the opposite open reading frame. When that happens, the opposite open reading frame is transcribed in the antisense direction. Antisense transcripts can hybridize with the sense transcript and dsRNA may be produced. This could serve as a basis for RNAi degradation. RNAi degradation of RNA will be handled in Appendix I of this book.

The larger (2.8 kb) transcript is essential for *Mutator*-associated activities, while the elimination of the smaller (1.0 kb) transcript still permits transposition and mutability but the suppression activity seems to be lost. The sequence of the fully spliced larger transcript (coding for *mudrA*) indicates that it contains a region with significant similarity to bacterial Insertion Sequences transposases, some similarity to retroviral integrase and a zinc-knuckle region. These findings strongly support the claim that *mudrA* is indeed the transposase of the *MuDR* system.

Bennetzen (1996) noted that the TIRs are rather unique to *Zea* and the related species *Tripsacum andersonii* but other sequences of the *MuDR* have been identified in rice. This led him to suggest that the *Mutator* elements have been present in plants for at least 60–70 million years, when the ancestors of maize and rice diverged.

When the sequence of the single regulator of the *Mutator* system, the *MuDR*, was established, its further identification in maize genomes was simplified. For example, the endonuclease *SstI* digests the TIRs, provided they are not methylated. Thus, when *SstI* digestion generates a 4.7 kb fragment, from genomic DNA of maize, this fragment indicates that it was cut out of the 4.9 kb *MuDR*. The identity of the *MuDR* can be verified by an appropriate hybridization probe. On the other hand, when the diagnostic restriction site (of *SstI*) is lost or the TIRs are methylated, the respective *MuDR* and the homologues of the *MuDR* (*hMuDR*) may be overlooked. It is now evident that inbred lines of maize that are not *Mutator* lines but do contain several such *hMuDR* elements. These *hMuDR* elements were apparently derived from *MuDR* by step-wise point mutations or by extensive changes, such as the addition of DNA sequences in their intergenic region, where the repetitive DNA could be extended by several hundred bp. In their review, Walbot and Rudenko (2002) updated the subject of *hMuDR*. The reader is reminded of the first few sentences in the Introduction of this book, where the "skeletons" of transposable elements in present-day genomes were mentioned.

Figure 52. Structure of MuDR, endogenous mudrA and mudrB transcripts, the CaMV 35S-mudrA construct in cA+ transgenic maize lines, and the probes used for RNA gel blots. A. Structure of an endogenous MuDR element. The element has two open reading frames, termed mudrA and mudrB, encoded in anti-parallel orientation. The intergenic region between the two genes is composed of diverse short repetitive elements. The promoters are located within the ~215-bp TIRs. The mudrA region with high similarity to bacterial transposases is shown. The DNA probes for RNA analysis are located above the element. B. The diversity of endogenous mudrA and mudrB transcripts in active Mutator seedlings (Hershberger et al., 1995). Intron sequences shown in solid black are in-frame with exons. Alternative mudrA transcription initiation sites (+169 and +252) produce transcripts with a short or long 5' leader sequence. [aa], amino acids. C. The structure of the CaMV 35S-mudrA cDNA in M13 transformed into maize to make cA lines. In construct phMR53, the native promoter, alternative start sites, 5' UTR, and introns were removed. The CaMV 35S promoter and 130-bp leader sequences were substituted. The mudrA 3' UTR (polymorphic region) was truncated and fused to the nopaline synthase (nos) terminator. (From Raizada and Walbot, 2000).

4.3. Transposition

As described above, the transposition of *Mu* elements requires the presence of an active *MuDR* element in the same maize genome. In the elements of the *Mutator* family, as in the T.E. of other families, transposition means two processes. We shall start with the second process, insertion (because it is the same process in all tissues and developmental stages of maize), in which the element is inserted in a new site in the genome. We already noted that there is no obvious preference of insertion sites. The first stage of transposition may proceed in either of two ways. The element may leave the donor-DNA site during the excision process and not leave a copy of itself behind (the "cut and paste" way), or the element undergoes replication and only the replicated element moves to the new site, where it is inserted, but a copy of this element stays at the original site.

Thus, in both ways, there is an element available for forward mutation (by an insertion into a new site) but only in the "cut and paste" way is there no element left in the site in which it was previously integrated. The examples of spots of wild-type cells in the aleurone are the results of the "cut and paste" method. Because in the aleurone this "cut and paste" transposition takes place at the last developmental stages of the endosperm differentiation, there is no way to trace the mobility of the element to the "paste" location. The "switch" from the common "cut and paste" transposition to a replicative transposition was already revealed in Tn7. In the part on bacterial Insertion Sequences we described Tn7 in detail but mentioned only the "cut and paste" transposition of Tn7. Now we would like to note that this is true for "wild-type" Tn7 but not for a specific mutant. May and Craig (1996) reported that a single amino acid alteration in one of the Tn7-encoded proteins (*tnsA*) may occur. In this alteration an alanine at position 114 is substituted for aspartic acid. In such a mutant there is a "switch" from "cut and paste" to replicative (cointegrate) transposition.

One should, obviously, be very careful in the comparison between the mutation induced "switch" in Tn7 and developmental/tissue induced "switch" in *MuDR*. The chemistry of transposition of Tn7 is known, while we know very little on the details of the chemistry of transposition in *MuDR*. Moreover, for the Tn7 transposition there is an *in vitro* system and the DNAs are not complexed with histones, as they are in *MuDR*. Nevertheless we should acknowledge that a single amino acid substitution in a protein involved in transposition can drastically change the mode of this process.

We indicated above that *MudrA* has sequence similarities with the transposases of other T.E. elements. This and genetic studies clearly indicate that *mudrA* is the transposase of the *MuDR* system. Further characterization of *mudrA* encountered a technical problem. The *mudrA* gene is toxic to *Escherichia coli* and to other bacteria that serve in genetic engineering of plasmids and transformation vectors. Obviously, the investigator would like to isolate the *mudrA* coding region, flank it with defined regulatory elements and then build a vector that can be used to produce transgenic plants. Transgenic plants that have a mutated gene in which a *Mu* element is inserted can serve as host to a transgene with an engineered *mudrA* gene. Recently it became possible to express *mudrA* in yeast (*Saccharomyces cerevisiae*), because yeast tolerates the fully spliced MURA-823 product. This protein could thus be produced in sufficient quantities to analyze its role in transposition.

The transcription of the two genes of *MuDR* (*mudrA* and *mudrB*) is regulated by the respective promoters that reside in the two TIRs. How much the transcription is also modulated by host sequences is not entirely clear, but it was recorded that individual *Mutator* lines exhibited slightly different transcript levels. Because the *MuDR* elements of these lines are in different sites of the maize genome, it suggests that host sequences affect transcription. In the same *Mutator* line, transcription levels differ in the various tissues of the host plants. Thus, the transcript was found to be highest in the cells surrounding the macrogametophyte. This is a clear indication that the TIR-located promoters are sensitive to developmental or tissue-specific clues from the host plant. The TIRs of non-autonomous *Mutator* elements (such as *Mu1*) are similar to those of the *MuDR* element but not identical to them. Clearly non-autonomous *Mu* elements are not independently transcribed. The promotion capability of *MuDR* TIRs was investigated by constructs in which such TIRs were fused to reporter genes. When this was performed with non-dividing maize protoplasts, the expression of the reporter genes was very low. However, when TIRB was fused with the β-glucuronidase (GUS) reporter and transferred to transgenic maize plants, there was a strong expression of GUS in all the plant tissues. This can be interpreted as a requirement of the TIRB promoter for a clue in the actively dividing maize cells; the levels of GUS was especially high in meristematic cells. The GUS expression was also very high in developing flowers, and particularly in the pollen.

Transcription starts at nucleotide number +169 for *mudrA* and at +163 for *mudrB*, when the nucleotides are counted from the left and the right terminal ends, respectively. Thus, the regulatory clues should have their impact at about 150 nucleotides upstream of the transcription-initiation point. We should keep in mind (see also Figure 52) that the *mudrA* has two transcription initiation points. One, as noted above, at +169 and another at +252. This causes two lengths of "leader". The translation initiation site for *mudrA* is at nucleotide 450 (i.e. an ATG codon). The fully spliced *mudrA* message of 823 amino acids codes for the putative transposase that we shall henceforth term *MURA*. *MURA* binds to a conserved 32 bp region in the TIRs of the *MuDR* element and interferes with the methylation of these sequences. It has characteristics that should enable it to enter the nucleus and was therefore defined as a nuclear protein.

The processing of *murbB* is more enigmatic than that of *murbA*. The former also has 3 introns, and accordingly, there seems to be more than one specific and canonical protein product (*MURB*). Moreover, the function of *MURB* is not yet clear. By *in situ* staining, *MURB* appears to be mainly a nuclear protein. But there is a paradox. *MURB* lacks elements that should lead it into the nucleus. Moreover, when *MURB*: GUS is bombarded into (the non-dividing cells of) the onion epidermis, *MURB* appears as a soluble cytoplasmic protein. This author suggests to consider a solution to this paradox: *MURB* "enters" the nucleus in the dividing cell, at the stage in which there is no nuclear envelope.

As already mentioned, *Mu* excision can be observed best in alleles affecting the pigmentation in the aleurone of the grain's endosperm. Each plant carries several hundred grains and spots resulting from excision of a *Mu* element from a gene such as *al-mum2* are easily scored. Experiments showed that the frequency of excision is not increased by the increase of copies of *MuDR* elements, although the latter do increase the abundance of *MuDR* transcripts. This increase in transcript levels is not accompanied by an increase of the levels of *MURA* and *MURB*. It seems that protein levels are regulated post-transcriptionally. How much of these proteins is required for excision is not clear. One additional note: *mudrA* and *mudrB* contain many codons that are rarely used in maize. Consequently, the translation of *mudrA* and *mudrB* is probably not efficient. This could lead to a lot of transcripts but a low level of translated proteins.

The excision performed by *MURA* seems to be a "dirty" job. While in *Ac/Ds* and *Spm* excision is "clean", leaving behind only the footprints, this is not so in the *MuDR* system. Various "scars" were detected after excision. For example, all or none of the 9 bp host sequence duplication ("footprints") may be retained. Also, there are common deletions that extend into the host sequences. There may also be base additions that are copies of the *Mu* elements. All these "scars" will not restore the wild-type. In short, reversion to wild-type (e.g. spots in the aleurone) constitutes only a minority of the total excisions of *Mu* elements, when the *Mu* resides in an exon. A higher rate of reversion to wild-type is expected when the *Mu* resides in introns, or in the promoter. Excision could also be abortive. One of the possibilities is that rather than causing two ds cuts on both sides of the element, there is only one ds cut.

To better understand the "cut and paste" transposition and especially the fate of the *Mu* element after excision, Walbot and associates used the *RescueMu* system (Raizada et al., 2001). The *RescueMu* had the following features. It contained the *Lc* (R) gene for anthocyanin production driven by a plant promoter (CaMV 35S). Near the 5' end of this gene a DNA sequence was inserted that was a modified *Mu1*. The latter had the TIRs of *Mu1* as well as an origin of replication for bacterial plasmids, and an antibiotic selectable marker (Amp^R), so that the whole construct could be propagated as a selectable plasmid in bacteria. Between the two TIRs there were also either of two tags for identification. The construct also had a poly (A) termination signal behind the *Lc* (R) gene. A small amount of maize tissue that contains this construct or even only the modified *Mu1* could be used to extract the genomic DNA and to identify the construct in appropriate bacterial plasmids (by plasmid rescue).

Moreover, as the modified *Mu1* had also either of two unique 400 bp tags for the positive identification of rescued plasmids, the follow-up could be on two *RescueMu* constructs. The genetic transformation of embryogenic maize was performed with embryogenic maize callus in a procedure by which the two *RescueMu* constructs, as well as a vector with a selectable marker (resistance to *Basta*), were co-bombarded (see Galun and Breiman, 1997, for transformation procedures), into the embryogenic maize callus. A high proportion of the *Basta*-resistant regenerants also contained the *RescueMu* construct. The embryogenic callus was from a line that did not contain *MuDR*. For analysis of the fate of *RescueMu* in the presence of *MuDR*, transgenic maize plants with *RescueMu* were crossed with a *MuDR* containing maize stock. Stocks with different numbers of *MuDR*s could be used and the fate of *RescueMu* could be followed in germinal as well as in somatic tissues. Spots of *Lc* would indicate excision and small tissue samples could serve for plasmid-rescue to follow the fate of the excised modified-*Mu1*.

The description above is merely a summary of the *RescueMu* procedure. The procedure is somewhat elaborate; the planning required a thorough knowledge of the *MuDR* system, proficiency in molecular genetics and familiarity with genetic transformation of maize. The procedure was followed in practise and yielded a wealth of information. Here we shall present only a summary of the results. However, if one is looking for an example of a study in molecular genetics for a class-course, this publication (Raizada et al., 2001) is recommended.

A summary follows. First the hoped-for transgenic plants were obtained among the *Basta*-resistant regenerants. Actually, the transgenic maize plants contained multiple linked copies of one or both of the *RescueMu* constructs. These plants were thus crossed with *MuDR*-containing stocks. Indeed, grains with spots were revealed in the progenies of these crosses, clearly indicating that the modified *Mu1* was excised in the presence of *MuDR*. Actually, the somatic excision (SE) happened at high frequency. In most cases these were single-cell spots but larger and complexed spots were also revealed and they were found in transgenic lines that harbored several copies of the *RescueMu*. This clearly indicated that the SE occurred most frequently at or after the last cell divisions of the aleurone layer. It thus seems that excision is correlated with the cessation of cell division. The excision of the modified *Mu1* from the *RescueMu* could be followed by appropriate PCR procedures. About half of the 45 analyzed cases showed a more or less "clean" excision, leaving intact the CaMV promoter and the *Lc* gene behind.

The other transgenic plants also had major deletions of the *Lc* gene. Some of the transgenic plants had "filler" DNA where the modified *Mu1* was removed. The results suggested that most broken DNA ends, during excision/repair are repaired by a non-homologous mechanism (as indicated in another system by Gorbunova and Levy, 1997a,b). Due to the existence of specific restriction sites in the *RescueMu* construct, it was possible to enrich digested leaf-tissue with new insertions of modified *Mu1*. Insertion of modified *Mu1* into new locations was thus verified. These took place late in leaf development in the F_1 seedlings that resulted from a cross between *MuDR* and transgenic plants with *RescueMu*. Insertion of *RescueMu* was also followed and found to happen frequently after the element left the original transgene array. Element's insertions during germinal development were accompanied by excision. This verified the previous notion that *MuDR* elements do not transpose in this tissue by the "cut and paste" mode.

Interestingly, 93% of *RescueMu* insertions were into maize genes. This substantiated previous information on the insertion of *MuDR* elements into genes, hence rendering this system useful for gene tagging, and indicating that *RescueMu* is useful for this purpose.

There were also notable exceptions. The *RescueMu* was inserted into the *mudrA* gene of a *MuDR* element. This indicates that *MuDR* are not immune to self-integration. Generally, the *RescueMu* system was rather useful to detect insertion in somatic tissue where genetic analysis cannot reveal them, as they are not transmitted into the next generation and do not necessarily cause visible mutated phenotypes. Moreover, the sequences around the inserted element can be analyzed. Briefly, such insertions in somatic tissue (e.g. leaves) are accompanied by 9 bp host sequence duplications, as was previously observed for insertion into germinal tissues.

While not revealed by the *RescueMu* methodology, *early* somatic excision (in the *Bz2* locus) and *early* somatic insertions (visible in a dominant mutation) were found; albeit they are probably rare.

4.3.1. Germinal Transposition in the MuDR System
We discussed above the possible increase in the copy number in the *Ac/Ds* and *Spm/dSpm* elements. These elements transpose by the "cut and paste" mode in the soma and germinal tissue. In the case of *Ac* it was suggested that the element excises from one of the already replicated DNA double helixes into an unreplicated site. Thus, with the replication of the new site, an additional *Ac* is formed. One

"daughter-cell" will thus have two *Ac* sites: an "old" and a "new" one, while the other "daughter-cell" will have only the "new" *Ac* site.

In the *MuDR* system there are many indications that the elements replicate and that germinal insertions must accumulate without loss of existing *Mu* locations. In other words: duplicative or replicative transposition does occur. But how "regular" this replication is and how much this rate of replication is affected by the number of active *MuDR* elements in the genome, is still under investigation.

4.3.2. The Functions of MURA and MURB

There is now good evidence for the claim that the full-size *MURA* protein (*MURA-823*) is essential and sufficient for the excision of *Mu* elements. As for *insertion* of *Mu* elements, the *MURA* is probably essential, but not sufficient. The subject is still being investigated.

The role of *MURB* in transposition is even less clear than the role of *MURA*. There are actually several *MURB* transcripts, which provide the different spliced mRNAs, and these proteins may have different roles. There are some indications (see Walbot and Rudenko, 2002) for a role of *MURB* in insertion (but not in excision). There is a reasonable argument for the role of *MURB* in the "switch" from "cut and paste" transposition to replicative transposition. This argument is based on the family of *Jittery* elements. In *Jittery* there is only "cut and paste" transposition and it produces a *MURA*-like protein but not *MURB*. While this possibility is worth further investigation, at present it is a weak argument, especially because no chemical interaction between *MURA* and *MURB* has yet been observed.

4.3.3. The Timing of Transposition

If we accept that the 5' TIR (or part of it) has the role of the promoter for *MURA*, then we should consider TIR as involved in the timing of transposition. We should recall that transpositions that cause somatic excision of *MuDR* elements (such as *Mu1*) are happening "late" in somatic differentiation, i.e. where cell division has stopped or is about to stop. It is thus possible that clues during active cell division restrict the activity of the promoter. There is a problem with such a claim: the transcript may be regulated but this does not necessarily reflect the level of the translated protein (*MURA*). Thus, the level of the protein and/or its activity may be developmentally regulated. The *increase* of transposition of *MuDR* elements, that parallels the *decrease* in cell division opens a regulatory possibility. TIRs contain histone promoter motifs, hence transcription factors that are abundant during active cell division (when histone mRNAs are actively transcribed) may bind to TIRs and interfere with the binding of functional transposase complexes to the TIRs.

There is an additional observation that concerns *MURB*. This protein is in high levels in tissues of actively dividing cells but in low levels during terminal tissue differentiation and in shoot apical meristems. A model was developed that attributes the *MURB* a role in gap repair. This gap repair drives the transposition in proliferating cells towards the replicative transposition and away from the "cut and paste" transposition that causes excision.

4.4. Epigenetic Regulation of Mutator

There is evidence that silencing of *Mutator* lines is the major basis of the loss of

Mutator activity in the progeny of such lines; loss of activity by segregation is rare. It seems that lines that have lost the *Mutator* activity have selective advantage. Because silencing is a common feature in *Mutator* lines, it was assumed to be of advantage to either the host or to the elements. Such possible advantages were therefore suggested.

4.4.1. Silencing in Somatic and Germinal Tissues

Silencing can happen in several types of tissues. Silencing can be imposed during somatic growth and it is clonal (e.g. in ear sectors). The silencing is manifested by several phenomena: cessation of germinal insertion activity, methylation of TIRs, or loss of somatic excision activity. Once silenced, reactivation is rare. Once a *MuDR* is silenced, it apparently changes to an *hMuDR* by accumulating mutations and loses its activity forever. Silenced *MuDR* lines that lost their activity only one or two generations ago can be reactivated by irradiation (UV-B or UV-C) and, to a lesser extent, by crossing with an active *MuDR* line.

4.4.2. Molecular Mechanisms of Silencing

Silencing is accompanied by increased methylation of the TIRs of *Mu* and *MuDR* elements. *MuDR* silencing is associated with *mudrA* and *mudrB* transcriptional down-regulation, reducing the production of functional *MURA* and *MURB*. It is not yet clear how the progressive methylation takes place, moreover, it is not yet assured whether or not methylation merely accompanies silencing or is the causal agent of silencing. We mentioned above that RNAi may be involved in the suppression of *Mutator* activity, but this possibility has also not yet been assured.

4.5. Applications of the MuDR/Mu System

A major use of the *MuDR/Mu* system is its application in gene tagging. The advantage of this system over other transposable element systems is that the *MuDR/Mu* elements can move to any chromosome, and have no defined target motifs, although they tend to insert into gene sequences rather than into repetitive DNA sequences. As noted above, Appendix III of this book is devoted to Gene Tagging.

5. THE TRANSPOSABLE ELEMENTS OF SNAPDRAGON

Chronologically, the variegations in snapdragon (*Antirrhinum majus*), caused by transposable elements (T.E.), were the first T.E.-induced variegations that attracted the attention of investigators. In their review on transpositions in *Antirrhinum*, Coen et al. (1989) traced the early reports on variegation in snapdragon. This phenomenon was already well known to C.R. Darwin. It was later investigated by H. de Vries and a full interpretation of floral-pigment variegation was provided by the pioneer of maize genetics and breeding, Rollins A. Emerson (1914). These masters were mentioned in the Historical Background of this book.

The snapdragon attracted the attention of European scientists, especially of German geneticists such as E. Baur, H. Kuckuck, and H. Stubbe. The interest in snapdragon variegation crossed the English Channel in a similar route to the

Figure 53. Transposable elements in flowers of snapdragon. When in white flowers, with mutated niv the T.E. is excised, the sites of excisions are visible as dark spots. (From Harrison and Carpenter, 1973).

crossings of the channel by the Saxons and the Angles in the 5th and the 6th centuries. Snapdragon variegation was the PhD thesis subject of K. Mather, who later became CBE and FRS, and one of the most prominent British geneticists.

Then, Harrison and Fincham (1964) continued the long tradition of snapdragon studies at the John Innes Institute. Snapdragon is a popular ornamental plant. It has a terminal inflorescence that bears 20 or more flowers. Each flower develops a capsule with over 200 seeds and three generations of snapdragon can be grown in one year. The snapdragon can also be propagated vegetatively and the biochemistry of the flower-color of snapdragon has been amply studied. One may be content with the aesthetic beauty of normal and variegated snapdragons but some were eager to have the intellectual satisfaction of solving the genetic mystery of this variegation.

One sometimes wonders, are the systems of transposable elements really complicated or do they appear complicated in the language of the scholars investigating them. For illustration, here is a sentence from Harrison and Fincham

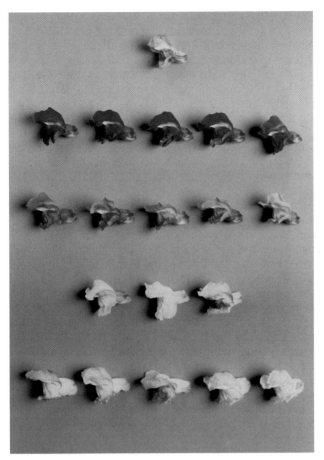

Figure 54. Pallida alleles arising from pal^rec. The top flower illustrates a pal^rec progenitor and the two rows below show alleles conferring various intensities of pigmentation. The three flowers in the second row from the bottom illustrate alleles conferring different stable patterns of pigmentation. In the bottom row flowers grown at 15°C show altered pigmentation. (From Coen et al., 1986).

(1964): "These properties (properties of mutable genes) include the occasional transfer of mutability from one locus to another, an extreme sensitivity of gene expression and mutation rate to the presence or absence of other elements in the genome, and often to common environmental variables, and sometimes an induction by the mutable alleles of the mutable gene in heterozygotes ("paramutation")." These introductory words may not have been very clear to all the readers of the journal *Heredity*, but surely made sense to the Chief Editor of this journal, K. Mather, because as stated by the authors, their study was "an extension, and in a part a repetition of experiments performed by Mather at the John Innes Horticultural Institution some 30 years ago".

In a subsequent publication from the John Innes Institute (after it moved from Hertford to Norwich) by Harrison and Carpenter (1973), the authors already discussed the possibility that *transposable elements* at the *pal* and *niv* loci are responsible for the instability. However, they noted that proof for the involvement of such elements is still lacking. Flower-color variegation is a very appropriate subject for the study of transposable elements. It has aesthetic appeal (Figure 53). Moreover, excisions in single cells, out of a mutated allele, are clearly visible, while excisions that would affect morphology cannot be scored unless they happen very early in the development of the floral initials.

By 1982 the pathway of anthocyanin synthesis became clear. The gene Niv^+ (*Nivea*) was found to encode the enzyme *chalcone synthase*, thus leading to dihydroquercetin and the gene Pal^+ (*Pallida*) was found to encode the enzyme *dihydroflavonol 4-reductase*, thus leading to the cyanidin and to red color. When both these genes are wild-type, namely Niv^+ and Pal^+, the flowers are red. But, if Niv^+ is expressed while Pal^+ is not expressed (null allele of *pal*) the flowers are ivory-colored. This is a simplification of the actual situation as the expression could be modified quantitatively and spatially, as presented in Figure 54 for the *Pallida* alleles. The year 1982 marked a notable German, English and US collaboration for the study of plant genes at the molecular level. The first target of these investigators was to study the chalcone synthase gene in several plant species, including snapdragon (e.g. Wienand et al., 1982). In the very same year the *Cin* T.E. of maize was cloned (Shepherd et al., 1982, 1984) and soon after that other maize transposable elements were sequenced (e.g. Döring et al., 1984; Fedoroff et al., 1984; Müller-Neumann et al., 1984; Schwarz-Sommer et al., 1984) as detailed above.

In 1984 the interesting transposable element system was ready for study in snapdragon; the molecular tools for the isolating and cloning of alleles were available and efficient DNA sequencing was at hand. Thus, the H. Saedler team (of the MPI in Cologne) with the collaboration of a snapdragon expert, B.J. Harrison (of the John Innes Institute in Norwich), sequenced the transposable elements of snapdragon (Bonas et al., 1984a,b; Upadhyaya et al., 1985; Sommer et al., 1990) as shall be detailed below.

5.1. The Structures of the Tam Elements

The transposable elements of snapdragon were termed *Tam* for transposable elements of *Antirrhinum majus*. The first of the *Tam* elements that was cloned and subsequently sequenced was termed *Tam1*. It was isolated from the T53 niv^{rec} mutant, in which white flowers are spotted with anthocyanin pigmentation. It was previously also shown to revert in germ cells to niv^+, resulting in normally pigmented plants. Using a probe from the chalcone synthase gene, DNA fragments were isolated from the niv^{rec} (T53) mutant that did not exist in the wild-type niv^+ (Bonas et al., 1984a,b). The additional DNA of a sequence in the niv^{rec} was first revealed by Southern blot hybridization and restriction endonucleases, as well as by heteroduplex analysis, and the length of this sequence was estimated to be 17 kb (it was later found to be about 15 kb). Subsequently, this DNA sequence, now termed *Tam1*, was partially sequenced (Bonas et al., 1984b). This sequencing showed that the *Tam1* from niv^{rec} (T53), then termed *niv-53*, had 13 bp of perfect inverted repeats and is flanked by 3 bp target-site duplications. The main characteristics of *niv-53* (as well as of other T.E. of snapdragon) were summarized by Coen and Carpenter (1986) and Coen et al. (1989) and are presented here in Table 7.

Table 7. Transposable elements of snapdragon.

Allele	Flower phenotype	Element designation	Size (kb)	Element position	Length of inverted repeat (bp)	Length of target site duplication (bp)
niv-53	Red sites on white background	*Tam1*	15	17 bp upstream of TATA box	13	3
niv-44	Stable white	*Tam2*	5	First exon/intron boundary	14	3
niv-98	Red sites on pale red background	*Tam3*	3.5	29 bp upstream of TATA box	12	8
pal-2	Red sites on ivory background	*Tam3*	3.5	41 bp upstream of TATA box	12	5

(Modified from Coen and Carpenter, 1986 and Coen et al., 1989).

Tam2 was isolated from another *niv* allele: *niv-44* (Upadhyaya et al., 1985). The *niv-44* causes white (stable) flowers. The location of the insertion of *Tam2* is at the first exon/intron boundary of the chalcone synthase gene. *Tam2* has several similarities to *Tam1*. The 14 bp inverted repeats of *Tam2* are almost identical to the 13 inverted repeats of *Tam1* and both have 3 bp target-site duplications. However, the total length of *Tam2* is only 5 kb, rather than the 15 kb size of *Tam1*. Both *Tam1* and *Tam2* have similarities with *Spm* elements and may be classified in the same superfamily of transposable elements. Notably in addition to the 14 bp terminal repeats, *Tam2* has several other repeats. Moreover, by Southern blot hybridization, it was found that there are several other DNA sequences in snapdragon that are identical or similar to *Tam2*. While Upadhyaya et al. (1985) assumed that *Tam2* is potentially mobile, there was no proof for this mobility.

The third snapdragon T.E., *Tam3*, was also first isolated and characterized from a *niv* allele: *niv-98* (Sommer et al., 1985). This is an unstable allele that causes red spots on a pale-red background. The allele also causes germinal instability (Table 7). The element *Tam3* has a length of 3.5 kb and perfect inverted repeats of 12 bp at its terminals. These inverted repeats have similarity to those of *Ac/Ds* elements and differ from the inverted repeats of *Tam1* and *Tam2*. *Tam3* has a 8 bp target-site duplication, which was also found in *Ac/Ds*. Consequently, *Tam3* and *Ac/Ds* probably belong to the same family. The *niv-98* can also revert germinally and a homozygous revertant with red flowers was isolated. It was maintained (as *niv-164*) and served to sequence the locus of excision of *Tam3* from *niv-98*. *Tam3* was also isolated from the *pal* locus (Martin et al., 1985). The phenotype of *pal-2* differs from that of *niv-98*. The former allele causes red spots on an ivory background, rather than on a pale red background. The *pal-2* allele is very sensitive to temperature. There are 1,000-fold more red spots on the flowers when the plants are kept at a low temperature (15°C) than at a higher temperature (25°C) (Figure 54); the germinal reversion also occurs at higher rates when the plants are maintained at lower temperatures. The level of *pal* transcript was lowered in *pal-2*, indicating that the *Tam3* insert interfered with normal transcription.

5.2. Transposition

Our knowledge of *Tam* transposition is derived primarily from the correlation of flower-pigmentation in the specific *pal* and *niv* alleles, with the analyses of DNA sequences at the sites of excision and integration. Because *Tam3* has great similarity to the *Ac/Ds* family of T.E., and *Tam1* and *Tam2* are similar to the *Spm* family, the maize T.E., where more information on transposition is available, are instrumental for the understanding of transposition in *Tam* elements. However, no direct chemical evidence on the detailed steps of transposition in *Tam* elements is available.

5.2.1. Excision

In many cases where a variegated mutant of either *pal* or *niv* alleles was reverted germinally, stable wild-type progeny was obtained. The DNA analyses then revealed that the *Tam* elements were removed from the loci. Whether or not the element was inserted elsewhere in the snapdragon genome could not be ascertained. There were many cases in which the reversion was not to full-color, wild-type flower-pigmentations. The stabilized revertant could have pale pigmentation or the pattern of pigmentation was altered, meaning, for example, that full-color was restored in the "tube", but much less in the "lobes" of the flower. As we indicated in some previous T.E. (e.g. the *Mu* excision) in *Tam* elements the excision is also not "clean". The site from which a *Tam* element is removed has a variety of changes, relative to the original sequence, before the insertion of the element.

Many such excision sites were analyzed and the phenotypic results were documented (see review of Coen et al., 1989). We shall give only a few examples to illustrate the excision consequences. A revertant of the *pal-2* allele is the stable *pal-518*. The *pal-518* did not contain any remnants of the *Tam3* that was inserted into the promoter of *pal* (41 bp upstream of the TATA box). Moreover, all the target-duplications were removed, but for 2 bp. In addition, there was also a deletion of 10 bp from the original promoter of the gene into which *pal-2* was inserted. The *pal-518* has pale red pigmentation with very much reduced transcript of the gene.

When observed more carefully, it was revealed that pigmentation of the "lobes" was much more reduced than in the "tube". This suggest that the promoter has spatial control and that the motif that controls "lobes" pigmentation was damaged, in the promoter.

On the other hand, revertants of *niv-44* where a *Tam2* element was excised from the first exon/intron boundary, restored pigmentation in the *niv-545*, even though the site of excision did not revert to the original sequence. Several revertants of *pal-2* were found to have deletions of various lengths (between 9 and 19 bp) upstream of the excision sites and their phenotype was pale red flowers. The investigators, who studied the phenotypes and the sequence changes after excision of *Tam3* from the promoter of *pal* (Coen, Carpenter and associates), suggested therefore that the promoter is composed of a set of sequences that respond to diverse regulatory signals, spatially arranged in the flower.

The situation seems to be much more straightforward when, during excision, an exon sequence is eliminated. This was found in a revertant of a mutant that had a *Tam2* in the *niv* gene. The revertant was a stable null-mutant with 32 bp missing in the first exon of the chalcone synthase gene. The great majority of the alleles with mutated pigmentation genes are recessive. However, exceptions were recorded when, after excision of a *Tam* element, the mutation was semidominant. In one such case an antisense RNA of 40 bases was probably transcribed. Thus, it was suggested that silencing could be imposed. Based on our present knowledge, we may speculate that RNAi could be involved by the formation of a ds RNA of 40 bp.

Models for the excision of *Tam* elements were proposed; some models were rather detailed (e.g. Figure 7 of Coen et al., 1989). The models were attractive and could explain the changes left at the excision site after the removal of the *Tam* element, but there is no detailed proof for the mechanism. Albeit, at least for the excision of *Tam3*, which is very close to the *Ac/Ds* of maize, the mechanism seems to be the "cut and paste" procedure.

The rate of excision can be influenced by environmental and genetic factors. As for environmental factors, a lot of information exists that shows that low temperature (15°C) induces more excision than higher temperature (25°C), but the effect of such thermal shifts is different in different elements. Thus, *Tam3* excisions are much more affected by temperature than the excision of *Tam2*. Also, some genetic repressors of excision were suggested for the excision of *Tam1* and *Tam2*, and a stabilizer for *Tam3* was also identified. The molecular bases of these changes in excision rates are not yet known.

5.2.2. Integration

The phenotypic result of a *de novo* insert of a *Tam* element into a gene could be, in the case of the *niv* and *pal*, that white and ivory spots, respectively, appear on a pigmented background. Such cases are very rare but at least one such case was recorded in a line designated TR.75 that involves an integration of a *Tam3* element into *pal*.

The transposition of an element from one gene location to another was also documented in the *niv* gene. The *niv-98* allele is an insert of *Tam3* into the promoter of *niv*. When *Tam3* is excised there are red spots on a pale red background. However, when the excised *Tam3* element reintegrates into the last exon of *niv*, all the background pigmentation is abolished.

As with excision there are models for integration, but no hard facts are available for the chemistry of integration. The process for *Tam3* integration is probably similar to that of *Ac/Ds*, meaning, that the integration is coupled with the excision process and there is no "free" element between the excision phase and the integration phase. The question of how the number of copies of the *Tam* elements are increased, could be answered in the snapdragon genome by the following claim. Transposition takes place primarily during the S phase of cell division. At this stage a *Tam* element on a ds DNA, which has already been replicated, is transposed to a ds DNA that has not been replicated.

As was reported for Ac/Ds *elements*, it seems that *Tam3* also integrates into nearby loci, and this may indicate that for transposition the donor DNA region and the new recipient DNA-region have to come into close association.

5.3. No Non-Autonomous and Mobile Elements in the Tam System

While in the "sister-families" of *Tam* elements, namely the *Ac/Ds* and the *Spm* transposable element families, there are two types of elements: autonomous (*Ac* and *Spm*, respectively) and non-autonomous elements (*Ds* and *dSpm*) – no non-autonomous mobile *Tam* elements were reported.

Why are non-autonomous, but mobile elements missing from the *Tam* system? The answer to this question did not come from either side of the English Channel. Investigators in the northern island of Japan, Hokkaido, went on to solve this riddle. Kishima et al. (1997) sequenced two elements of *Tam* that were not mobile from a variegating (*niv^{rec}*) snapdragon. They found that while the middle regions of these two elements had identical regions to that of a mobile *Tam3* element, their terminals, and especially their terminal repeats, were different from the mobile *Tam3*. This is the converse situation to that of the *Spm* and *Ac/Ds* elements. In these two systems the non-autonomous (but mobile) elements have terminals that are identical to the autonomous elements, but the middle regions are very different from those of the autonomous elements.

In a further study by this Japanese team (Kishima et al., 1999) it was found that a single variegated snapdragon can have 40 *Tam3*-like elements, and all of them had intact and homologous internal regions, without deletions. However, the terminals of many of these elements were aberrated, indicating that they had lost their mobility.

A further study of Yamashita et al. (1999) investigated, in detail, the terminals and sub-terminals of *Tam3*. When these were fully sequenced 11 hairpin structures were revealed at the 5' side of the element and 5 hairpin structures at the 3' side of the element. It was also found that the gap repair, which is normally associated with synthesis-dependent strand annealing, is inhibited during *Tam3* excision. In effect, these Japanese investigators provided evidence for the arrest in gap repair in *de novo* somatic excision of *Tam3*. Moreover, the gap repair was halted within the hairpin structure region (Figure 55). These authors therefore suggested that the structural homogenity (i.e. the lack of non-autonomous but mobile elements) in the *Tam* system was caused by the immunity to gap repair at the hairpins.

This suggestion was based on the knowledge that gap repair is a mistake-prone process and can lead to deletions and aberrations of the internal region. This latter region encodes the transposase, thus, in T.E. systems where gap repair occurs, autonomous elements may lose their autonomous mobility. When the terminal ends of these non-autonomous elements remain intact, they can still be mobilized by the transposases of autonomous elements in the same genome.

Such a process was indeed reported previously for the *Ac/Ds* system by Rubin and Levy (1997). These authors transferred *Ac* into transgenic plants and studied the results of excision of single *Ac* elements and found that there were frequent *de novo* alterations in the excised *Ac*. These alterations consisted of internal deletions with breakpoints usually occurring at short repeats and in some cases of duplications of *Ac* sequences or insertion of *Ac*-unrelated fragments. Such alterations were actually revealed in *Ds* elements. The authors suggested that abortive *Ac*-induced gap repair, through synthesis-dependent strand annealing, is the underlying mechanism for the derivation of *Ds* elements from *Ac* elements.

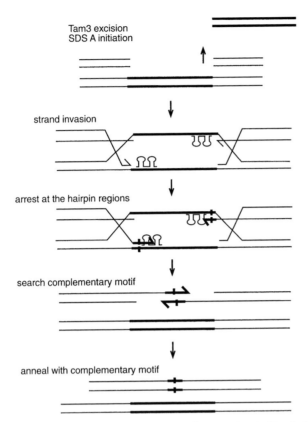

Tam3 excision
SDS A initiation

strand invasion

arrest at the hairpin regions

search complementary motif

anneal with complementary motif

Figure 55. Model for the processes of the arrested gap repair in Tam3. Each thin line represents a single-stranded DNA. The gap repair process (the SDSA, synthesis-dependent strand annealing pathway) is initiated by transposon (thick line) excision from a sister chromatid in a homozygous allele. The ends of the donor site then invade the sister chromatid. The DNA synthesis stops within the hairpin regions located in the terminals of Tam3. These short stretches search for complementary motifs (short vertical bar) and are annealed with these motifs. (From Yamashita et al., 1999).

5.4. Gene Tagging

As other T.E. of plants, the *Tam* elements and especially *Tam3* can be useful in gene tagging. True, *Tam3* elements do not have the advantage of *MuDR/Mu* elements of integrating into gene sequences that are unlinked to the previous site of the element. However, snapdragon has an advantage of its own: its flowers may have homeotic alterations that are useful to study the genetics of structuring morphologies. For example, the recessive allele *deficiens*^{globifera} causes such an homeotic change: sepals grow in place of petals and stigmatic tissue instead of stamens.

Another homeotic allele is *cycloidea^radialis* that causes a radial symmetry rather than the normal bilateral symmetry in snapdragon flowers. Such mutated alleles may be the results of *Tam* insertion. For screening for insertional mutants, that result from a cross in which one parent carries an actively transposing *Tam* element, one needs a very large F$_2$ population because only germinal inserts will be commonly revealed. On the other hand, once a *Tam3* is inserted into a homeotic gene, it may move to a nearby location, so that the gene and its controlling region can be "dissected".

Indeed, by 1990 Carpenter and Coen moved into the study of homeotic mutations by transposon-mutagenesis in snapdragon (Carpenter and Coen, 1990). It should be noted that during the "transition" of attention of the John Innes team, the homeotic changes in *Arabidopsis* flowers were studied by E.M. Meyerowitz at the California Institute of Technology, in Pasadena, California. A collaborative publication thus appeared in *Nature* that carried the title: "The war of the whorls" (Coen and Meyerowitz, 1991). This bridged the common interests of the investigators in England and in California. Interestingly, not only the investigators at John Innes shifted their attention from the *Tam* elements towards floral morphogenesis in snapdragon. A similar shift happened at the Max-Planck-Institute in Cologne (e.g. Schwarz-Sommer et al., 1990). However, the joint Cologne/Norwich publications were discontinued.

6. CLASS II TRANSPOSABLE ELEMENTS IN *ARABIDOPSIS THALIANA*

6.1. Class II Elements that Populate A. thaliana

After many years of promotion by George P. Redei, of the University of Missouri, the cruciferous plant *Arabidopsis thaliana* was adopted by molecular geneticists as their model-plant of choice. No wonder, *A. thaliana* is a tiny plant with modest culture requirements. It has a life cycle of about six weeks, at the end of which one plant may produce 50,000 seeds. A thousand or more seeds can be germinated in a single petri dish to observe the seed germination. A few thousand plants can be grown in a medium-size tray and screened for a variety of mutations. The entire nuclear DNA of *A. thaliana* has been sequenced and the sequencing-information is freely available to the public. Finally, while up to about five years ago genetic transformation of *A. thaliana* was more difficult than transformation of other plants such as tobacco, potato and tomato; in recent years the *in planta* transformation has been developed (see Breiman and Galun, 1997 for review), which renders transformation of *A. thaliana* amenable even in the hands of investigators who tend to avoid tissue culture procedures.

Until 1992 investigators did not turn to *A. thaliana* as a source of endogeneous transposable elements, probably because its genome was considered to be densely populated with genes and controlling sequences of genes. Moreover, because investigators assumed there are no endogenous Class II T.E. in *A. thaliana*, they transferred to it single elements, such as *Ac*, to investigate the behavior of such "lone" elements in an alien plant genome. Then, in 1993 came the surprise, and as shall be detailed below, the first endogeneous Class II T.E. was found in *A. thaliana*. The finding of other Class II transposable elements followed. In summary, there are three Class II types of transposable elements in *A. thaliana*: (1) elements that are endogeneous and autonomous; (2) small endogeneous elements that apparently do not encode transposases to drive their mobility; (3) alien transposable elements such as *Ac/Ds*, *Spm* and *MuDR* that were transformed by investigators into the *A. thaliana* genome. In the following we shall detail one element of the first type. Small, non-autonomous elements such as *MULE*s and *MITE*s will be handled subsequently. Information on the third type of alien transposable elements in *A. thaliana* shall be provided briefly. More information will be given on these elements when we deal with Gene Tagging in Appendix III of this book.

6.2. The Tag1 Transposable Element

A team of N.M. Crawford and associates (Tsay et al., 1993) from the University of California at San Diego and the John Innes Centre in Norwich intended to use the *Ac* element of maize to produce nitrate-assimilation mutants in *A. thaliana*. Like the biblical King Saul they were "looking for asses and found a crown" (First Book of Samuel, chapter 9). More specifically, Tsay et al. (1993) used the chlorite-toxicity system in their search for nitrate-assimilation mutants. This system was developed several decades ago and utilized in *Aspergillus* genetics. It is based on the fact that chlorate is not very toxic to cells but chlorite is very toxic. Normal cells with functional nitrate assimilation take-up chlorate and reduce it to the toxic chlorite. This kills normal cells. However, cells that are defective in either chlorate transport or in nitrate reductase (the enzyme that reduces chlorate to chlorite) are not killed. Tsay et al. (1993) transformed *A. thaliana* with a vector that contained *Ac*. They were looking for the progeny of transformed plants that were resistant to chlorate due to the insertion of *Ac* into either a gene for chlorate transport (CHL1) or for nitrate reductase. They found resistant *A. thaliana* progeny-plants and focused on one of them (*chl-6*). When *chl-6* was analyzed, the investigators were surprised to find in the *Chl* gene not an *Ac* element but another insert of 3.3. kbp. This latter insert did not hybridize with an *Ac* probe and was found to be a novel transposable element that the investigators termed *Tag1*. The genomic sequence flanking *Tag1* was duplicated and the new element was found to have terminal inverted repeats (IR) of 22 bp. The *Tag1* was inserted into an intron of *Chl1*. It was assumed that excision of *Tag1* should restore the normal expression of this gene (rendering the revertant chlorate-sensitive). Indeed, such revertants were found among the progeny of plants that had the *Tag1* in *Chl1*. Analysis of the revertants showed that *Tag1* had been excised, leaving behind a small insert. The genome was then searched for additional *Tag* elements. Only one additional *Tag1* element was found in the Landsberg *erecta* ecotype. There were also two *Tag*-like elements in the genomes of this ecotype but not in other ecotypes of *A. thaliana*.

Once found, the *Tag1* element was thoroughly investigated by Crawford and his associates (e.g. Frank et al., 1997; Liu and Crawford, 1997, 1998; Liu et al., 1999, 2001). First, the search by probing digested *A. thaliana* DNA with the appropriate probe identified only two *Tag1*. When the *Ac* was removed, by outcrossing, the *Tag1* was still mobile, indicating that the presence of *Ac* is not required for the transposition of *Tag1*. Still, *Tag1* could have been mobilized by another element in *A. thaliana* that is not *Ac*. To eliminate this possibility, a construct that contained *Tag1* between a plant promoter and a marker gene (GUS), was transformed into tobacco. Tobacco cells that by appropriate staining will be pigmented (blue), should have the *Tag1* excised from this construct.

On the other hand, a defective *Tag1* that had a deletion in the internal region was utilized, this Tag1 should not be excised from the GUS containing construct. Such experiments indicated that the *Tag1* is indeed autonomous, even in an alien plant species. It was also found that there are transpositions in germinal as well as in somatic tissues. The rate of germinal insertions had a positive correlation with the number of *Tag1* elements. Somatic excision usually ocurred late in organ differentiation.

In further studies Liu and Crawford (1998) focused on the transcription of *Tag1*. For that the *Tag1* was transferred by *in planta, Agrobacterium*-mediated transformation into *A. thaliana* ecotypes that had no endogeneous *Tag1*. A major transcript of 2.3 kb and several minor transcripts were revealed. The major transcript encoded a putative transposase of 84.2 kD, with two nuclear localization-signal sequences and a region that is conserved among plant Class II transposases (e.g. *Ac*). The level of transcript varied among the transgenic *A. thaliana* plants but it did not correlate with the frequency of excisions in these lines, indicating that not only transcription activity controls transposition rates. Another observation was that even defective *Tag1* elements could cause the activation of non-active *Tag1* elements.

The next question of Crawford and his associates (Liu et al., 1999) was how *Tag1* would behave in a monocot plant. They chose rice, which is relatively easy to transform and has a smaller genome than other Gramineae species (such as wheat, maize and barley). It was found that indeed *Tag1* is also active in rice. Moreover, the timing of transposition in rice is similar to that in *A. thaliana*. The transcript patterns and the footprints left after excision in rice are similar to those in *A. thaliana*. This indicated that *Tag1* could be useful for gene tagging in rice.

In a more recent study (Liu et al., 2001) Crawford and associates "dissected" the transcript of *Tag1*. For that the investigators used two vectors. One had a CaMV promoter upstream of the cDNA of the transposase and the other was a "target-defective" vector, where a defective *Tag1* was engineered between a promoter and a code for GUS. The latter construct should indicate the excision activity imposed by the transposase encoded by the first construct. It was found that the 5' intron for the major 2.3 kb *Tag1* transcript is essential for the accumulation of transcript that will lead to a high rate of somatic excision. The 3' introns (there are three such introns) were not essential for somatic excision in primary transgenic plants, but were important to maintain high levels of somatic excision in subsequent plant generations. When the endogeneous promoter of *Tag1* was exchanged with the CaMV 35S promoter, only germinal excision was inhibited. These and additional results indicated that somatic and germinal excisions of *Tag1* are differentially controlled.

Mack and Crawford (2001) studied *in vitro* the recognition of specific DNA motifs by the transposase of *Tag1* (a 729 a.a. protein). They found that the transposase has a DNA-binding domain near its N' terminal. This domain binds to hexamers at the sub-terminal repeats of *Tag1*. The binding is metal-dependent, it is enhanced by zinc and requires key cysteine residues. These properties suggest a zinc-finger DNA-binding domain.

6.3. The "Hidden" Class II Elements of A. thaliana

The "wind" that is able to resurrect the "skeletons" (or the "dry bones", Ezekiel 37) of transposable elements in *A. thaliana* was identified. There are two main approaches to reveal the "hidden" (or "skeletons" of) transposable elements. One approach is to "resurrect" these elements and cause them to transpose. The other approach is to "mine" these elements either by computer search of sequence data or by a PCR procedure. We shall look at examples of these approaches in *Arabidopsis thaliana* and start with the resurrection approach.

6.3.1. The Resurrection by Demethylation

The correlation between methylation state and transposition activity was detected by several investigators and this correlation was mentioned above, when T.E. as the *Ac/Ds*, *Spm* and *MuDR* were detailed. Thus, promoter sequences of active *Ac* and *MuDR* transposase-codes, are usually hypomethylated. Also, *in vitro* transposases bind to the respective DNA (e.g. in the TIRs) more efficiently when this DNA is hemimethylated. It was found that in *Mutator* strains, which show a high rate of transposition, the TIRs of *Mu* and of *MuDR* are demethylated. It was also revealed that when some transposable elements spontaneously lose activity, this process is accompanied by methylation. Most of the *Arabidopsis thaliana* genome is "dense" with genes. Like the "knobs" in maize, there is also a "knob" in *A. thaliana*. This latter "knob" was sequenced and found to contain transposable elements of various kinds (e.g. retrotransposons and Class II transposons). These elements were found to be highly methylated (Singer et al., 2001). Singer et al. (2001) intended to investigate the effect of methylation and demethylation on the transposition of *A. thaliana* T.E. For that they used a gene previously identified in this plant that is involved in methylation: *DDM1*. Vongs et al. (1993) found that *DDM1* is required for DNA methylation in *A. thaliana*. In the presence of the null mutation, *ddm1*, there is an immediate loss of DNA methylation from the heterochromatin, as well as from the "knob", and a gradual loss of methylation also took place from euchromatic DNA over successive generations of inbreeding. Available sequence data enabled these investigators to focus on the TIRs of several *MuDR*-like elements in *A. thaliana*. They found that while in *DDM1* plants, these TIRs were methylated, in *ddm1* plants, the TIRs were demethylated and the respective *MULEs*, which were inactive in *DDM1* plants, were activated (transposing) in the presence of *ddm1*. The *MuDR*-like elements were termed *MULEs*.

While Singer et al. (2001) found that demethylation resurrected *MULEs*, a novel family of Class II elements was revealed after demethylation of *A. thaliana* by a Japanese team, Miura et al. (2001). They also utilized the *ddm1* mutation, as one of this team participated in its discovery (Vongs et al., 1993). However, Miura et al. (2001) started from a different angle. They found that in an unstable mutant of *A. thaliana*, impaired in the synthesis of brassinosteroids (regulators of cell elongation) there was an insert of several thousand nucleotides in the DNA that encodes the respective enzyme. The sequences of terminals of this insert indicated that it belonged to the CACTA family of transposons, to which *Spm* and *Tam1* also belong. The conserved TIR-sequences of these elements are CACTACAA. Several additional CACTA type elements were revealed in *A. thaliana* and were termed *CAC1*, *CAC2*, *CAC3*, etc. It was found that in the unstable dwarf mutant (with the *CAC1* insert) the *CAC1* transposed to another location (in another chromosome). Moreover, in the presence of the *ddm1* gene, the *CAC* elements started to become mobile. Indeed, the demethylation of *CAC* elements paralleled their mobility. Still, there is as yet an open question. While in the presence of *ddm1* there is reduced DNA methylation; this gene is probably affecting the chromatin state. Is it possible that demethylation is a secondary effect to chromatin modulation? This open question stresses again the notion that a better understanding of the relation between chromatin modulation and the epigenetic changes in T.E. activity in plants is required.

6.3.2. Fishing for T.E. in the DNA Sequence Data of A. thaliana

The entire *A. thaliana* genome is now sequenced and the tools to "fish" (or "mine") for specific sequences that have the characteristics of transposable elements, are now available. The investigator can now feed his computer with the data and with the questions and let the computer come up with answers regarding the T.E. that inhabit the *A. thaliana* genome. Unless obviously truncated, the question whether or not the revealed elements are active or "skeletons" cannot be answered by such a computer search. Methylation has to be evaluated in each element by appropriate methods (e.g. digestion by endonucleases that do not cut methylated restriction-sites). An extensive study of the diversity of transposons in *A. thaliana* was performed by Le et al. (2000). This work was executed before the full length of the *A. thaliana* DNA was sequenced. These authors therefore focused on 17 Mb of this DNA. The concluding sentence in their summary indicated the essence of the findings: "... these findings further underscore the complexity of transposons within the compact genome of *A. thaliana*".

Beyond this final conclusion there are several interesting outcomes of this study. The investigators used "RESites" that are sequences similar to the empty sites of insertions. This RESite mining identified past insertions that were converted to repetitive sequences. The authors estimated that most of the repetitive sequences that exist in *A. thaliana* are remnants after excision of T.E. The search revealed T.E. of various kinds such as *CACTA* elements and *MULEs*, as well as *Ac*-like elements, *Mariner*-like (*MLE*) and miniature transposable elements (*MITE*). The surprise was that over one quarter of the mined elements were novel, not belonging to any previously described T.E. family.

The authors termed this novel family of T.E. *Basho*, after the nomadic Japanese haiku poet of the 17th century, Matsuo Munefusa (Pseudonym: Matsuo Basho). The *Basho* elements have defined termini (5'-CHH...CTAG-, where H=A, T or C); their target site preference is for T such as 5'-AT-3'. In total, Le et al. (2000) revealed 632 transposons in the 17 Mb sequence of *A. thaliana*. Of these, 134 elements were of Class I (e.g. retrotransposons, *SINEs* and *LINEs*), 310 elements were of Class II (e.g. *Ac*-like, *CACTA*-like, *MULEs*, *MITEs*) and 179 were *Basho* elements. Hence, even in the "compact" genome of *A. thaliana* there exists a diverse arsenal of transposable elements.

In a subsequent study of the Bureau team (Wright et al., 2001), the investigators focused on one group of T.E. termed *Ac-III* elements. The purpose was actually to study the selection pressure imposed by T.E. in a mainly self-pollinating species (*A. thaliana*) and in a related out-crossing species (*A. lyrata*). The study suggested that the *Ac-III* elements became recently active in both *Arabidopsis* species; the insertions were mostly into non-coding regions.

We shall avoid further details on the detection of T.E. in *A. thaliana*. With the analysis of the now available entire genomic DNA sequence it is probable that more T.E. elements will be revealed, possibly even "novel" families. Though it should be mentioned that T.E. were found in *A. thaliana* already over a decade ago (Peleman et al., 1991). These investigators found a 431 bp insert in a gene coding for Sadenosylmethionine synthase 1 (*Sam1*, in the ethylene synthesis pathway). There were several copies of this element in the *A. thaliana* genome; they had 13 bp in their TIRs and a 5-bp target-site duplication. Evidence was found for the transposition of this element.

A much more abundant element was discovered in the promoter of the arginine decareboxylase gene. The insert was characterized as *MITE* (miniature inverted-repeat transposable element). The full-length versions of these elements were mostly found in the immediate vicinity of genes. These elements were termed *AtATE* (El Amrani et al., 2002). *MITE*s were also revealed in *A. thaliana* by another team of investigators (Feschotte and Mouches, 2000). While the mode of transposition of *MITE*s, which are widespread among eukaryotic organisms, is unknown, the authors suggest they are mobilized by a full-size element in *A. thaliana*: *Lemi1*. These authors suggested that the *MITE*s of *A. thaliana* are derived from *Lemi1* elements by deletions. *Lemi1* is related to the *pogo* family of transposable elements (see below). This suggests that the *Lemi1* is either of very ancient origin or that there was horizontal transfer of *Lemi1* into *A. thaliana*.

"Fishing" (or "mining") for T.E. elements can also be performed in a more specific manner, meaning, fishing for specific T.E. in specific genes. For that, two PCR primers are used with the genomic DNA as template. One primer should have a sequence typical of a specific T.E. and the other primer should have a sequence from the specific gene. The DNA fragment that will be multiplied by the PCR process will be homologous to a fragment of the genome that spans from the T.E. to part of the gene. The PCR-derived fragment can then be used to identify endonuclease-digested DNA fragments on Southern blots of genomic DNA. By this procedure, DNA fragments of genomic DNA that contain a known T.E. inserted into, or adjacent to a known gene, can be cloned. The clones can be sequenced to verify the T.E. and to identify the exact location of the insertion.

6.4. Alien Transposable Elements Inserted into A. thaliana

The third group of Class II transposable elements that inhabit *A. thaliana* are those T.E. that were transferred to this species by genetic transformation. There are numerous cases of this kind and several will be mentioned in the Appendix on Gene Tagging (Appendix III). Here, we shall provide only a few examples.

An early mobilization of the *Ac* element into *A. thaliana* was performed by Van Sluys et al. (1987). At that period the genetic transformation of *A. thaliana* by *Agrobacterium tumefaciens* was still problematic, but *A. rhizogenes*-mediated transformation yielded transgenic plants with active *Ac* elements.

To increase the transposition of *Ds* that is transferred to *A. thaliana*, Long et al. (1993) added to the transgenic *A. thaliana* the code for the *Ac/Ds* transposase, downstream of a strong promoter (CaMV 35S). To test whether or not the excision of *Ds* was increased in *A. thaliana* that was furnished with additional transposase, the investigators used a genetic marker. A vector was engineered that contained a gene for resistance to "hygromycin". It also contained a *Ds* inside a gene for streptomycin resistance. The progeny plants of the transformed *A. thaliana* that were resistant to hygromycin could be tested. Those that were resistant to streptomycin, "lost" the *Ds* from the streptomycin-resistant gene. When this vector was introduced together with the construct of the transposase code behind the CaMV promoter, there was much more excision of *Ds* than when the transposase code was behind the natural (*Ac*) promoter.

A very extensive study was performed by investigators of the Max-Planck-Institute in Cologne, Germany (Wisman et al., 1998). These authors introduced the *Spm* element into *A. thaliana* plants and came up with 8,000 transgenic *A. thaliana* plants that had a total of 48,000 insertions of *Spm* (they termed it *En/I*, since a senior participant in this team, H. Saedler, collaborated with P. Peterson in the identification of *En* and Peterson always kept *his* designation for this T.E.). This work was part of a study of genes involved in flavonoid biosynthesis. Using a PCR-based screening procedure, the *Spm* insertions were found in 55 genes. There was no preference to either transcribed or untranscribed regions. After the initial insertion the *Spm* also moved into new locations in the *A. thaliana* genome. The authors recommended their procedure as an efficient gene-disruption strategy for assigning functions to genes defined only by their sequence.

7. THE *TC1/MARINER* SUPERFAMILY OF TRANSPOSABLE ELEMENTS

In the late nineteen seventies transposable elements were well established in plants and bacteria. Both types had terminal inverted repeats. The retrotransposons were then revealed in animals. This caused the simplistic notion that animals have Class I T.E. (i.e. retrotransposons), while bacteria and plants contain Class II T.E. This notion was refuted in the early nineteen eighties. A tiny worm, the nematode *Caenorhabditis elegans*, gave this notion a decisive blow. *C. elegans* is an inconspicuous animal that is abundant in soil and was raised to genetic fame by Sidney Brenner of the MRC of Cambridge, England (e.g. Brenner, 1974; Sulston and Brenner, 1974).

In the middle of the twentieth century there was a burst of bacterial genetics and biochemistry. Some scientists studying bacteria even claimed that all the principal genetics and biochemical genetic features of higher animals could be studied with much more efficiency in bacteria. S. Brenner, who was already a famed geneticist (he was a pioneer of molecular genetics), did not accept this idea. He claimed that one can use a tiny organism that really represents animals and study with it genetics, biochemistry and developmental genetics. Indeed, *C. elegans* is only about 1 mm long and has about 1,000 cells but it contains features of animals such as a pseudocelom, ectoderm, endoderm, mesorderm and sex organs. The fate of individual cells of *C. elegans* can be traced during differentiation, and a single worm can produce hundreds of eggs. Its life cycle is about four days. While commonly hermaphrodite, it can be crossed and its genome contains about 8×10^7 bp.

7.1. Discovery of Transposable Elements in C. elegans

The first clue about T.E. in *C. elegans* came probably from the University of Colorado, in Denver, rather than from the MRC in Cambridge, England. S.W. Emmons et al. (1980), in Denver, studied the general structure of the genome of *C. elegans*. Their methods were based on biophysical procedures such as DNA reassociation kinetics and electron microscopy of heteroduplexes, rather than on DNA sequencing. The main conclusion of Emmons and associates was that *C.*

elegans DNA is highly interspersed with repetitive sequences, most of them only a few hundred bp long. While a detailed discussion of the meaning of their results was presented by Emmons et al. (1980), transposable elements were not mentioned. Emmons then moved to the Bronx (NY), while his coauthors D. Hirsch and B. Rosenzweig stayed in Boulder.

After about three years a transposable element in *C. elegans* was indeed reported, first from the Bronx (Emmons et al., 1983) and then from Boulder, Colorado (Liao et al., 1983). Emmons et al. (1983) used mainly cloning of DNA fragments from *C. elegans* strains and Southern blot hybridizations and found that the genome of this worm contains a 1.7 kbp repeated DNA sequence that they termed *Tc1* (for transposable element of *C. elegans*, number 1). *Tc1* was present in different numbers in the various *C. elegans* stains that were analyzed. For example, they found about 20 copies in the Bristol strain and about 200 copies in the Bergerac strain. Studies of several genomic sites of the Bergerac strains indicated that the *Tc1* element underwent excisions from their points of insertion at a high frequency. These and other features strongly suggested that *Tc1* is a transposable element. The report from Denver (Liao et al., 1983) was very similar to that from the Bronx. The former report also concluded that *C. elegans* contains copies of a 1.7 kbp T.E. and they also used the same name, *Tc1*. They also indicated that the terminal inverted repeats of *Tc1* are shorter than 100 bp. The authors looked for hybrid dysgenesis after interstrain crosses but found none. Neither Emmons et al. (1983) nor Liao et al. (1983) sequenced the *Tc1*. The full DNA sequence of *Tc1* was reported soon afterwards from Denver (Rosenzweig et al., 1983). A typical Class II T.E. emerged. The total length of *Tc1* was found to be 1,610 bp (not exactly 1.7 kbp as based on biophysical procedures). The element has terminal inverted repeats of 54 bp, and it is flanked by 2 bp target duplications. The elements have two open reading frames (ORFs) coding for polypeptides of 273 and 112 amino acids, respectively. Now the *Tc1* was established as a T.E.

Further studies drifted into the finding of extrachromosomal *Tc1* (e.g. circular) elements in *C. elegans* (e.g. Ruan and Emmons, 1984; Rose and Snutch, 1984) but the biological significance of such extrachromosomal elements in *Tc1* transposition was not revealed.

The *Tc1* of *C. elegans* has another interesting feature. This element exists probably in all *C. elegans* strains but in different copies per genome. In all these strains there is somatic transposition of *Tc1*, but germ-line transposition was observed only in the Bergerac strain. Genetic loci ("mutators") in the Bergerac strain are probably enabling germ-line transposition.

After the *Tc1* of *C. elegans* was established as an authentic Class II transposable element, reports appeared that it is not a "lonely" element. Harris et al. (1988) found a very similar element in another species of the same genus, *C. briggsae* (*TCb1*) and also a family of *Drosophila melanogaster* elements, *HB*, had a reading frame that was very similar to the reading frames of *Tc1*. The similarity between *Tc1* and HB1 was also noted in a short (one page) report by Henikoff and Plasterk (1988). The submission and acceptance dates of these two publications (both in *Nucl. Acid Research*) are interesting. Harris et al. (1988) was submitted on February 26, 1988; Henikoff and Plasterk (1988) was submitted (and accepted) on May 30, 1988; then on May 31 Harris et al. (1988) was accepted. We shall see that these publications were only the swallows that heralded the spring. As detailed below, *Tc1* is a member of a huge superfamily of Class II transposons found in many phyla of organisms. In fact, it belongs to the most widespread superfamily of Class II transposable elements.

7.2. The Transposition of Tc1

Several interesting features of the *Tc1* transposition and the *Tc1* transposase (TcA) were revealed (e.g. Vos et al., 1993). First, structurally the TcA is encoded by an RNA that has an intron that is spliced to an mRNA that encodes a protein that is 343 amino acids long. It was subsequently found that the Tc1A transposase is the limiting factor for *Tc1* transposition. When the code for TcA was engineered behind a heat-shock promoter and the construct was introduced into nematodes, heat-shock substantially increased the rate of transposition of the transgenic nematodes. The Tc1A is specific for *Tc1* because Tc1A did not affect the transposition of the related T.E., *Tc3* that was also found in *C. elegans*. The transposase (Tc1A) binds to the terminals of the *Tc1*. There are two binding regions in the Tc1A protein: one is in the 63 residues of the amino-terminal of the protein and the other is between amino acids 71 and 207. The binding of Tc1A is to the inverted terminals of *Tc1*, between nucleotides 5 and 26. Some additional information on the transposition of *Tc1* will be provided below, after other elements of the *Tc1/Mariner* superfamily of T.E. are briefly reviewed.

7.3. Discovery of the Mariner Elements in Drosophila

Hartl (1989, 2001) detailed the interesting story of how the transposable element *Mariner* was discovered in 1981. D.L. Hartl and his laboratory (including assistants, post-graduate students and *Drosophila* flies) moved from Purdue University in West Lafayette, Indiana to the Washington University, St. Louis, Missouri. The lab had a project to look for hybrid dysgenesis in the progeny of interspecific crosses between *D. simulans* and *D. mauritiana*. Therefore, before packing, they prepared milk bottles with interspecific matings. As careful investigators, they also made intraspecific matings, as controls. Nothing of interest resulted from the bottles of interspecific matings, but a bottle of intraspecific mating of *D. mauritiana* flies resulted in an interesting eye-mutation that was revealed after arrival in St. Louis. One male fly had peach-colored, rather than (wild-type) brick-red eyes. The new mutant, *white-peach* (w^{pch}), was found to be X-linked and an allele of *white*. The (w^{pch}) mutant was unstable: occasionally resulting in mosaic eye-color of single or several pigmented omatidia in an otherwise peach-colored eye. The frequency of mosaic-eyed flies was about 4×10^{-3}. There were also germ-line mutations that were higher in males ($\sim 3 \times 10^{-3}$) than in females. It became evident that a transposable element was inserted into the w^{pch} allele of *D. mauritiana*. This T.E. was designated *Mariner*. As Hartl (2001) pointed out, this name was not in honor of the ballad of Samuel Taylor Coleridge ("*The Rime of the Ancient Mariner*"). *Mariner* was a choice of an associate of Hartl, who had a new-born baby called Marin.

7.4. The Main Features of Mariner

To verify that w^{pch} contains a (*Mariner*) transposable element, the relevant DNAs

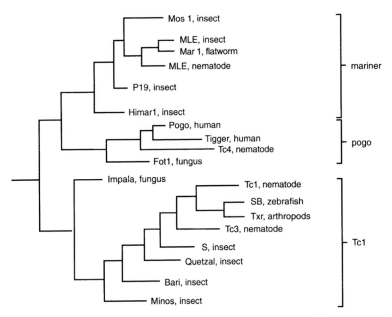

Figure 56. Phylogeny of the Tc1/Mariner family. (From Plasterk and Van Luenen, 2002).

were probed. It was first found that w^{pch} contained an "extra" DNA fragment of about 1.3 kpb that was inserted into the 5' of the first exon and 131 bp upstream of the transcription start of the eye-pigmentation gene. This fragment was lost when pigmentation reverted to wild-type. It was then found that this fragment exists, in several copies, in additional sites of the *D. mauritiana* genome. Moreover, different strains of this fly species contained different numbers of this DNA fragment, and in different chromosomal sites. These data clearly verified that *Mariner* is a transposable element. The complete nucleotide sequence of *Mariner* was published five years after its discovery (Jacobson et al., 1986). The *Mariner* then received its first special review by Hartl (1989). The sequencing showed that *Mariner* has a total length of 1,286 bp. There are two terminal inverted repeats (TIRs) of 28 bp each. The two TIRs do not precisely match. There are four mismatching nucleotides on each terminal. It is therefore one of the smaller Class II T.E. *Mariner* has a single, uninterrupted reading frame of 1,035 nucleotides. The codon usage is typical of *Drosophila*, and so is the promoter sequence. There is also the common AUG translational initiation codon. The reading frame ends with GAATAA, which are codons for glutanic acid and for termination, respectively. There is also a regular poly (A) signal. The insertion into w^{pch} is at the untranslated leader, right after a TA dinucleotide.

The *Mariner* was revealed, in different numbers of copies, in several other species of the *D. melanogaster* complex, but apparently not in *D. melanogaster* itself. In several cases the *Mariner* element was found to be truncated in the middle or at the 5' end. Sometimes, the *Mariner* of *D. mauritiana* caused a high rate of mutability in the soma (mosaic eyes) and in the germ-line. This high rate was

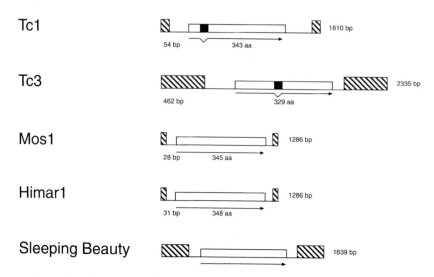

Figure 57. Structure of the most studied members of the Tc1/Mariner family.
The terminal inverted repeats (dashed boxes) and the open reading frames (open boxes)
are indicated for Tc1, Tc3, Mos1, Himar1 and Sleeping Beauty.
(From Plasterk and van Luenen, 2002).

correlated with a locus in chromosome 3. This was termed *Mos. Mos* turned up to
be a *Mariner*-type transposon that can also transpose, so that it can move to other
loci.

As in *Tc1*, the transposase of *Mariner* also has a motif of the D.D(35)E kind.
We met this motif (or *signature*) above and it is shared by a very large group of
transposable elements and even bacteriophages and viruses (such as HIV, the Human
Immunodeficiency Virus). The D.D(35)E stands for two aspartic acid residues (D, in
the single-letter designation of amino acids) that are separated from each other by
more than 90 amino acids, and a glutamic acid (E) separated from the second
aspartic acid by 35 amino acids. The distance can also be 34 amino acids and in
Mariner there is no glutanic acid as the third acidic amino acid, but another aspartic
acid.

The nomenclature of this motif is therefore D.D(34)D for *Mariner*. The
transposases of many transposable elements have been sequenced in recent years.
This has enabled the construction of a phylogenetic tree of these elements as shown
in Figure 56. Based mainly on the transposase sequences the *Mariner*-like elements
that were included in a "family" of *MLE* (*Mariner*-like elements) and separated from
two other, related families, of transposable elements: the *Pogo* family and the *Tc1*
family. These three families, as well as several additional families, were grouped in
one superfamily of Class II T.E.: The *Tc1/Mariner* superfamily.

A schematic representation of the members of the *Tc1/Mariner* superfamily is
given in Figure 57. In this figure, *Mos1* actually stands for the full size *Mariner*
elements. The element *Himar1* is from the horn fly, *Haematobia irritans*, and this
name is according to the nomenclature suggested by H.M. Robertson (Lampe et al.,
1998). *Sleeping Beauty* is a synthetic, *Mariner*-like element based on fossil

elements found in fishes. It will be mentioned again below. The latter element is therefore man-made and not a natural element.

In the following sections, the elements that are grouped in the *Tc1/Mariner* will be handled together as has been done in recent reviews (e.g. Plasterk, 1996; Hartl et al., 1997; Plasterk et al., 1999; Plasterk and van Lueren, 2002; Feschotte et al., 2002).

7.5. Structure and Function of the Tc1/Mariner Superfamily

The *Tc1/Mariner* elements have several common or similar features that justify their inclusion in one superfamily. We mentioned above the similarity in their transposases and that they all have the D.D(35)E (or D.D(34)D) motif. They also have conserved sequences in their terminal inverted repeats (e.g. 5'-CAGT in *Tc1*-like elements). All integrate into TA. Because of greater similarity within two groups of the *Tc1/Mariner* superfamily, these groups were frequently designated *Mariner*-like elements (*MLE*) and *Tc1*-like elements (*TLE*), respectively.

7.5.1. Transposition

A brief description of transposition of *Tc1* was provided above. A more general model was suggested by van Luenen et al. (1994), based on the transposition of *Tc3*. The latter model is still valid and probably can be applied to the whole *Tc1/Mariner* superfamily, although it may be amended by future research. In this model the transposase binds to the termini of the transposon and makes double-strand DNA breaks at the two ends. There is a hydrolysis of the phosphodiester bond at the 3' end of the transposon. At the 5' end the break is two nucleotides within the transposon. This leaves a GT protruding at the "left" side of the transposon and a TG at its right side, both with (active) hydroxyl groups. While this happens, the transposase forms a complex with the termini of the transposon. A staggered cut is made at the target DNA, at both sides of the TA dinucleotide. The active hydroxylated nucleotide of the cut transposon then "attacks" the target nucleotide and single strands of four nucleotides (ATGT and TGTA) are formed at both ends that connect the transposon with the target DNA.

The second strand is then filled, probably by the host repair enzymes. The stagger, at the target TA, generates a TA-target duplication at the ends of the integrated element. The target sequence is not completely cut during the transposition; the cleavages of the phosphodiester bonds, at the target site, are carried out in one step with the fusion to the 3' ends of the incoming transposon. The gap, left after the transposon is excised from the donor site, is also filled, probably by the host machinery. This latter repair may use different DNA sequences as templates and one of the results could be that the transposon is copied back into the "old" site. In hetero-allelic mutants, virtually all the templated repair that uses the homologous chromosome as template, will result in exact restoration of the "empty" site.

The excision and strand transfer described above were observed in cell-free systems when the transposases of *Tc1*, *Himar1* or *Mos1* were the only proteins in the reaction. Such experiments suggested that only transposase is required for the *in vitro* reaction, but they did not prove that other proteins are not involved in the *in vivo* process.

7.5.2. The Transposases of the Tc1/Mariner Superfamily

A schematic representation of the transposases in representatives of the *Tc1/Mariner* elements has been provided in Figure 57. There are two domains in these transposases. One is the N-terminal domain. This consists of two "helix-turn-helix" (HTH) regions that bind specifically to the DNA ends of the transposon. The other domain is the C-terminal, catalytic, domain that consists of the DDE (or DDD) motif that was mentioned above. This motif is considered to serve as a binding domain to Mg^{2+} or Mn^{2+}, that are required for the catalysis. In *Tc1* and *Tc3* the most N-terminal HTH is sufficient for binding to the 20 bp that are located 5–30 bp from the terminal end of the element. There are some differences between the *Tc1* and *Tc3*. For example, in *Tc1* it was found that the second (less N-terminal) HTH renders the transposase binding more selective. It is noteworthy that while the most terminal 4 bp of the elements are conserved and essential for transposition, they are *not* a binding site for the transposase.

Despite the similarity of the transposases within the *Tc1/Mariner* superfamily, and even more so within *TLE* and *MLE*, there is a remarkable specificity. Thus, the *Tc1* transposase will not bind to the DNA of the *Tc3* transposon, and *vice versa*.

Mutants with altered transposases served to improve the understanding of the functionality of these complexed proteins. It was found that transposase mutants of *Tc1*, in which the 153 N-terminal amino acids were missing, lost binding capability to the DNA of the element. Several transposase mutants were not precisely characterized, but they were useful to derive the efficient T.E. *Sleeping Beauty*. Such screening of mutants for more efficient transposition was performed with the transposase of the *Himar1* element when it was mobilized into bacteria. Indeed, this resulted in mutants with altered transposase that had increased transposition in *E. coli*. A similar approach was also used to alter the *Mariner* transposase for enhanced transposition in cultured human cells. Obviously, it seems that the natural transposases are not the most efficient ones for transposition. Such a situation should not surprise us. For regular enzymes, such as those active in the household metabolism of organisms, it is reasonable that there is a selective pressure for higher efficiency. Not so for transposable elements. A too efficient transposition will kill the host or at least reduce its vitality. Therefore, natural selection will not necessarily lead to the most effective transposases.

While, as mentioned above, the insertion is always in a TA dinucleotide of the host DNA, there are probably "hot-spots" and "cold-spots", where *Tc1* and *Tc3* insert frequently or rarely, respectively. For example, the sequence CAYATA**TA**TRTG is a preferred sequence for insertion. In certain cases, as in plant elements of the *Tc1/Mariner* family, it appeared that transposition to nearby sites is preferred. This led to speculation for a requirement of proximity between the donor site and the host site which are attractive and logical, but have no real proof. As mentioned elsewhere in this book, the investigators dealing with eukaryotic T.E. tended to ignore the fact that the DNA of eukaryotes is not "nude" but rather complexed, in chromatin, with histones.

7.5.3. Regulation of Transposition

There are several indications that transposition in the *Tc1/Mariner* superfamily is highly regulated. For example, already in early studies with *Tc1* (Emmons and Yesner, 1984) it was noted that this element transposed frequently in somatic cells but not in the germ-line cells. This is so when *Tc1* is in the Bristol N2 strain. It was then found that in another strain of *C. elegans*, the Bergerac BO strain, there is also transposition in the germ-line. Indeed ample mutations were derived in Bergerac from these germ-line transpositions. Due crosses and genetic analyses revealed *mut* genes in Bergerac that permitted the germ-line transposition of *Tc1*. Moreover, a special *mut* gene was induced by a mutagen (EMS) that further enhances transposition of *Tc1* in Bergerac. Similarly, EMS-derived mutants were also isolated from the Bristol N2 strain, and these enhanced not only *Tc1*, but also *Tc3* and similar *Tc* elements. Could it be that at least some of these *mut* mutants are loss-of-function mutants in which the *silencing* of transposition is not functioning? Indeed, one *mut* gene seems to involve lack of RNAi production. An appendix on RNAi is provided at the end of this book.

There is also regulation of *MLE* transposition. There are numerous *MLE* in various organisms but there are only two active *Mariner* elements in flies. One is the *Mos1* in *D. mauritiana* and the other is the *Himar1* of *Haematobia irritans*. All others are inactive and were characterized only by their DNA sequences. Both of these active elements require no additional host factors for *in vitro* transposition. The *Mos1* element served Hartl and collaborators (see Hartl et al., 1997, for details and references) in a two-component experimental assay to explore the transposition of *MLE*. The *Mos1* is a full-length *MLE* element of 1,286 bp with 28 bp of TIRs and an ORF that encodes a functional 345 amino acid transposase. The latter, as indicated above, has the D.D(34)D catalytic motif. *Mos1* was introduced into *Drosophila* that contained a *peach* element that is not capable of autonomous mobility. The *peach* was inserted into the *white* eye locus (in the 5' non-coding region of the X-linked *white* gene) and caused peach-colored eyes. In the presence of *Mos1*, the *peach* was transposed very frequently, causing mosaic eyes by soma transposition and reversal to wild-type eyes in the progeny, by germ-line transposition. As the frequency of excision in this assay is very high, it served to explore various features of the transposition. One of these is the formation of footprints. The typical footprints of *Mos1* were found to be 5'-TACCATA-3' or 5'-TATGATA-3'.

There were two proposals for the regulation (*in vivo*) of *MLE* elements, which are not necessarily mutually exclusive. One proposal was that there is "overproduction inhibition". For example, high expression of the *Mos1* transposase (e.g. by using a very potent promoter upstream of the code for the transposase) decreased the transposition of this element. The exact mechanism for this inhibition is not known. Another phenomenon was termed "dominant-negative complementation". This term stands for the inhibition of the wild-type transposase by defective transposases. The latter probably competes with the binding sites on the DNA of the element with the wild-type transposase. Several such defective transposases were isolated that indeed reduced the transposition. It is reasonable to assume that in many organisms there are defective *MLE* elements in addition to the active ones, and that the former reduce the transposition. In summary, while it is evident that the transpositions of the *Tc1/Mariner* superfamily are not maximal, but rather regulated, the exact means of regulation are not yet clear.

7.6. The Tc1/Mariner Elements as Genetic and Biotechnological Tools

The use of *Tc1/Mariner* elements for gene tagging is straightforward. The actual procedures for gene tagging are reviewed in Appendix III. Basically, for efficient tagging, there are three requirements. One is that the mutated organism does not already have many natural elements as the element that is used for tagging. Another requirement is that the transposition is sufficiently high. The third requirement is that there is a low preference of integration into specific sites. The *P* element of *Drosophila* has these three characteristics and has been amply used for gene tagging. Among the TLE the *Tc1* is also useful for gene tagging in *C. elegans*, but there are two problems that should be considered. One is that there are already *Tc1* elements in this species and the other is that the transposition should be enhanced. The latter issue can be handled by adding genes such as *mut-2* or *mut-7* that will facilitate the transposition. The interference by endogeneous *Tc1* may be handled by adding to the T.E. vector, a marker as the GFP (green fluorescence protein), that will mark the integration of the incoming T.E. However, such methods still await actual application.

Transposable elements can be used as vectors to introduce alien DNA into host organisms. Such vectors should contain a coding sequence (i.e. a gene) that is flanked with suitable controlling sequences (e.g. promoter and terminator). The alien genes with the controlling sequences have to be engineered into the transposable element. The modified T.E. should then be used to infect the host organism. The *Tc1/Mariner* elements could be useful as vectors because these will integrate into a wide range of host organisms. This is probably so because integration is not dependent on element-specific host factors. In fact, many cases were recorded in which *Tc1, Tc3, Sleeping Beauty, Mos1, Himar1* and *Minos* were transferred to other organisms. In a number of these transfers the integration was to organisms of other kingdoms. Such "horizontal transfers" are listed in Plasterk and van Luenen (2002) and summarized in Table 8. While such horizontal transfers with *Tc1/Mariner* elements were shown to be possible, the question remains how much additional DNA (genes with controlling sequences) can be added to these "Trojan Horses". It is expected that in the future, investigators will explore, and even put into practise, this novel means of genetic transformation for crop improvement, gene-therapy and the manufacture of medical products. Using the *Tc1/Mariner* elements for plant transformation and crop improvement is less attractive because, for angiosperm plants, efficient means for genetic transformation are already available (Galun and Breiman, 1997; Galun and Galun, 2001). In most cases, *Agrobacterium*-mediated transformation, with or without microprojectal bombardment, can serve for such transformations. In specific cases (e.g. in *Arabidopsis thaliana*) *in planta* genetic transformation was readily applied.

7.7. Examples of Tc1/Mariner and Other Class II T.E. in Diverse Organisms

As already remarked above, the *Tc1/Mariner* elements are the most widespread Class II T.E. Since these elements were found in virtually all plant families and animal phyla in which they were investigated, we shall not try to list them all. The *Tc1/Mariner* elements are considered to be transferred vertically (e.g. to progeny through sexual and non-sexual propagation) as well as by "horizontal" transmission (new insertion into the genome from outside). Direct proof for horizontal transmission of a T.E. is very rare, but there are cases in which the evidence for this phenomenon is very strong. For example, when the DNA sequences of a fly *MLE* and a mammal *MLE* are very similar, while the coding sequences for common metabolic enzymes are very different in these two organisms, the conclusion that

Table 8. Experimental horizontal transfer of Tc1/Mariner elements.

Transposon	Endogenous host	Heterologous host
Tc1	Nematode, *Caenorhabditis elegans*	Human embryo retinal cell line Mammalian cells lines
Tc3	Nematode, *Caenorhabditis elegans*	Zebrafish, *Danio rerio*
Sleeping Beauty	Fish, reconstructed from salmonid Fish species	Human Hela cell line Mouse embryonic stem cells
Mos1	Insect, *Drosophila mauritiana*	*Drosophila melanogaster* Queensland fruit fly, *Bactrocera tryoni* blowfly, *Lucilia cuprina,* Yellow Fever mosquito, *Aedes aegypti,* *Drosophila virilis,* Zebrafish, Chicken embryonic cells Protozoan, *Leishmania major*
Himar1	Insect, reconstructed from *Haematobia irritans*	Bacteria, *Escherichia coli,* *Mycobacterium smegmatis,* Human embryonic kidney cells
Minos	Insect, *Drosophila hydei*	Mosquito, *Anopheles gambiae,* *Anopheles stephensi,* *Drosophila melanogaster,* Medfly, *Certitis capitata*

(From Plasterk and van Luenen, 2002,
where references to the horizontal transfers are provided).

there was horizontal transmission of this *MLE* is justified. There is a claim that T.E. have selective/evolutionary advantages. Therefore, when T.E. are lost from an organism, they can be replenished by horizontal transmission. The evolutionary loss of T.E. has good evidence but how exactly this happens is a matter of speculation. D.J. Lampe and H.M. Robertson (Lampe et al., 2001) elaborated this subject and with some fantasy the lines of "Also sprach Zarathustra" come to mind. We indicated above that for transposition it is necessary that certain regions of the transposase bind to specific regions in the terminals of *Tc1/Mariner* elements. Even slight changes in either the DNA sequence of the terminals or in the amino acid sequence of the transposase will abolish transposition. An intimate matching of these sequences is thus a prerequisite of transposition.

On the other hand, a T.E. that has lost the transposition capability is doomed. This will happen because there is a "biological clock" of nucleotide mutations. One estimate is that one nucleotide change happens in each site, each 10^6 years. However, if there is no selection for retaining transposition capability (when this capability does not exist any more), the sequence of the T.E. will drift away until there are so many nucleotide changes that a search for T.E. will not recognize it. The element will then be lost forever. Restoration of activity is impossible after a certain number of nucleotide changes have taken place. If we associate the *Leben* (~female vitality) and *Zarathustra*, in "*Also sprach Zarathustra*", to the transposase and termini of T.E., and the *Liebe* (love) to the intimate association between the transposase and termini, the following words of *Leben* to *Zarathustra* will ring a bell.

"Wenn dir deine Weisheit einmal davonliefe, ach!
Da liefe dir schnell auch meine Liebe noch davon ... .
... ich weiss es; dass du mich bald verlassen willst."

(From: F.W. Nietzsche *Also sprach Zarathustra*; the meaning of *Leben's* words is: once your wisdom will be lost then my love to you will be lost as well ... . I know; that you shall soon desert me).

Based on the nucleotide sequences that encode the transposase, and especially the region surrounding the D.D(34/35)E/D motif, the *Tc1/Mariner* elements can be divided into three monophyletic groups: 1. *Tc1*-like elements (*TLE*); 2. *Mariner*-like elements (*MLE*), and 3. *Pogo*-like elements (*PLE*) (Figure 56). In the following sections I shall provide information on these groups according to the organisms in which they reside.

7.7.1. MITEs and Tc1/Mariner Elements in Angiosperm Plants

Tc1/Mariner elements have been recently reviewed (Feschotte and Wessler, 2002). Using several procedures for "fishing" (or "mining") for *Tc1/Mariner* elements of the *MLE*, *TLE* and *PLE* groups, these authors "found" almost a hundred such elements. Most of these were "fossil" (or "skeleton") sequences with similarity to *MLE*. Only two real *MLE* elements were previously reported to exist in angiosperm plants: one of these was found in soybean and was accordingly termed *Soymar1*; the other was revealed in the rice (*Oryza sativa*) genome and termed *Osmar1*. These two elements have similarity to *Mariner* by their transposase code but they are also similar to *Stowaway* elements. The *Stowaway* elements (that shall be mentioned again below) have no coding sequences but their similarity to *Mariner* could be based on their mobilization, in *trans*, by *MLE*. *Stowaway* elements are widespread in plants. Thus, if these elements are considered to belong to the *Tc1/Mariner* superfamily, the number of plant elements that can be grouped with this superfamily is substantially increased. Feschotte and Wessler (2002) analyzed the genomic DNA of only a very few plant species, but even in these the authors could identify sufficient kinds of elements so that a phylogenetic tree of 91 plant *MLE* transposase fragments could be constructed. By using certain degenerate primers these authors were able to derive, by PCR, sequences that are between the two first aspartic acids (D) of a D.D(34)D motif (that exist in the rice *Tc1/Mariner* element, *Osmar1* and the soybean *Soymar1* element). Such sequences were found in many monocot and dicot species. This provides strong support to the claim that plants harbored, in the past, *Tc1/Mariner* elements. Sequencing and conceptually translated amino acid sequences provided further interesting information. It became evident that most plant elements become inactive. Once inactive, mutational drifts may cause gradual changes so that what were once coding sequences (for transposase) cannot be recognized by the "mining" procedures. In summary, we can only be sure that there were "many" *Tc1/Mariner* elements in "most" plant species. However, what is "many" and "most" cannot be determined.

While in recent years, the search for T.E. was mostly based on DNA sequences, in the past, T.E. were revealed in genetic studies. Once an allele was suspected, by genetic means, to contain a T.E., this element could then be further studied by molecular techniques. For example, a transposable element was found in the DFR-C gene of petunia. The element was revealed by the instability of flower pigmentation. Molecular methods then verified (Gerats et al., 1990) that an insert existed in the *An1* locus. This insert was 283 bp long and included perfectly inverted terminal repeats of 12 bp. The insertion was flanked by a 8 bp target site duplication. It became evident that this was a small non-autonomous transposable element. It was termed *dTph1*. *dTph1* was mobile only in the presence (in *trans*) of an autonomous T.E. Because there was no coding sequence in *dTph1*, it could not be grouped into one of the existing Class II T.E. groups. Thus, this element and subsequently revealed short T.E. with inverted terminal repeats were given the acronym *MITE* (for Miniature Inverted-repeat Transposable Elements).

Since *MITE*s have no transposase-coding sequences, they cannot be categorized to any of the full size Class II T.E. However, because they appear to be very common and of wide distribution among monocots and dicots, they will be mentioned briefly below.

An element that is even smaller than *dTph1* was found in maize. It was termed *Tourist*. The *Tourist* elements constitute a family of *MITE*s with an average of 132 bp, containing terminal inverted repeats. *Tourist* elements are flanked by 3 bp direct repeats and are very abundant in the maize genome (Bureau and Wessler, 1992). In a subsequent study (Bureau and Wessler, 1994a) an additional *MITE* family was announced: *Stowaway*. While the *Stowaway* family has similarities to the *Tourist* family they differ sufficiently to assign them different names. *Stowaway* elements were revealed in many plant species (in over 40 monocot and dicot plant genes that were deposited at the time in the EMBL GenBank). The search for *Tourists* went on to several Gramineae species (e.g. sorghum and rice). Many such elements were detected in introns or flanking sequences of genes and it appeared that the *Tourist* elements were only recently mobile in this grass family (Bureau and Wessler, 1994b). Other investigators found a *MITE* in bell pepper (*Capsicum annuum*) with 28 bp TIRs (Pozueta-Romero et al., 1995) and termed it *Alien*. *Aliens* were then revealed in additional dicots and in monocots (Pozueta-Romero et al., 1996).

The availability of genomic sequences in rice and *Arabidopsis* allowed additional "fishing" for *MITE*s (Bureau et al., 1996) and indeed, new families of *MITE*s were discovered. The *MITE* families were given names: *Gaijin*, *Castaway*, *Ditto*, *Wanderer* and *Explorer*. The *Arabidopsis* genome was used not only by the Wessler team to fish for *MITE*s. Other investigators (Casacuberta et al., 1998) revealed, in the same genome, another *MITE* family: *Emigrant*. In fact, *Emigrant* elements were also found in the genomes of other Cruciferae genomes.

At this point the reader can conclude that the discovery of additional *MITE* is not limited by their existence in plant genomes. They most probably exist, and even in abundance, in most (or even all) plant species. More can be found by improved techniques to reveal them. Indeed, one such improved technique came from the Lowlands, from the University of Gent (Van der Broeck et al., 1998) to where T. Gerats, who published the discovery of *dTph1* of petunia (Gerats et al., 1990) had moved (from Amsterdam). This technique was termed *Transposon Display*. It was instrumental to analyze further the behavior of the *dTph1 MITE* during ontogeny and breeding of the ornamental species *Petunia hybrida* (De Keukeleire et al., 2001). This technique enabled spatial and temporal detection of the *MITE*s in specific cells.

An indication for the origin of one *MITE* element in maize was recently reported (Zhang et al., 2000). This study started with *PIF*. *PIF* is a P-instability factor, an active DNA transposable element family from maize that was identified after mutagenic insertions into the same site: intron 2 of the *R* gene of maize (that regulates anthocyanin synthesis). Based on *PIF*, the authors isolated a family of *Tourist*-like *MITE*s and termed it: miniature *PIF* (*mPIF*). It was found that *mPIF* had several similarities to *PIF*. Both had identical TIR and they shared similar subterminal sequences. Both also had a striking similar preference of insertion into a 9-bp target site. This strongly suggested that *PIF* and *mPIF* are mobilized by the same transposase. Indeed, one *PIF*, termed *PIFa*, was isolated and its putative transposase sequence could be analyzed and compared with the transposases of other full-size Class II elements. The results indicated that *mPIF* belongs to a new family of Class II elements and is distantly related to the bacterial element IS5.

From the latter study we can predict a future trend. Although *MITE*s cannot be categorized because they lack transposase, there may be a way to put them in a specific T.E. group. This will be possible if sufficient similarities are found in the existing *MITE* sequences, to sequences in full-size T.E.

The *Heartbreaker* of maize constitutes a clear indication for the talents of investigators to find exotic names to transposable elements and that further "fishing" will probably yield additional T.E. elements, even in organisms where there has been an extensive search for such elements. The first element of the family termed *Heartbreaker* (*Hbr*) was recently found in the 3' untranslated region of a mutated allele of the maize disease-resistance gene *HM1* (Zhang et al., 2000). Further search indicated that this element has the typical features of a *MITE* and is a member of a large family of similar elements. The similarity is so great that the investigators concluded that it spread only recently in the maize genome. About 4,000 *Hbr* elements were estimated to exist in maize. The first *Hbr* was 314 bp long, has TIRs and seems to insert preferentially into or near genes. Although the evidence for the latter characteristic is not conclusive because the search is based on sequence data and these come mostly from gene sequences, rather than from random DNA sequences of maize. It is also not clear how many of the *Hbr* elements are still mobile, nor was the autonomous element that can mobilize *Hbr* found. The genome size of maize is huge, thus the full sequence of this genome will not be available in the near future. Therefore, elusive autonomous elements may easily hide from database screening.

7.7.2. Class II Transposable Elements in Hyphal Fungi

Ever since the pioneering studies of G. Pontecorvo on the genetics of fungal organisms (e.g. *Aspergillus nidulans*) at the University of Glasgow, Scotland, the fungal gene for nitrate reductase (NR) has been a genetic "treasure trove". I have already mentioned this two-component gene when I described a method to screen for NR mutants: exposure to chlorate. The mutants are resistant because they will not toxify themselves with the chlorite that is produced by fungi with wild-type NR that are exposed to chlorate. The NR gene was also a starting point for the search for T.E. in fungal organisms by M.-J. Daboussi (CNRS-Université Paris-Sud, Orsay, France) and her associates. Prior to the search of Daboussi there was only one report on a transposable element in the hyphal fungus *Neurospora crassa* (Kinsey and Helber, 1989). The latter authors revealed transposable elements in a specific strain of *N. crassa* from the Ivory Coast and termed them *Tad*. *Tad*s had a length of 7 kbp but it was not sequenced. Actually "preliminary results" suggested that *Tad* had neither short inverted repeats nor long direct repeats at its terminals. Therefore, *Tad* could not be classified to any specific T.E. group.

Daboussi and associates (Daboussi et al., 1992) focused on the plant pathogen *Fusarium oxysporum* (there are many strains and *formae specialis* of this species). The structural gene for NR (*nia*) was the locus where T.E. were searched. As indicated above, NR⁻ (*nia⁻*) mutants can be isolated in great numbers on chlorate-containing medium. Hundreds of such mutants were produced, but attention was directed to those that were unstable and reverted to wild-type. One mutated strain yielded up to 10% unstable mutants. However, as the investigators intended to analyze the *nia* mutation at the molecular level, and this was at the time possible with the *nia* of *Aspergillus* but not with *nia* of *F. oxysporum*, the investigators transferred a single *nia* gene from *Aspergillus* into *F. oxysporum*. Indeed, unstable *nia*D mutants were then found in the transgenic *F. oxysporum*. Furthermore, an insert was found in the unstable mutated gene. The insert was of 1,928 bp with TIRs of 44 bp. The two TIRs differed by one base. There was an ORF coding for 512 amino acids. These were clear indications of a transposable element and the element was coined *Fot1*. The excision of *Fot1* was found not to be "clean" – remnants were left behind, but because the original insert was into an intron, the excision still allowed wild-type expression of NR. The *Fot1* element was found to have features that are similar to *Tc1* and other *Tc1/Mariner* elements, but its DNA sequence differed from the then known elements.

The same approach to study an unstable mutant of NR after a *nia* gene was transferred from *A. nidulans* into a specific strain of *F. oxysporum*, was used to isolate a different insert into the *nia* gene (Langin et al., 1995). The new insert was called *impala*. *Impala* is 1,280 bp long, has TIRs of 37 bp and inserts into a TA site. These features and similarity in ORF sequences to those of other Class II elements put *impala* into the *Tc1/Mariner* superfamily. In fact, as indicated above, the *Mariner*, full-size element *Mos1* is 1,286 bp long has TIRs of 286 bp and an ORF that encodes 345 amino acids. There is actually a family of *impala* elements in some strains of *F. oxysporum*.

By 1997 genomic sequences of *F. oxysporum* became available. Therefore, a computer search could be performed to "fish" for transposable elements in this fungal organism. Indeed, Okada et al. (1998) of Nagoya, Japan used sequence analysis and found another T.E. that they called *Tfo1*. However, *Tfo1* was not a member of the *Tc1/Mariner* family. Its length was 2,673 bp, it had TIRs of 15 bp and a long ORF that coded for 777 amino acids. These amino acids indicated a transposase with similarities to the *hAT* superfamily (that include the *Ac/Ds* elements). There were several *Tfo1* elements in the genome of *F. oxysporum*.

Subsequently, another T.E. of the *hAT* family was revealed in a different fungal organism: *Cryphonectria parasitica* (Linder-Basso et al., 2001) and was termed *Crypt1*. *Crypt1* is 3,563 bp long and contains TIRs of 21 bp. It has an ORF that encodes a 946 amino acid putative transposase.

During further studies of Daboussi and associates (Hua-Van et al., 2001a) on the *impala* elements of *F. oxysporum*, these authors found that the *impala* elements are composed of five sub-families in this fungus. These sub-families differ from each other by about 20% at the nucleotide sequence level, and *impala* elements were found in most strains of *F. oxysporum*. These authors (Hua-Van et al., 2001b) also found that some *impala* elements are autonomous and they could trace the mobility of the *impala* elements. Trans-activation was possible even when the two elements diverted considerably in the nucleotide sequences. The autonomous version of *impala* could transpose into a different species of *Fusarium*: *F. moniliforme*. By using an autonomous copy of *impala* and a defective one, the authors could establish a two-component transposable element system that is mobile in *Fusarium*.

The *F. oxysporum* complex is posing an enigma to fungal geneticists. The variability between strains, pathovars etc. is extensive. On the other hand, the variability cannot be based on sexual crosses because there is no sexual reproduction in *F. oxysporum*. Thus, Daboussi et al. (2002) recently attempted to use transposable elements (*Fot1* elements) to resolve this enigma. One of the conclusions of these authors was that because *Fot1* elements with very similar nucleotide sequences reside in *F. oxysporum* strains and even in other species, the transposable elements moved horizontally between strains and species. Moreover, it was suggested that the existence and mobility of transposable elements causes ample genetic changes in the genomes of *F. oxysporum*.

The subject of transposable elements in (hyphal) fungal organisms was recently approached by a combined effort of an *Aspergillus* geneticist (C. Scazzocchio), M.-J. Daboussi and collaborators in Orsay, France (Li Destri Nicosia et al., 2001). First, they looked at the genome of the model ascomycete *Aspergillus nidulans* and found ample *Fot1*-like elements. This was a surprise because *A. nidulans* is one of the most studied hyphal fungi since the studies of G. Pontecorvo in the early nineteen fifties (e.g. Pontecorvo et al., 1953), and no T.E. were reported previously in *A. nidulans*. These *Fot1*-like elements were found to be transcribed in *A. nidulans*. However, there was a difference between *F. oxysporum* and *A. nidulans*. In the former species the *niaD* (NR) gene could serve as a transposon trap; not so in *A. nidulans*. However, the authors used several clever manipulations in order to render *impala* and *Fot1* elements useful in gene tagging in *A. nidulans*. For example, they caused an *in vivo* insertion of a *F. oxysporum* transposable element into the *niaD* gene of *A. nidulans*. The mutated *niaD* was then reintroduced into an *A. nidulans* fungus line that had a deletion in its endogeneous *niaD*. In this way it was expected that only in the case that the transposable element was excised from the mutated *niaD* would the fungus be able to grow on nitrate. In practice, in one case *Fot1* was introduced into an intron of *niaD*, and in the other case *impala* was inserted into the *niaD* promoter. The results showed that the alien elements were transposing at high frequency (causing nitrate utilization). The excised transposons tended to reinsert at random. Actually, the authors tagged the *impala* transposon so that it could be traced. Several indications suggested that *impala* is transposed by a cut and paste system as was actually suggested for another fungal transposon, *Restless*, which belongs to the *hAT* superfamily (Kempken and Kuck, 1998). One can now claim that if there were not sufficient means to cause genetic mutations in *A. nidulans*, this is now amended by *F. oxysporum* transposable elements. Moreover, an efficient gene tagging system can now be utilized in *A. nidulans*.

7.7.3. Tc1/Mariner-like Elements in Protozoa
In the field of transposable elements we encounter paradoxical terms. One of these is "transposons that do not transpose". Well, the investigators may use a "softer" characterization: "elements (that) have not been observed to transpose" (Doak et al., 1994). This is how some sequences of ciliated protozoa were characterized. Nevertheless, names were given: *TBE*, *Tec1* and *Tec2*. *TBE* was found in *Oxytricha fallax* and *Tec1* as well as *Tec2* were found in *Euplotes crassus* (Klobutcher and

Jahn, 1991). All had ORF that encoded putative transposases and the C-terminal side of the computer-translated amino acid sequence had a DD(35)E motif that brought them close to the transposases of the *Tc1/Mariner* superfamily. They also seem to have been targeted into the TA dinucleotide (Jahn et al., 1993; Doak et al., 1994). Ciliated protozoa undergo, during their life-cycle, an extensive genome rearrangement. Still, they are able to transmit their genetic load to the next generation in an intact form. This happens because these protozoa have two nuclei. The *micronucleus* stays intact with a normal chromosome complement. These micronuclei are the "germline" nuclei and transfer the genes to the next generation. The other nucleus is the *macronucleus*, from which the genes are transcribed. In the latter nuclei, the DNA undergoes extensive replication and fragmentation, and 10,000 or more replications of some chromosomal sequences may be produced. *Tc1*-like sequences are found in the micronuclei of these ciliated protozoa but not in their macronuclei. During DNA processing in the macronuclei, the *Tc1*-like elements are excised and then form circles. Klobutcher and Jahn (1991) studied putative transposable elements in ciliate protozoa of the genera *Euplotes* and *Oxytricha* and found indications that these putative transposable elements play a role in the DNA rearrangements of the macronuclei, possibly by their contribution to the rejoining of DNA sequences.

The putative transposable elements of ciliate protozoa (e.g. *TBE*, *Tec1* and *Tec2*) are champions in excision: they excise "cleanly" from the DNA of the macronucleus, leaving no remnants behind. However, in the DNA of the micronucleus these elements are not excised. The micronucleus produces two nuclei in the next generation: one micronucleus and one macronucleus. The elements are replicated with the rest of the DNA of the micronucleus. In this way the elements are transmitted from generation to generation. Doak et al. (1994) found a similarity between the phenomenon in the "somatic" nucleus (macronucleus) of ciliate protozoa to the preferential transposition of some transposons, in metazoa, in the somatic cell line.

Finally, a trans-kingdom transfer of an insect transposon (*Mos1*) into the pathogenic protozoa *Leishmania major* was successfully undertaken (Gueiros-Filho and Beverley, 1997). The insect *Tc1/Mariner* element *Mos1* was found to undergo frequent transpositions in its trans-kingdom host.

7.7.4. A Note on Minos, Hobo, Pogo and Other Class II T.E. that Reside in Insects and in Other Invertebrate Metazoa

Estimates for the number of insect species range from one to thirty million. There is no wonder that the estimates for the total number of insect species varies so extensively. This number is based on very rough estimations. One method is to "isolate" one tree and to count all the insect species that inhabit this tree. This is repeated with several trees. This leads to a calculation of how many "unique" species are in each tree. From this the calculation is made how many species of insects are in a given forest. From one forest it is extrapolated further, until a worldwide estimate is made. When a "final" number of insects is calculated, there is no chance that anyone will be able to prove that the number is wrong ... who is going to count? This reminds one of the mathematical theorem of Gödel (1906–1978). However, even the extrapolation requires a great effort, because the number of insect species on one tree in the Amazon rain-forest may exceed the total number of mammalian species in the world. A very small fraction of these species were searched for T.E. and several of these elements (or element families) were already mentioned above.

A special section will be devoted to the *P* element of *D. melanogaster*, where this element was found by genetic studies and in which the hybrid dysgenesis phenomenon (also noted above) was revealed. However, insects harbor other Class II elements. We shall mention some of them such as the *Pogo* (e.g. Tudor et al., 1992) and the *Hobo* (e.g. Lim, 1988; Atkinson et al., 1993) of *D. melanogaster*. Another element, *Minos*, was found in *Drosophila hydei* (e.g. Franz et al., 1994). Several families of Class II elements (grouped as *MITEs*) were found in mosquitos (*Aedes aegypti*), where they were suggested to be associated with genes for Yellow Fever (Tu, 1997, 2000). The *Mariner*-like elements (*MLE*) of many insects were surveyed by Robertson (1993).

An interesting case of the same transposable element in two very different insects was reported by investigators in Tsukuba, Japan (Yoshiyama et al., 2001). These investigators found a *MLE* in a moth (*Adoxophyes honmai*) and the same *MLE* also resides in the parasite of this moth, the parasitoid wasp *Ascogaster reticulatus*. The moth *MLE* was found in two other moths of the same genus but not in other parasitoid wasps. It was thus suggested that during this parasitic relationship the *MLE* were transferred horizontally from host to parasite. Nature is playing its own games: the *A. reticulatus* "used" the *A. honmai* as a substrate for its eggs but the *A. honmai* reciprocated by transferring a "parasitic DNA" to the wasp.

We detailed above the Class II elements in plants, insects and round worms. This was due to the available information. Such elements are probably abundant in additional organisms. One indication for this abundance is the survey of Robertson (1997) on the many *MLEs* in flatworms and hydras.

We shall leave now the world of Class II elements of protozoa, insects and other invertebrate metazoa (we shall return to the *P* elements of *Drosophila* towards the end of this book), and in subsequent sections we shall deal with Class II elements of higher metazoa such as non-mammal vertebrates and man.

7.7.5. Class II T.E. in Non-Mammalian Vertebrates: Tc1-like Transposable Elements in Amphibia

The discovery of transposable elements in amphibia showed that these elements may jump in jumping frogs (e.g. Garrett and Carroll, 1986; Garrett et al., 1989; Lam et al., 1996a). Well, this is an exaggeration: the DNA sequences in frogs resemble sequences of known transposable elements of the *Tc1* family, but as for jumping – all the frog sequences probably lost this capability many millions of years ago.

During a few years after the discovery of retrotransposons (Class I T.E.) in mammals and other vertebrate metazoa, investigators assumed that there was a division: vertebrates contain Class I but not Class II T.E. As we shall see below, the finding of Class II elements in fishes showed that this assumption was wrong. *Tc1*-like elements in frogs were found in relatively recent years (Garrett and Carroll, 1986; Garrett et al., 1989; Lam et al., 1996a). The frog elements were not revealed by genetic studies but rather by "fishing" for these elements from specific DNA sequences or from the GenBank database. Sequences coding for the transposase of T.E. in zebrafish (*Tzf*) were utilized as "bait". Lam et al. (1996b) found several *Xenopus laevis* (clawed frog) sequences that matched *Tc1*-like sequences. Two of the frog DNA sequences matched well with regions of the transposase-coding sequences of the *Minos* element of *Drosophila* and the *Tzf* element of zebrafish, respectively.

The frog sequences contained defective transposase codes and 220 bp inverted repeats. Two of the *Xenopus* elements differed substantially: they were less than 50% identical. Also, the inverted repeats, though being of about the same length, differed substantially in their sequences. Further search, by PCR and specific probes, revealed additional sequences with homologies to *Tc1*-like transposases. The putative elements were termed *IXz* and *TXr*. The transposase-like sequences were used to construct a possible phylogenetic tree that included *Tc1*-like families of hagfish, zebrafish, frog and salmon. The authors came up with a suggestion that there was horizontal transfer of *Tc1*-like elements among these vertebrates.

Pontecorvo et al. (2000) started their report on a novel *Tc1*-like transposable element in the genome of the water green frog (*Rana esculenta*), with a rather extensive review on the relevance of transposable elements for understanding how genomes function and evolve. They mentioned several phenomena such as the maintenance of telomers in *Drosophila* and gene-expression silencing by RNAi, before they described the elements found in the "edible" (*esculenta*) frog. There is actually a family of elements, termed *R.e./Tc1*. The family consists of tandemly repeated sequences, located at the centromeric regions of the frog's chromosomes. The repeat unit contains a residue of a *Tc1*-like transposon of the fly *Haematobia irritans* (that was mentioned above as *Himar*) bordered by two short direct (target) repeats of 9 bp. The *Tc1*-like remnants are located near a sequence that is identical to the human Werner Syndrome gene. The *Tc1*-like "remnant" is short, its length is about 650 bp. Clearly, these *R.e./Tc1* elements have lost the ability to transpose and cannot jump in the frogs' genome.

7.7.6. Fishing for Transposable Elements in Fishes

The immunoglobin (*IgM*) heavy chain constant region gene of the channel catfish, *Ictalurus punctatus*, was studied in detail (Wilson et al., 1990). During processing the transcript of this gene is undergoing splicing that is different from the splicing found in mammals and in some vertebrates (a shark and an amphibian). The main point for our deliberations is that the sequence of this gene became available. Henikoff and associates (Henikoff, 1992 and see references in this publication) developed a method to search for distant relationships between genome sequences by using multiple protein blocks for the detection of similarities. The method was developed in order to find similarities even between sequences that had "drifted" away during evolution, so that many bases were exchanged and the sequences are no longer recognized by regular search procedures. Hence, Henikoff (1992) used his method to search for new members of the *Tc1* family in several groups of organisms. The above-mentioned *IgM* gene sequence was available, and indeed, such a similarity to the transposase code of the *Tc1* elements was found in the gene of the catfish.

Another sequence with similarity to *Tc1* sequences was revealed by Heierhorst et al. (1992) in the Pacific hagfish (*Eptatretus stouti*). The existence of *Tc1*-like transposons in several fishes was undertaken by Radice et al. (1994). These authors used mainly PCR procedures and thus isolated clones with such *Tc1*-like sequences, with putative ORFs for the transposase. The clones were one from a rainbow trout (*Salmo gardneri*), one from a zebrafish (*Brachydanio rerio*) and three from the Atlantic salmon (*Salmo salar*). They had inverted terminal repeats with conserved terminal nucleotides and a length of about 1.7 kbp. All were "fished" out by the similarity of the derived amino acid sequences to the amino acid sequences of the transposases of the *Tc1* family of transposons. The *Tc1*-like elements were sequenced and a phylogenetic tree of the derived amino acids was constructed. All these *Tc1*-like sequences were derived from one branch. This "tree" and its relation with the "trees", of fruit flies, nematodes and some arthropods, suggested to the authors that these fish elements are very ancient (i.e. possibly rooted in the early protozoa) and were transmitted in vertebrates vertically rather than horizontally. In the "tree", all the salmon-types were on one branch, while the zebrafish element was on a branch that had earlier split-off.

Even more diverse was the branch that led to the catfish sequence. A similar search for *Tc1*-like elements in fishes was concluded by Goodier and Davidson (1994) and at about the same time as the Radice et al. (1994) study. The former authors found such *Tc1*-like elements in salmon, trout and charr. A complete element was sequenced from the Atlantic salmon (*SALT1*). This element was 1,535 bp long and included inverted terminal repeats of 35 bp. The *SALT1* had a degenerated ORF. The derived amino acids of this ORF provided a sequence that contained a D.D(35)E-like motif, similar to the respective motif found in the transposases of *Tc1*-like elements. Lam et al. (1996b) of Harvard's Biological Laboratories in Cambridge, MA, surveyed briefly the *Tc1*-like transposable elements and noted that all these reported elements were degenerated, and none of them was a full-size element with certain transposition capability. These authors assumed that such intact (and active) elements exist but they are a very small minority among defective elements. In order to locate an intact element one has to screen thousands of such elements. They thus devised an efficient screening method to look for intact *Tc1*-like elements in zebrafish (*Brachydanio rerio*, or *Danio rerio*). The search started with a degenerate oligonucleotide probe, corresponding to conserved blocks of amino acids in the *Tc1* transposase. After cloning suspected DNA sequences from the zebrafish genome, they devised a "Two Dimensional Transposon Display" (a special two-dimensional electrophoresis). Finally, *Tzf* was revealed. The *Tzfs* constitute a family of elements (one of them *Tzf-41* seems to be intact). There are many copies of *Tzf* in each zebrafish genome. By identifying *Tzf* elements in a site of an offspring DNA where there was no *Tzf* in the parents, these authors identified putatively mobile *Tzf* elements. Several such cases were recorded and the authors also suggested that *Tzf* transposition causes mutations in zebrafish.

An additional report on "new" zebrafish transposons also came from Cambridge, but this time from Cambridge, UK (where the zebrafish was brought to fame) rather than from Cambridge, MA (Gottgens et al., 1999). I put the "new" in quotation marks because it is new to the investigators but ancient for the zebrafish. This *Tdr2* element was also "fished" from the genome of the zebrafish by using a derived DNA sequence, from conserved *Tc1* transposase regions, as "bait". It should be noted that before the isolation of *Tdr2*, Izsvak et al. (1995) identified a similar element from zebrafish and termed it *Tdr1*. While *Tdr1* is close to the *Tzf* family, mentioned above, the *Tdr2* is on the same phylogenetic branch as the nematode *Tc3A*. The *Tc1*-like elements of fishes served Ivics et al. (1996) to conduct a detailed analysis of these elements. They revealed functional domains in these elements. One of these is a bipartite nuclear localization signal (*NLS*) that is located in the ORF, upstream of the code for the D.D(35)E catalytic domain. A second motif is a "novel" combination of a paired domain-related protein motif, juxtaposed to a leucine zipper-like domain located in the putative DNA-binding regions of the transposases. An additional element with inverted repeat terminals was found in the medaka fish

(*Oryzias latipes*) by Koga et al. (1996) and termed *Tol2*. Whether or not it is autonomous and mobile was not clarified.

Further studies with the transposable elements of the zebrafish (e.g. Ivics et al., 1996) led to a multinational collaboration that can be called the Hungarian Quartet: Z. Ivics and P.B. Hackett of St. Paul, Minnesota; R.H. Plasterk of Amsterdam and Zsuzsanna Izsvak of Szeged, Hungary (Ivics et al., 1997) to "build" an autonomous "fishy" transposon. The idea was to use this transposon in various mammals for experiments such as gene tagging mutagenesis and genetic transformation. It was already known to the authors that *Tc1/Mariner* elements were very modest with respect to host factors. This was made clear by the "trans-kingdom" transfer of the insect *Mos1* element into the protozoa *Leishmania major*, mentioned above. Thus, an element of this superfamily of fishes could probably also be mobile in other vertebrates. The authors therefore chose a zebrafish element (*Tdr1*) as well as several other related elements and started a "molecular reconstruction." In this reconstruction all the suspected mutations were amended until a fully active transposase was reconstructed.

The reconstructed element was capable of catalyzing transposition of engineered, non-autonomous, salmonid elements in fishes and mammals. The artificial element was termed "*Sleeping Beauty*". The "resurrection" of "skeletons" was mentioned in the Introduction of this book. The *Sleeping Beauty* (*SB*) may be considered a resurrected skeleton. The efficiency of the *SB* in mobilizing defective elements was verified in cell culture systems of fish (carp), mouse and man into which the appropriate DNA sequences were introduced. To facilitate the detection of transposition, a selective marker was used: a gene for resistance to the neomycin-analogue G-418. Resistance to G-418 was an indication for transposition. Indeed, such transpositions were verified by molecular means in G-418 resistant colonies of human HeLa cells. Ivics et al. (1997) thus showed that *SB* is a fully functional transposon system that can perform all the complex steps of cut and paste DNA transposition. The nuclear-localized transposase is able to recognize and to excise its specific DNA substrate from an ectopic plasmid and to insert it into chromosomes. Because of its simplicity and apparent ability to function in diverse organisms, these authors suggested that *SB* should prove useful as an efficient vector for transposon tagging, enhancing trapping and transgenesis in species in which DNA-transposon-technology is currently not available. These suggestions were substantiated by a later publication of the European collaborators of the aforementioned study (Izsvak et al., 2000).

7.7.7. Tc1/Mariner-like Transposons in Man and Other Mammals

The existence of *Tc1/Mariner*-like transposons in mammals other than man has not been researched intensively. A relatively early and rather brief report on such elements was provided by Auge-Gouillou et al. (1995) of France. These authors detected the respective elements by PCR-amplification of genomic DNA, using degenerate primers that were based on conserved regions of the transposase code (of *Tc1/Mariner*). They obtained *Mariner*-like element (*MLE*) clones of man. The 5' terminal inverted repeats were further utilized in PCR procedures to obtain full-size and deleted *MLE* elements from mammalian genomes. In this way *MLE*s were detected in man (*Homo sapiens*), mouse (*Mus musculus*), rat (*Rattus norvegicus*), Chinese hamster (*Mesocricetus auratus*), sheep (*Ovis aries*) and cattle (*Bos taurus*). Since these authors did not mention in which mammals such *MLE* were *absent*, we may assume the *MLEs* exist in all, or at least in nearly all, mammal species. Two additional short reports on *MLEs* in man were published in 1995. One was of Oosumi et al. (1995) of the USA, who revealed in the DNA-base of man inverted repeats of 45 bp that had similarity to *Tc1/Mariner* inverted repeats. The human elements were termed *hum1* and they were present in multiple loci of the human genome. Using sequences from these inverted repeats as primers, PCR multiplication of human DNA, specific DNA regions were identified. The latter had a length of 1.3 kbp and were similar to *Mariner* elements. They were termed *humar* (Figure 57). A second family of small, putative non-autonomous elements (*hum7*) led to another full-size (?) human *MLE*.

Two families of Class II elements were detected in the genome of man by Morgan (1995) of Nottingham. Morgan based his findings on computer search procedures in GenBank sequences. One of the families of elements was closely related to the *Mariner* elements, on the basis of the putative code for the transposase. These elements had terminal inverted repeats of 31 or 32 bp. The other family of human Class II elements revealed by Morgan's search, consisted of *SINEs* (i.e. short interspersed repetitive elements). The latter were composed of two inverted repeats of 37 bp that were surrounded by 6 unique bp. The inverted repeats of the latter *SINEs* were very similar to the inverted repeats of the former elements.

With the accumulation of additional DNA sequences from the human genome the computer search for human transposable elements became more fruitful. Smit and Riggs (1996) of Duarte, CA, utilized the sequencing data from the human genome to reveal "*Tiggers*" (*Tigers*?) in this genome. Well, there are various kinds of virtual *Tigers*, such as Paper *Tigers*. The *Tiggers* of Smit and Riggs may be termed *Computer Tigers*, at least in reference to how they were hunted. These authors did not reveal what inspired them to choose this name for these interspersed human repeats. Possibly, they saw a jungle of repeated sequences in the human genome. Anyway, Smit and Riggs focused on two *MERs* (medium reiterated frequency repeats) and named them *Tiggers*. These were found to be related to the *pogo* transposons of *Drosophila* (e.g. Tudor et al., 1992). These transposons were found by database searches. In their publication these authors compiled a comparative list, that included *Ac* (of maize), a frog transposon, *Tc1*, *pogo*, a *Mariner*-like human transposon as well as 14 human *MERs*. The latter could be divided into two main groups. One group of *MERs* (7 *MERs*) had similarity to the *Ac* of maize. This *MER1* group had a target site duplication of 8 bp and TIRs of 12 to 18 bp. None of the *MERs* of this group was a full-size transposon and their length ranged from 190 to 527 bp. The other group (*MER2*) consisted of the two *Tiggers* and 7 additional *MERs*. In all the members of the second group, *MER2*, the target site was TA. It should be recalled that the TA target site duplication is a property of *IS630* and *Tc1*. The TIRs of *MER2* were considerably longer than those of *MER1* (23 bp or more). The length of *Tigger1* and *Tigger2* was found to be 2,417 and 2,708 bp, respectively. The other elements of this group (the *MERs*) were much shorter (234 to 434 bp with three exceptionally longer sequences: some had versions of the same element that were 726 to 1,205 bp long). Various degrees of similarity were revealed, in sequences belonging to *MER2*, to known Class II transposons of insects and other metazoan animals.

Further analyses of the human genome revealed that one of the *MER2* sequences had an ORF that codes for an amino acid sequence that is similar to that of the *pogo* transposase. This was the one termed *Tigger1*. By using the DNA sequence of

Tigger1, Smit and Riggs (1996) found *Tigger2*. *Tigger2* also had similarity (in its code for transposase) to the *pogo* element. The authors made an interesting suggestion that concerns *CENP-B*, the major mammalian centromeric protein. One region of the ORFs of *Tigger1* and *Tigger2* (as well as the ORF of *pogo*) codes for an amino acid sequence that is similar to the sequence of *CENP-B*. Smit and Riggs thus suggested that this is a very ancient sequence and that *CENP-B* is derived from a *pogo*-like transposase. This may be another example of the acquisition of a cellular function by a transposable element. A similar acquisition was mentioned above: the contribution of transposons to the telomeres of flies.

The two *pogo*-like *Tiggers* are really "*Paper Tigers*" or "*Computer Tigers*" rather than real biological entities. They were constructed from databases of human DNA sequences and were not isolated as discrete DNA fragments. Robertson (1996) further analyzed (by computer) the *pogo*-like transposons of man. He made various calculations based on accepted assumptions. One of these assumptions was that in primates the rate of evolution (nucleotide changes) per site is 0.16% every 10^6 years. Robertson thus suggested that *Tiggers* were active 80–90 million years ago, in the genome of an early primate or primate ancestor. One cannot be sure whether or not Robertson is right, but his arguments were nicely presented.

While there was a "flow" of information on "fossil" Class II T.E. in the human genome during the years 1995 and 1996, this flow has subsided in more recent years. Now that the DNA sequences of the human genome are known and are becoming available to the public, we can guess that many more such "fossil" elements will be revealed. Are there also active *Tc1/Mariner* elements in man? One cannot exclude this possibility. What is now sure is that such elements can be introduced into the human genome and become mobile in it. This was shown by the introduction of the *Sleeping Beauty* into this genome, as noted above (Ivics et al., 1997).

The *Sleeping Beauty* (*SB*) could also be used as an *in vivo* mutagen in mice (Luo et al., 1998). This was shown in a study in which mice embryonic stem cells and engineered cassettes for introducing *SB* into the mouse genome, were used. The cassettes were engineered so that cells in which there was an excision of a non-autonomous *SB* (*dSB*), could be selected. This strategy was similar to the strategy used in previous studies on the excision of T.E. The *dSB* was engineered into a gene for puromycin resistance (*puro^R*). When the cassette was integrated into the mouse chromosome, the respective cells were sensitive to the drug – unless the *dSB* was excised. Cell colonies that are resistant to the drug are assumed to be those in which the *dSB* was excised. Such cells can thus be selected and analyzed. When a cassette also contains a sequence for site-specific recombination with a given mouse gene (e.g. *Hpr*), there can be a targeting of the engineered sequence to a required locus. The *dSB* will not be excised unless an *SB* transposase is furnished, in *trans*, to the same cells. The excised *dSB* may reintegrate into another site of the genome. Using molecular analysis of selected mouse cells, Luo et al. (1998) could verify the transposition of *dSB* and look into the details of the excision and the insertion of the *SB* elements. The transposition of *SB* was of the cut and paste type and was similar to the transposition of *Tc1/.Mariner* elements. The frequency of *dSB* transposition in mice was low (3.5×10^{-5} events/cell, per generation), but the authors claimed that this frequency may be increased in the future. In principle, this study clearly indicated that *SB* may be used to establish a general transposon-tagged mutagenesis scheme in mammals. It can thus be claimed that *Sleeping Beauty* is a fairy tale that came true.

We shall depart from the Class II transposable elements of man and other mammals with a question: why are *active* Class II elements abundant in insects and other low metazoa, while such *active* elements are very rare or non-existent in man and other mammals?

8. THE *P* ELEMENT OF *DROSOPHILA*

8.1. Introduction

We mentioned the *P* element of *Drosophila* above when we mentioned *hybrid dysgenesis*. Below, we shall describe this element in greater detail. In the above heading the *P* element is put in the singular rather than in plural. This is by intent. While, when dealing with other elements such as *Tc1* and *Mariner*, I used the plural (e.g. elements or even families of elements), the *P* represents one specific (full-size) element. Although, in addition of the full-size *P*, there are deletion versions of the intact *P*. Moreover, *P* was studied exclusively in *Drosophila* and mainly in *D. melanogaster*, the humble fruit fly that served geneticists for about 100 years. There are probably no other "family-members" of the *P* element in other organisms. As indicated in the Historical Background of this book, this tiny, but mighty, fruit fly was a key factor in the conversion of T.H. Morgan from a disbeliever to a believer in Mendel's laws of inheritance. In this chapter, on the *P* element, we shall witness that this fly retained its major role in genetic studies through the end of the 20th century and the beginning of the 21st century. This fly has even its own website: http://www.fruitfly.org that is maintained at the Department of Molecular and Cell Biology of the University of California, Berkeley, CA. The Bishop/philosopher George Berkeley (1685–1753), in honor of whom the town in California was named, probably never imagined that there would be a *Berkeley Drosophila Project*, that carries his name. G. Berkeley, who authored the "*A Treatise Concerning the Principles of Human Knowledge*" (1710) and the fruit fly, each contributed to human knowledge in very different ways, but now they both name the same project.

The *P* element, like several other important transposable elements, was revealed by genetic studies, rather than by "mining" the databases of sequenced genomes. Soon after its discovery, molecular approaches were joined to the genetic studies, leading to outstanding molecular genetic research and to most interesting discoveries (see review of Rio, 2002).

8.2. The Discovery of the P Element

Genetic instability in the progeny of the matings of *Drosophila melanogaster* flies from different strains was already observed in the first half of the 20th century. One such observation was made by Sturtevant (see Sturtevant and Beadle, 1939), who was a pioneer of *D. melanogaster* genetics. Incidentally, this author (E.G.) witnessed A.H. Sturtevant, in 1961, at the California Institute of Technology, scoring flower pigmentation, in a flower field, together with his wife. Sturtevant was color-blind, therefore, the help of his wife was essential. Genetical peculiarities resulting from unidirectional mating between *D. melanogaster* flies of different strains were observed many years ago. To give just one example, the study of

Hiraizumi (1971) will be mentioned. Usually there is a lack of genetic recombination in *D. melanogaster* males. Hiraizumi found that after mating male flies from natural populations with specific laboratory female flies, there was abundant recombination in the male progeny. While this observation was worthy of a publication in the *Proceedings of the National Academy of Sciences (USA)*, no possible involvement of a transposable element was suggested by Hiraizumi. Such a suggestion had to wait 6 more years, when Kidwell and associates and Picard and associates (e.g. Kidwell et al., 1977; Picard et al., 1977) published on the phenomenon termed *Hybrid Dysgenesis*.

The phenomenon of *hybrid dysgenesis*, that was previously noted in this book, shall be described in more detail below. Kidwell et al. (1977) who coined this term, described it as a *syndrome*, which includes mutations, sterility, recombinations, transmission-ratio distortions, chromosomal aberrations, local increase of recombination in females and non-disjunction. To describe hybrid dysgenesis, Kidwell et al. (1977) partitioned the *Drosophila* strains into two types: P and M, according to the paternal and maternal contribution that is required to lead to this syndrome. The paternal P was derived from male flies that were collected in recent years from the wild, while the M maternal parents were flies maintained for many years as laboratory stocks. The hybrid dysgenesis syndrome was typically observed in progeny resulting from the mating of P fathers and M mothers. Before the P was defined as a specific sequence of DNA it was described as the P factor. The laboratory stocks were said to have an M cytotype and the natural stocks were said to have the P cytotype. The real meaning of these cytotypes came to light only after molecular-genetic studies were applied to the hybrid dysgenesis phenomenon. While the cross $P_{females} \times M_{males}$ resulted in normal progeny, the reciprocal cross, $M_{females} \times P_{males}$ resulted in progeny that showed a range of surprising phenotypes that were collectively called hybrid dysgenesis. Dysgenic flies were often sterile. When fertile, their offspring showed abnormalities, such as high levels of mutation and chromosomal rearrangements, that were passed to successive generations. Moreover, some of the mutations were quite unstable in dysgenic flies and reverted to wild-type, or to altered alleles, at high rates. Also, the peculiar chromosomal recombinations appeared in dysgenic male flies. When these genetic observations were recorded, there was already an awareness of transposable elements. Therefore, a search for the involvement of a "mobile element", in an allele affected by hybrid dysgenesis, was initiated. The *while* (w^+) locus was a good target for genetic studies. Hence, molecular cloning of an allele that was mutated by hybrid dysgenesis was initiated (e.g. Bingham et al., 1982; Rubin et al., 1982). The intensive genetic studies on hybrid dysgenesis during the years 1977 to 1983 were reviewed by Engels (1983; 1996) and by Rio (2002).

8.3. The Structure of P and its Truncated Versions

The detailed structure of P and its truncated versions were investigated by O'Hare and Rubin (1983). These authors used the library of random DNA fragments of a P strain (π_2) in the bacteriophage λ vector (Charon 28), and screened this library with fragments of DNA that contained the mutated *white* ($w^{\#6}$ and $w^{\#12}$). The positive phages were counterscreened with fragments containing the wild-type *white*. The results indicated that this P strain contained about 50 P elements. The full-size P elements had a length of 2.9 kbp and most of the smaller P elements were 0.5–1.6

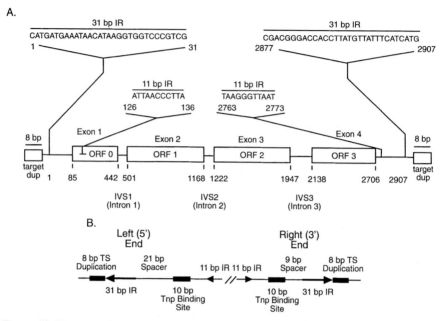

Figure 58. Features of the complete 2.9 kb P element. (A) Sequence features of the 2.9 kb
P element. The four coding exons (ORF O, 1, 2 and 3) are indicated by boxes with
nucleotide numbers shown. The positions of the three introns (IVS 1, 2, 3) are indicated
below. The DNA sequences of the terminal inverted 31-bp repeats and the internal 11-bp
inverted repeats are shown, with corresponding nucleotide numbers shown above.
The 8-bp duplications (dup) of target site DNA are shown by boxes at the ends of the
element. (B) cis-acting elements of P element transposition. Key sequence features of the
left (5') and right (3') end are indicated. The terminal 31-bp inverted repeats and the
11-bp internal inverted repeats are indicated by arrows. The transposase binding sites
and 8-bp target site duplications are indicated by boxes. The distinct spacer lengths
between the 31-bp repeats and the transposase binding sites, 21 bp at the 5' end and
9 bp at the 3' end, are indicated above. (From Rio, 2002).

kbp. The smaller *P* elements all appeared as the full-size element from which
middle-sequences of different sizes were eliminated. The full-size 2.9 kbp was first
identified by comparison of the normal *white* locus with a mutated *w*, in restriction-
site maps (using several endonucleases). It became evident that the mutant allele was
2.9 kbp longer than the wild-type allele due to an insert in the mutant. It also
became evident that the *P* strain contained multiple copies of this 2.9 kbp insert. At
the level of restriction enzyme mapping, it seemed that all the full-size inserts (*P*
elements) were either identical or highly conserved.

One of the clones that contained the insert (Pπ 25.1) was propagated and fully
sequenced. The total length was 2,907 bp. At the terminals there were inverted
repeats of 31 bp. A more detailed scheme of the full-size P elements resulted from
several additional studies and is presented in Figure 58. The *P* has four ORFs
(termed, ORF 0, ORF 1, ORF 2 and ORF 3) and the derived RNA has four exons:
Exon 1, 2, 3, & 4 (why ORF 0 is for exon 1 was not made clear by *P* element

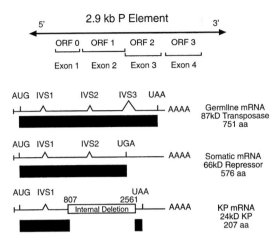

Figure 59. P element mRNAs and proteins. The 2.9-kb P element and four exons (ORF O, 1, 2 and 3) are shown at the top. The germ-line mRNA in which all three introns are removed, encodes the 87-kDa transposase mRNA. The somatic mRNA, from which only the first two introns are removed (and which is also expressed in germ-line and somatic cells), encodes the 66-kDa repressor mRNA. Shown at the bottom is a KP element, which contains an internal deletion. This truncated element encodes a 24-kDa repressor protein. (From Rio, 2002).

scholars). Flanking the elements on both the left (5') and the right (3') terminals (i.e. the 31 bp inverted repeats), there are 8 bp of target duplications. Inside the element, in ORF 0 and downstream of ORF 3, respectively, there are 11 bp that are termed internal inverted repeats. These internal inverted repeats are located about 100 bp inside of the terminals of the element (Figure 58).

From the sequencing results of O'Hare and Rubin (1983) it became evident that *P* is different from all the previously identified Class II transposable elements. The full-size elements are all highly conserved and those that are smaller are variable in size (e.g. 501 bp or larger) and were considered to have been produced by deletion of internal sequences of the 2.9 kbp elements; but the end-sequences of the smaller elements were identical to those of the full-size element. About one third of the 50 to 60 *P* elements of the fly genome were full-size elements.

8.4. Transcription and Translation of P

The coding sequence of the *P* element spans from bp no. 85 to bp no. 2,706 (Figure 59). It is composed of four ORFs, as noted above. The three introns are termed, respectively IVS 1, IVS 2 and IVS 3. When all three introns are spliced, an mRNA is derived that encodes an 87 kD protein, which is the *P transposase*. However, when IVS 1 and IVS 2 are spliced and IVS 3 is not spliced, the resulting mRNA contains a stop codon (UGA, in IVS 3). Thus, a shorter protein, of 66 kD, is translated. This 66 kD protein is the *repressor* of transposition. The transcript, splicing sites and translation products are shown schematically in Figure 59. The splicing of IVS 3 is thus of prime importance: when this splicing takes place, the transposase (87 kD protein) will be synthesized, while without this splicing the repressor (66 kD protein) will be synthesized. Some additional details about the regulation of this splicing will be provided below, but it should be noted here that the splicing of IVS 3 is confined to the germ-line nuclei; hence, the transposase is produced only in the germ-line cells. In somatic cells (as well as to a large extent in germ-line cells) the primary transcript of *P* is not spliced at IVS 3. Consequently, the mRNA in which only IVS 1 and IVS 2 were spliced will be exported from the nuclei to the cytoplasm, and there cause the synthesis of the repressor. The latter is termed *Type I repressor*. Other truncated proteins were also revealed and these may also act as repressors of transposition (e.g. *Type II repressor*). One of the latter proteins is the KP protein that was found in some natural fly populations that had an internally deleted *P* element.

Informatics analyses indicated that insertion sites of *P* elements into the target DNA tended to be CG rich, and that there is a symmetric palidromic pattern, of 14 bp, centered on the 8 bp target duplication. The consensus structure of the favored insertion site is 5'-GGCCAGAC-3'.

8.5. Transposition of P

8.5.1. Additional Information on Hybrid Dysgenesis

Before handling the process of *P* transposition, we should consider some additional aspects of hybrid dysgenesis. As noted above, the syndrome of hybrid dysgenesis is caused by the introduction of *P* elements that are carried on the chromosomes of the paternal parent into the environment of an *M* strain egg that is devoid of *P* elements. Thus, both the *P* elements in the sperm and a *P*-free maternal environment are required to yield hybrid dysgenesis. This already suggested that the presence of *P* in maternal chromosomes will repress *P* transposition. Without *P* transposition there will be no hybrid dysgenesis.

Indeed, the high rates of mutations and reversions are largely the results of insertion and excision of *P* elements. The insertions appear to be confined to germ-line cells. The rate of insertion can be high – about one insertion per chromosome in each generation of flies. Insertions and excisions of *P* in progeny resulting from mating *P* males with *P* females (non-dysgenic progeny) are much less frequent. For the induction of hybrid dysgenesis at least one full-size *P* element (in the male parent) is required. This *P* element can mobilize deleted *P* elements that are in the same genome by *trans* activation.

A not yet fully understood aspect of hybrid dysgenesis is the transmission of the P cytotype. For example, there is a *pre-P cytotype* state. This state results from crosses that involve *Lk-P* (1A); consequently there is a single generation of repression even in the absence of the inheritance of a full-size *P* element DNA. How much repressor (66 kD) protein and repressor mRNA can enter the eggs? Is the elimination of transposase mRNA affected by an RNAi mechanism? Such questions are still not fully settled. In the *Lk-P* (1A) strain there are two full-size *P* elements inserted into the sub-telomeric telomere-associated sequences (*TAS* elements). The *TAS* repeats are 0.8 to 1.8 kbp sequence elements consisting of shorter internal repeat motifs that are often found near the ends of all *Drosophila* chromosomes. Mutations in the heterochromatin protein 1 reduce the ability of the *P* elements of *Lk-P* (1A) to mediate *P* cytotype. This suggests a role of the sub-telomeric heterochromatin in the cytotype effect.

8.5.2. Structural and Biochemical Aspects of P Transposition

Several experimental approaches showed that the two 31 bp terminal inverted repeats are required for transposition, and that the internal 11 bp inverted repeats function as transpositional enhancer motifs. Moreover, the 5' and 3' 31 bp at the termini are not precisely inverted repeats. Consequently, when an artificial *P* element with two 5' termini rather than a 5' terminus and a 3' terminus was tested, it did not transpose. Nevertheless, for transposition there should be a stage of pairing of the two (not fully perfect) inverted repeats. However, it must be assumed that the transposase of *P* can distinguish between the two 31 bp terminal inverted repeats.

Biochemical, genetic and molecular approaches revealed several domains in the transposase. At the amino-terminus end there is a C_2 HC putative zinc-knuckle motif that, in conjunction with an adjacent basic region, constitutes the site-specific DNA-binding domain. A little towards the carboxyl-end there is a dimerization region (with a "leucine-zipper" motif). Between the latter region and the DNA-binding motif are potential phosphorylation sites (S/TQ). Changes in the sequence of these sites affect the transposition activity. *In vivo* the transposase protein may be phosphorylated on the amino acids serine, threonine and tyrosine. It seems that the removal of phosphates from the transposase reduces its activity. Indeed, when the genomic DNA sequence of *Drosophila* (Adams et al., 2000) was computer-analyzed, two PI3-related protein kinases were located on this genome.

The transposase has also a GTP binding region. The role of GTP is probably that of a co-factor for *P* element transposition (rather than a contribute of phosphate). Alterations in the GTP binding region were found to abolish transposition. The carboxyl-end of the transposase (e.g. that coded by exon 4 (ORF 3) contains a high proportion of acidic amino acids. This is most probably the catalytic region. It belongs to the superfamily of polynucleotidyltransferases (e.g. transposase of bacteriophage *Mu* and bacterial Insertion Sequences, retroviral integrases, certain RNases and certain recombinases). All these enzymes use metal ions (usually magnesium) for the catalysis of phosphodiester bond hydrolysis. Although sequence-wise there is no similarity between the carboxyl-end of the *P* transposase and catalytic enzymes of the D.D(35)E superfamily, there are indications of similar functions. Mutagenesis of acidic amino acids in the carboxyl-end of *P* transposase in which these amino acids were replaced by alanine, markedly reduced transposase activity, both *in vitro* and *in vivo*. Albeit lack of sequence identity between the catalytic region of *P* transposase and the D.D(35)E superfamily of enzymes does not exclude overall three-dimensional structure similarity. Even within the D.D(35)E superfamily the amino acid sequence similarity can be rather low (e.g. 18% identity).

When the *P* transposase was purified, its binding sites to the *P* element could be revealed: it was found to bind near the ends of this element, between the terminal 31 bp and the internal (11 bp) inverted repeats. It is noteworthy that while the terminal 31 bp are essential for transposition, they do not constitute binding sites of the transposase. The binding site at the 5' end of *P* overlaps the TATA box of the minimal *P* element promoter. This strongly suggests that transposition and transcription of the *P* element are mutually exclusive. For further acquaintance with the transposition, Kaufman and Rio (1992) developed an *in vitro* system in which transposition events between two specific plasmids could be followed in bacteria. One plasmid (the donor DNA plasmid) contained a *P* element carrying the bacterial tetracycline-resistant gene and was unable to replicate in normal *E. coli* strains. The recipient plasmid was a standard ampicillin-resistant plasmid. Transposition events were scored after transformation of the reaction products into *E. coli* and selection for tetracycline and ampicillin-resistant colonies. Indeed, the transfer of *P* from one plasmid (donor) into the other (recipient) was revealed. Other aspects of the transposition could be analyzed, such as GTP, serving as co-factor but not as supplier of high energy. It was also found that 3'-hydroxyl groups, at the ends of cut *P* elements, are required for transposition, meaning that these groups are used (as in other transposition systems) as a nucleophile during strand transfer, indicating that *P* transposition proceeds by a cut and paste mechanism. The cleavage of the *P* from the donor DNA is rather unique. The 3' ends of the transposon DNA are cleaved (as in other T.E.) at the junction of the element and the flanking target repeats. This cleavage yields the aforementioned 3'-hydroxyl group. However, the cleavage at the 5' ends is different. At these ends the cleavage is 17 nucleotides into the 31 bp inverted repeats. This generates novel 17 nucleotide 3' extensions on both the excised transposon and the flanking donor DNA. The generation of these unique 3' extensions may facilitate non-homologous end-joining (NHEJ) of the donor DNA after *P* excision. Furthermore, by further *in vitro* studies it became evident that the full-length 17 nucleotide (single-stranded) extensions are essential for *P* element integration. Shortening these extensions drastically reduced the integration (e.g. Beall and Rio, 1998).

Transposition of *P* is preferentially to nearby sites, meaning to other sites on the same chromosome and commonly within 100 kbp of the original site. After the development of homologous recombination in *Drosophila*, and based on the now known full-sequence of the genome of this fly, one can use the *P* and direct it to any required location rather than the previously required "walking" with "steps" of about 100 kbp from one insertion to another. On the other hand, there are some "hot-spots" for *P* in the fly's genome, such as the *singed* gene (on the X chromosome). What the characteristics of such "hot-spots" are is not yet clear. It is also not clear why the transposition of *P* in the germ-line cells of females is more frequent than in germ-line cells of male flies. Whenever *P* is inserted into a gene it is commonly located upstream of the transcriptional initiation site. Again, there is no definite explanation for this location but the assumption was made that the chromatin modulation (see Appendix II) at the promoters of genes is involved in this localization.

We noted that the 66 kD protein and the KP proteins repress transposition of *P*. What is the mechanism of this repression? While the full answer to this question is not yet available, one plausible answer came from the KP protein. This protein does contain regions for binding to the DNA of *P* but lacks functional catalytic properties. It is assumed that the binding of KP competes with the binding of the full-size (87 kD) transposase, thus blocking the cleavage of the *P* elements from its donor DNA.

8.5.3. Transposition and Gap Repair

Experiments were conducted in which the transposition from a defined locus (*white*) and the subsequent repair of the excision were followed. Such experiments showed that when the chromosome that contained the *P* had a normal homologous chromosome in the same genome, then the "precise" excision was elevated 100-fold. This strongly suggested that the wild-type homologous chromosome served as template during the gap repair. This mechanism fits the cut and paste process that was analyzed in detail in some bacterial Insertion Sequences (Tn*10* and Tn*7*). On the other hand, if templates are required for gap repair, then applying an ectopic (transgenic) template should enable the insertion of alien sequences into the gap. This seems to be the case. About 30 bp of homology are required for template use in such "alien" gap repairs. Actually, the gap repair can proceed by either the NHEJ process or by another process of strand-invasion and extension. The mode of gap repair can lead to several chromosomal aberrations and even to the restoration of a new copy of *P* at the donor site. The latter phenomenon can result from the SDSA pathway of repair. In the SDSA pathway, the cleaved P elements ends "search" for a homologous sequence. If found, this sequence serves as template for repair. When the "found" sequence is a sister chromatid or a homologous chromosome with a *P* element and the repair synthesis is interrupted, this could result in an internally deleted *P* element. Such deleted *P* elements were frequently found in wild fly populations. If the repair is not interrupted, a copy of *P* will be produced. The latter phenomenon could explain the multiplication of *P* elements.

8.6. Regulation of Transposition

8.6.1. P and M Cytotypes
The interpretation of early genetic studies on the inheritance of *P* and *M* cytotypes was complicated because the strains contributing the *P* elements had many such elements in their genomes. As indicated above, flies collected from the wild contain about 30–50 full-size and truncated *P* elements. The interpretation was made much easier when fly strains with only one or a very few *P* elements were isolated and used in such studies. One strain of the latter type is *Lk-P* (1A). This strain contains two full-size *P* elements at the tip of chromosome X, at a location termed 1A. When transposed, the *P* elements of *Lk-P* (1A) tend to insert into sub-telomeric heterochromatin, a region where there are *TAS* repeats (mentioned above). Although there are only two *P* elements in this strain, these elements manifest a strong *P* cytotype (repressive) effect, similar to the effect in strains having many *P* elements. The *Lk-P* (1A) produces very little transposase. When one of the two *P* elements excises, there is a loss of *P* cytotype strength. Either gonadal dysgenic sterility (GD) or *singed-weak* (sn^W) hypermutability were used to assay *P* cytotype regulation.

It was revealed that *P* cytotype repression in *Lk-P* (1A) was strong in the germline tissue, and this repression was transmitted maternally. In the somatic tissue, the repression was weak. In normal *P* strains, the repression is strong in both the germline tissue and the soma. Maternally-derived cytoplasm from a *P* strain female will confer repression of *P* mobility to offspring for a single generation. For the inheritance of the *P* cytotype in subsequent generations, the presence of maternally-derived *P* elements are required.

However, when the cytoplasm of a *P* strain female is transmitted without the chromosomal *P* element, P cytotype can be maintained, provided the chromosomal *P* is contributed by the paternal parent. It appears that eggs derived from *P* strain females carry an extrachromosomal determinant that was termed pre-*P* cytotype. This pre-*P* cytotype is functional only in the presence of a chromosomal *P* element (provided by either the male or the female parent). The exact nature of this pre-*P* cytotype is still a mystery but it does not correlate with the 66 kD repressor protein. Maternal genomic imprinting was suggested as a possibility; but what exactly is this imprinting?

There is experimental evidence that in *P* cytotype there is transcriptional repression of the *P* element promoter in the germ-line cells and this repression reduces the splicing of IVS 3. Once not spliced, the 66 kD rather than the 87 kD transposase will be produced. The 66 kD repressor could then be transmitted from the germ-line cells, where it was produced, to the unfertilized egg cells and there inhibit the transposition (by occupying the transposase binding sites on the *P* element) of a *P* element that enters the egg with the sperm. When the repressor binds to the binding regions of the DNA of *P*, it will compete with the binding of the transposase to these regions and thus reduce or eliminate transposition. This is post-transcriptional repression. On top of these elaborate interactions there may be an involvement of interaction of chromatin modulation. The small repressor element, KP (207 amino acids), contains a DNA binding domain and a dimerization domain that are the same domains found in the 66 kD repressor. Hence, KP can also bind to the *P* element and inhibit transposition. There are additional truncated *P*-like elements that may interfere with transposition.

8.6.2. Transposition Control by Antisense and RNAi

The control of transposition of the *P* element by antisense and/or by RNAi eluded several studies. It is possible that certain double-stranded RNAs that are homologous to the mRNA will cause (via the RNAi system) a specific degradation of the *P* element mRNA. A guide to recent literature on RNAi is provided below in Appendix I.

8.6.3. Pre-mRNA Splicing

How introns are recognized, and how splicing at specific nucleotides of the primary transcript (the pre-mRNA) is processed, are central issues in molecular biology. Several components of the splicing process are already clear. Thus, the existence of spliceosomes is well known and they are rather conserved in eukaryotic cells. However, by observing a given nucleotide sequence, it cannot always be predicted where exactly the RNA will be cut and spliced. The *P* element system was useful in solving some of the riddles of splicing control. Nucleotides in the exon that is upstream of the IVS 3 were identified as affecting the splicing. It was also reported that a "splicing extract" from mammalian cells did splice the IVS 3, but a "splicing extract" from somatic *Drosophila* cells was unable to perform this splicing. As shown schematically in Figure 60, the process of IVS 3 splicing (and most probably of other introns) is an elaborate process that requires the binding of several specific proteins to the pre-mRNA. Moreover, the splicing complex has to be assembled in a proper sequential order.

A summary of the presently known information on the involvement of the different proteins in the splicing of IVS 3 is provided by Rio (2002) and the control of pre-mRNA splicing was reviewed by Smith and Valcarcel (2000). Obviously, the availability of appropriate proteins in the nuclei of specific cells should have a decisive impact on the efficiency of the splicing. We should therefore not be surprised that the same splicing is effective in one type of cells (e.g. germ-line), while not effective or even missing in other cells (e.g. somatic cells). Conversely, in nuclei of certain cells there may be inhibitory factors that interfere with the assembly of the splicing complex, and thus reduce the splicing.

In summary, the *P* element provides a very useful system to study the control 5'

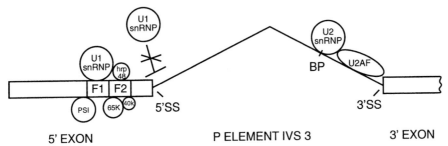

5' EXON   P ELEMENT IVS 3   3' EXON

Figure 60. Model for somatic inhibition of IVS splicing. U1 snRNP usually interacts with the IVS3 5' splice site during the early steps of intron recognition and spliceosome assembly. In somatic cells (and in vitro) this site is blocked. Mutations in the upstream negative regulatory element lead to activation of IVS3 splicing in vivo and in vitro. The F1 site binds U1 snRNP and the F2 site binds the hnRNP protein, hrp48. An RNA-binding protein containing four KH domains that is expressed highly in somatic cells, called PSI, has also been implicated an IVS3 splicing control. Several other uncharacterized proteins from Drosophila extracts have been identified by UV cross-linking to the IVS3 5' exon. (From Rio, 2002).

splice site selection and, in more general terms, to reveal the control mechanisms for alternative pre-mRNA splicing.

8.7. The Hobo Elements

There are similarities between the *P* element and *hobo*. *Hobo* is also a transposable element of *Drosophila*, and as revealed in *P*, old laboratory strains of *D. melanogaster* do not host *hobo* elements but fruit flies collected from nature, in recent years, do contain *hobo* elements. Also, like in *P*, there is hybrid dysgenesis in *hobo*. The first isolation and molecular characterization of *hobo* was reported at about the same time as the molecular characterization of *P*, but the genetics of *hobo* received much less attention. *Hobo* was regarded in the *Drosophila* genetics literature as a kind of humble relative of *P*. Thus, the *hobo* elements are described here, briefly, as a section under the chapter on the *P* element. An early comprehensive review on *hobo* was provided by Blackman and Gelbart (1989).

As in *P*, the *Drosophila* strains are divided into H strains that contain *hobo* and E strains that lack intact *hobo* elements. The full-size *hobo* has a length of 3 kbp. There are also (as in *P*) truncated versions of the full-size element.

There is a variability in the number of *hobo* elements in the genomes of different *Drosophila* strains. In some strains the number ranges from 2 to 10, while in other strains there may be between 30 and 50 *hobo* elements. Truncated *hobo* elements are apparently frequently amplified, probably by the action of transposase encoded in full-size *hobo* elements that reside in the same genome. Not only *D. melanogaster*, but also several other *Drosophila* species contain *hobo* elements.

One *hobo* element that was fully sequenced (*hobo108*) has 3,016 bp, including 12 bp of terminal inverted repeats. As in *P*, *hobo* produces an 8 bp target-site duplication, but the 8 bp of *hobo* differ from those of *P*.

Crosses of H-males with E-females caused high rates of mutations, but (unlike in *P*), the progeny of crosses between E-males and H-females also induced mutations in the germ-line, although at a lower level. There are also indications that *hobo* can be used as a vehicle for transforming germ-line cells, in analogy to the findings in *P*.

Hobo was included in the *hAT* superfamily of Class II transposable elements (e.g. Galindo et al., 2001). The *hAT* stands for three groups of transposable elements: the *hobo* of *Drosophila*, the *Ac/Ds* of maize and the *Tam3* of *Antirrhinum majus*.

As was found for the *P* element, the tracing of *D. melanogaster* strains that were collected from nature on different dates differ in their content of full-size ("canonical") *hobo* elements. Hence, all the strains collected before 1960 did not contain full-size *hobo* elements, while strains collected in later years did contain canonical *hobo* elements.

8.8. Use of the P Element in Drosophila Genetics

Relative to other eukaryotic transposable elements, the *P* element undergoes frequent excision and insertions. The awareness of this character of the *P* element initiated interest in the utilization of *P* for genetic manipulation in *Drosophila* since the early years after the *P* was identified. On the other hand, unlike the *Sleeping Beauty* (artificial) transposon that is mobile in many organisms, the mobility of *P* is restricted to *Drosophila*.

One important utilization of *P* is to apply it for germ-line transformation. In this procedure an alien gene is integrated into an internally deleted *P* element and this construct is engineered into a vector plasmid. In parallel, another (helper) plasmid is engineered that contains the DNA of a *P* element that encodes the transposase but lacks the terminal ends of the *P* element. The two plasmids are injected into the preblastoderm of a fly embryo at the posterior pole, where the germ-line cells will form. The recombinant *P* elements will be mobilized from the injected plasmids to the *Drosophila* chromosomes in the pole cells. In the next (sexual) generation, the DNA sequence that has P termini and an alien gene between these termini, will start to move, if the fusion-gene that encodes the transposase, co-segregates with the alien gene. However, in the absence of the transposase-coding fusion-gene, the integrated sequence with the alien gene will be stable at the locus of integration. If one intends to cause the alien gene to change its location, the fly that carries this gene should be crossed with another fly that produces *P* transposase.

Obviously, when a *Drosophila* line with a single (full-size) *P* is available, this line can be crossed to another line. Insertions into specific genes can then be phenotypically recognized. In fact, many lethal mutations were induced and characterized in this manner. Further "hopping" of an inserted *P* can mutagenize nearby genes in a process of "walking" or "local hopping" along the chromosome. Moreover, PCR methods based on the known sequence of *P* on the one hand, and the DNA sequence of a chromosomal site on the other, can be used to identify a mutated gene. Again, this can be combined with DNA hybridization and plasmid rescue for further analysis. Obviously, all these approaches became even more efficient with the completion of the sequencing of the genomic DNA of *Drosophila*. As indicated above, *P* frequently inserts upstream of the coding region of genes. This may cause, in some cases, only mild phenotypic changes. However, further transpositions may reveal more severe effects on the same gene.

When a *P* element transposase promoter (that is a weak promoter) is fused to a marker gene (e.g. the *lacZ* of bacteria), insertions downstream of a strong promoter, that is functional in specific tissues, at specific developmental stages, or under specific conditions, can be revealed. This method was termed *Enhancer Trapping*. The enhancer trapping can be combined with the use of a pigmenting gene and then specific "illuminated" cells can be isolated by a cell-sorter.

When appropriate tools (from yeast) were used, the integration of *P* could be aimed to specific "homologous" sites, by homologous recombination. These and several other utilizations of *P* elements, in the advanced studies of *Drosophila* genetics, were reviewed by Rio (2002). In fact, the *Berkeley Drosophila Genome Project* uses the above-mentioned methods to further analyze the genetics of this fruit fly model. This effort at Berkeley is a fine example of the combination of a transposable element, acquaintance with the genomic sequence and use of advanced molecular genetic methodologies, that led and is continuously leading to increased *Human Knowledge*. We return to George Berkeley and to his book on the *Principles of Human Knowledge*.

CHAPTER 7

EPILOGUE

We now have a rather broad and detailed knowledge of the phenomenon termed Transposable Elements (T.E.), but a full understanding of T.E. has not yet been achieved. Philosophically, whether or not we shall ever reach a full understanding of the "real", is under debate. Plato (~428 to 347 BC), in his *Republic* (Book VII) provided the *Analogy of the Cave*. The philosophically unenlightened are represented in this analogy as prisoners, chained from birth in an underground cave, their heads cannot turn and they are able to see nothing but moving shadows that are silhouetted by a fire behind them. The prisoners take the moving shadows as the reality. Unless the prisoners find a way to virtually escape from this cave, they will not be enlightened. There are "steps" in this "escape". The prisoners may use empirical experience in order to understand that their vision does not represent "reality". There is a further level of escape based on pure logic. Can we ever truly perceive an entity that is sensed by us? For example, can we perceive additional dimensions beyond the three dimensions in which we experience our "daily" life? George Berkeley (1685–1753) strongly denied the existence of any "real-matter". An entity may be regarded by us as having attributes such as colors, scents, tastes, etc. that we can perceive with our senses. Is this complex of perceptions a full manifestation of "real-matter"? Berkeley insisted that all that there is, is merely the compilation of our senses. Let us assume that the sense of vision did not exist, would we be able to have the same perception of our surroundings as we have *with* vision? The answer is probably: no. Now, by analogy, let us assume that there is an additional sense that we do not possess. If we can use this additional sense, will the perception of our surrounding world be different? The answer is probably: yes!

From such philosophical elaborations let us come "down to earth" – to the understanding of transposable elements. Instead of "senses" we shall deal with scientific "tools" such as genetics, cytogenetics and molecular biology. When Gregor Mendel started his investigations on inheritance, he possessed only a keen capacity for observation, an ability to formulate thought in mathematical equations and basic botanical knowledge. With these "modest" tools he was able to release himself from the "chains" of Plato's cave, come up with the basic rules of genetics and make sense of seemingly chaotic data. On the next level, Barbara McClintock started with baffling phenomena and with only the "tools" of genetics and cytogenetics came to the revolutionary conclusion on the existence of "Controlling Elements" that were later termed T.E. She was also capable of "freeing herself from Plato's cave" and her conclusions were later verified by molecular genetics methodologies. With the passing years, additional biological "tools" became available and molecular techniques were rendered more efficient. Additional systems and families of T.E. were revealed and the understanding of their mechanisms was increased, but we still surely do not possess a full understanding of T.E.

In the preceding text, we noted repeatedly that some phenomena are still enigmatic. To illustrate this point, let us deal with one issue: the epigenetic regulation of transposition of one system: the *MuDR/Mu* family. Clearly, there was a correlation between certain DNA methylations and transposition frequency. But does this provide a full explanation? In recent years, two mechanisms have been revealed: *RNA silencing* and the *histone code*. It emerges that organisms possess an efficient way to destroy RNAs having defined sequences of about 23 (or more) nucleotides. I briefly describe this RNA silencing in Appendix I.

The *histone code* has also been revealed only in recent years. This subject is summarized in Appendix II. It is plausible that both these mechanisms are involved in the epigenetic regulation of T.E. and, with the elucidation of the involvement of these mechanisms, the understanding of the biology and genetics of T.E. will rise to a higher level. We can visualize increased biological understanding by a ladder composed of rungs of increasingly new and better methods ("tools"); the scientist is using this ladder to reach greater heights of understanding, the exceptionally enlightened scientists may "fly" to higher levels even without some of the rungs. Clearly, the biblical ladder of Jacob comes to mind. However, there is a difference, in the biblical ladder angels were descending from heaven while in "our" ladder the scientists are ascending to greater heights. We expect improved rungs and hope for enlightened investigators to cause a better understanding of Transposable Elements.

CHAPTER 8

APPENDICES

All the three subjects of the following appendices, *RNA Silencing*, *Chromatin Remodeling* and *Gene Tagging* are related to transposable elements. These subjects are presented as appendices because a full description of these subjects will render this book too voluminous. Moreover, these are either rather complicated issues (e.g. RNA silencing and chromatin remodeling) or an issue (gene tagging) that is of great interest to a limited group of investigators. Furthermore, RNA silencing and chromatin remodeling are recent fields of endeavor and new reports on these subjects are constantly added. Thus, updating these subjects in a book is not feasible. The appendices will therefore lead the interested readers to the relevant literature, rather than provide all the pertinent information. The subjects of RNA silencing and chromatin remodeling are covered in two separate appendices. Indeed, up to 2001, these were handled in the literature as separate issues, but recently, there has been an abrupt change. RNA interference (RNA silencing) and chromatin silencing (e.g. by the methylation of lysine 9 in histone H3) were found to be interconnected in the fission yeast, *Schizosaccharomyces pombe* (Allshire, 2002; Volpe et al., 2002). It is plausible that the model of chromatin silencing via RNA silencing developed for fission yeast will be applicable to other eukaryotes. While readers should keep this interrelationship in mind, I have decided to present RNA silencing and chromatin remodeling as two distinct appendices.

1. APPENDIX I: RNA SILENCING

Sequence-specific RNA silencing is a process that was revealed, since 1998, in many eukaryotes and probably exists in all these organisms (unless special mutations inhibit this process). It was termed in plants *post-transcriptional gene silencing* (*PTGS*), in animals *RNA interference* (*RNAi*) and in fungi *quelling*. It was first described in detail by Fire et al. (1998) for the nematode *Caenorhabditis elegans* and subsequently the "gospel" was spread like "wildfire" to other organisms that serve as models for molecular-genetic studies.

In retrospect, "gene silencing" and "RNA silencing" was a phenomenon that was reported repeatedly by investigators long before the publication of Fire et al. (1998), but the molecular mechanism(s) behind this phenomenon were obscure. In plants there were indications for it following virus infection (e.g. Lindbo et al., 1993) and following the introduction of alien genes (especially alien promoters) by genetic transformation. There were also abundant reports on silencing involved in the transposition of transposons and retrotransposons. We shall provide here only references to investigations that followed the study of Fire et al. (1998) – with one exception: the transmission of the silencing across graft connections in plants.

Palauqui et al. (1997) reported on the transmission of a systemic acquired silencing that is transgene specific, from a stock to a scion (i.e. across the graft) in tobacco. The "silencing" moved great distances without losing its efficiency. We may even go back much further. About 50 years ago, Frankel (1956) reported the transmission of cytoplasmic male sterility (CMS) in *Petunia* from a stock to a scion. This long-range transmission of "silencing" may be much more prevalent than is presently assumed. This author dares to make a guess that changes in morphogenesis, in reaction to environmental clues, could also involve long-distance transmission of silencing specific genes. Can the clues of photoperiodism, perceived by leaves, be transmitted to meristems by mobile silencing complexes?

The "historic" paper of Fire et al. (1998) deserves special attention. The authors observed that when single-stranded RNA, consisting of either an mRNA (sense) or the antisense of this RNA, were injected into the nematode *Caenorhabditis elegans*, there could be a mild reduction in the expression of the respective gene. However, when the two single-stranded RNAs were simultaneously injected, or the dsRNA was injected, the expression of the respective gene was specifically silenced. Moreover, the silencing effect of the dsRNA was mobile! The effect moved quickly from cell to cell and the silencing was transferred to the next generation. The authors suggested that a stage of "propagation" of the silencing was involved: a few dsRNA sequences affected many cells and their impact "spread" quickly.

1.1. Early Studies on RNAi

During the years 1998 and 1999 a few reports substantiated the finding of Fire et al. (1998). First, the same team added some extra information. Timmons and Fire (1998) found that rather than injecting "pure" dsRNA into the intestine of the nematode, the animals can be fed with bacteria that express such dsRNA, and the same RNAi will result. Tabara et al. (1998) suggested the RNAi system as a genetic tool. Moreover, Tabara et al. (1999) found in *C. elegans* two loci (*rde-1* and *rde-4*) that cause resistance to the effect of RNAi.

A further study of RNAi in *C. elegans* was performed in Amsterdam. Ketting et al. (1999) found that activating the *Tc1* transposition in the germ cells of some *C. elegans* was initiated in some mutants. In w.t. worms there is usually no *Tc1* transposition in the germ cells. These specific mutants were found to be resistant to RNAi. Hence, these authors assumed that the w.t. RNAi prevents transposition of *Tc1* in the germ cells.

In a study with *Drosophila*, Jensen et al. (1999) attempted to reveal the reason for the reduced transposition rates that follow the initial high rate of transposition, when a "new" *I* element was introduced into a fly strain that did not harbor *I* (*I* is a T.E. with similarity to *LINEs*). This gradual reduction of transposition was stabilized at a low rate. The authors concluded that a process akin to RNAi caused the degradation of the mRNA that is required for *I* transposition.

The transgene-induced gene silencing was termed in fungi *quelling*. Cogoni and Macino (1999) worked on quelling in the fungus *Neurospora crassa*. They found that for effective quelling a gene termed *qde-1* was required. The product of *qde-1* is a kind of RNA-dependent RNA polymerase (RdRP). Such an enzyme is involved in the synthesis of dsRNA and is a component of RNAi. This was an obvious clue that RNAi can be involved in quelling of genes in fungal organisms. Another clue that dsRNA is a mediator of both post-transcriptional gene silencing (PTGS) and virus-resistance in plants was furnished by Waterhouse and associates (Waterhouse et al., 1998) of the CSIRO, in Canberra, Australia. These authors found that when a plant is manipulated to produce sense and antisense RNA for a specific gene or plant virus, then the respective gene is silenced and the virus is repressed. The authors even came up with a model to show how specific dsRNA can have the PTGS and/or the antiviral effect.

An interesting case of gene silencing was revealed in the oomycete *Phytophthora infestans* by Van West and Kamoun (1999). *P. infestans* is an important plant pathogen. It was regarded as a fungus in the past but now, by DNA phylogeny, seems to be closer to algae (that have lost their chloroplasts). These authors transformed *P. infestans*. They introduced into it constructs of sense or antisense sequences of a gene (*inf1*) that produces elicitin (a component of the infection process). They also added the *inf1* without its promoter. In all these cases there was silencing of the endogenous *inf1* as well as of the transgene. Moreover, because in *P. infestans* one can produce heterokaryons, the silencing could be transferred from one hypha to another. It was observed that once the silencing took place, the homokaryotic derivatives of the heterokaryons were "silenced" and they did not express the transgene.

1.2. PTGS and RNA Silencing in Plants

David Baulcombe and associates of the John Innes Centre (Norwich, UK) are pioneers in the investigations of PTGS and defense of plants against viruses (Voinnet et al., 1998, 1999). A 25 nt antisense RNA was revealed by this team (Hamilton and Baulcombe, 1999). This RNA sequence was involved with silencing of genes and viruses. This short RNA had homology with the target gene (or virus). However, it was found that the viruses may fight back! Voinnet et al. (1999) found that viruses developed means to counteract the plant anti-viruses defense, which is based on PTGS. During a short period of time the Baulcombe team came up with a series of interesting publications on gene silencing and viral defense in plants (e.g. Voinnet et al., 1998, 1999; Hamilton and Baulcombe, 1999; Dalmay et al., 2000, 2001; Thomas et al., 2001; Vaistij et al., 2002). Baulcombe (2001) provided an update on the silencing process with the short title: *Diced defense*. Baulcombe stated that it was clear that cells can block invading viruses or mobile DNAs by the use of an RNA-cleaving enzyme, found in *Drosophila* (Bernstein et al., 2001) and termed it *Dicer*. He suggested a model in which the *Dicer* cleaves dsRNA into fragments of 22 nt. In a second stage, the *Dicer* associates, through its PAZ domain with a component related to ARGONAUTE, which is also a PAZ domain, containing protein. The complex joins an RNase and the 22 nt fragment. The complex can now police mRNA and degrade any RNA that has homology to the 22 nt. While this model later went into further refinement, it was a plausible one.

The RNA silencing in plants was further explored by a team in Canberra, Australia (e.g. Wesley et al., 2001; Waterhouse et al., 2001). These investigators even came up with technical innovations. They provided information on how to construct tools for silencing specific plant genes and gave them nice names: *pHANNIBAL* and *pHELLSGATE*. The RNA silencing constructs were based on the hairpin RNA, preferably containing an intron. The construct named after the Punic/Phoenician leader (Hannibal) could be adopted to silence specific genes as the gene chalcone synthase (CHS) and the construct was termed *CHGS-HANNIBAL*.

Why *HANNIBAL* was not made clear by the authors; but Hannibal (247–183 B.C.) could talk to me; he spoke Phoenician, a West-Semitic language and a dialect of Hebrew. A further report on genes of plant viruses that suppress the plant-defence (PTGS) mechanism came from Singapore (Li et al., 1999).

Studies with *Arabidopsis* supplemented the information of PTGS in plants. Mourrain et al. (2000) found two genes in this plant that are required for the PTGS process: *sgs2* and *sgs3*. Of these, *sgs2* encodes a protein that is similar to proteins in other organisms (*N. crassa* and *C. elegans*) that are required for quelling and RNAi. The other gene, *sgs3*, seems to be unique to *Arabidopsis*. A study by Chuang and Meyerowitz (2000) paved the way for the use of RNA silencing techniques in the study of plant morphogenesis. These authors produced dsRNAs that had homologies to the genes *AGAMOUS, CLAVATA3, APETALA* and *PERIANTHIA*. The respective mRNAs were then repressed, causing specific morphogenetic changes.

At the beginning of 2001, it became clear that, at least in plants, specific genes (especially transgenes) could be silenced by PTGS. This silencing was manifested through "dicing" (Bernstein et al., 2001; Baulcombe, 2001) and the production of small-interfering RNA (siRNA) as well as policing and cleaving particles that were termed *RISC* for RNA-induced silencing complex. RISCs have a remarkable mobility and target specificity. Plants also utilize this system to repress invading RNA viruses, as well as to repress transposable elements. It was also frequently reported that silencing was associated with DNA methylation of specific sequences, in the inactivation of transposable elements that were integrated into the genome. Are the phenomena of PTGS and DNA methylation related? Wassenegger (2000) handled the relation between RNA silencing and the methylation of cytosine in DNA. A detailed discussion of the relationship between DNA methylation and RNA silencing in plants was provided by Bender (2001) of the John Hopkins University School of Public Health (in Baltimore, MD). The summary answer is that siRNA can indeed trigger the methylation of specific DNA sequences that have homology to this siRNA. Hence, the siRNA, at least in plants, seems to have a dual role: it causes the degradation of specific mRNA and also causes heterochromatization of specific DNA sequences, reducing or preventing transcription from these sequences.

The assumption was put forward that the siRNA (and other "aberrant" RNA species) actually make contact with specific DNA sequences and provide a signal for methylation. This can be termed " RNA-directed DNA methylation". Is there a feedback? Meaning, that methylated genes will transcribe "aberrant" RNA and that the latter is a trigger for additional methylation of the same sequence? It is a possibility that requires further verification. There are specific genes (some of them were already mentioned above) that are required for this kind of silencing: *sgs2*, *sde1* and *sde3*. Mutations in these genes will impair specific silencing. All this may not be the full picture. The protein HC-Pro drastically reduces the levels of siRNA but it does not reduce DNA methylation (Mallory et al., 2001). Loss of DNA methylation was recorded when two genes, involved in DNA methylation and chromatin remodeling, *DDM1* and *MET1*, were mutated. Although these mutations reduced DNA methylation, they did not affect the RNA silencing. It thus appeared that methylation is necessary to reinforce the RNA silencing signal. The subject became even more complicated when it became evident that, at least in plants, there may be the symmetrical methylation of cytosine of CG or CNG, but there is also a non-symmetrical methylation. These two methylations may have different control mechanisms. For example, the MET1 protein does not affect the asymmetrical methylation of virus-infected plants, while it is important to maintain CG methylation.

The question of the relationship between siRNA and DNA methylation in animals is still not resolved. In this connection it should be noted that the nematode *C. elegans*, where RNAi was first revealed, lacks genomic methylation.

Back to plants, the Australian-New Zealand team (Waterhouse et al., 2001) claimed that plants have a sophisticated adaptive defence mechanism against viruses and transposable elements, and that this plant mechanism has a number of parallels with the immune systems of mammals. The same conclusion was provided by Dangl and Jones (2001).

A more recent study by Foster et al. (2002) concerning PTGS of plants revealed an interesting issue. It concerns the trafficking of RNA in the phloem. This study explored the selective traffic of RNA into the shoot apex. Earlier observations indicated that RNA viruses are excluded from the shoot apex. This exclusion is actually the basis for utilizing shoot-tip-culture to clean plants (e.g. potato) from viral infection. However, the RNA of PTGS is also normally excluded from the shoot tip. The authors revealed the presence of an RNA signal surveillance system that acts to allow the selective entry of RNA into the shoot apex.

From a "practical" point of view, the RNAi system is now a developing tool in genetic studies. Genes containing known sequences can be explored by using these sequences for RNAi knock-out. The consequences of such knock-outs can then be explored (e.g. Gupta et al., 2002a,b; Han and Grierson, 2002).

The PTGS is not limited to angiosperm plants. A report of Wu-Scharf et al. (2000) on the alga *Chlamydomonas reinhardtii* is noteworthy. This alga has a gene (*Mut6*) that is required for the silencing of transgenes and transposon transposition. This gene encodes a protein that is highly homologous to RNA helicases of the DEAH-box family. The gene is thus required for the degradation of certain aberrant RNAs. The degraded RNAs can then initiate the PTGS mechanism. If this is verified, then not only dsRNA, but also some aberrant ssRNA can initiate the PTGS mechanism.

1.3. RNAi in Animals

By now, the RNAi system was reported in protozoa as well as in metazoa and, in the latter animals, in invertebrates as well as vertebrates. Thus, RNAi probably exists in all or in almost all animals.

RNAi was already reported in the protozoan parasite *Trypanosoma brucei* in the same year as the discovery of RNAi in nematodes (Ngo et al., 1998). The inheritance of RNAi was further studied in the nematode *C. elegans* by Grishok et al. (2000), and the link between co-suppression and RNAi in this worm was reported by Ketting and Plasterk (2000). A more recent study with *C. elegans* revealed several proteins that have a role in the RNAi process (Tabara et al., 2002).

Several studies have handled the RNAi system in the fruit fly *Drosophila melanogaster*. For example, Zamore et al. (2000) reported on the cleavage of mRNA at 21 to 23 nt intervals. Hammond et al. (2000) transfected cell cultures of *D. melanogaster* with specific dsRNA species and found that it caused the degradation of homologous RNA sequences in these cells. Aravin et al. (2001) followed the silencing of one specific gene in *D. melanogaster*.

Scadden and Smith (2001) found in *D. melanogaster* that the deamination of adenosines, rendering them to inosines, antagonizes the RNAi process. A more detailed study of the components of RNAi in *D. melanogaster*, such as the role of short interfering RNAs (siRNA) and the RNase III processing of long dsRNA, was performed by Elbashir et al. (2001b).

In mammals, Wianny and Zernicka-Goetz (2000) reported that dsRNA can be used to specifically inhibit three mice genes. Hence, the RNAi system should be useful in the study of development and gene-regulation in mammals and mammalian cells. Elbashir et al. (2001a) used the RNAi system to specifically suppress genes in different mammalian cells, including human embryonic kidney and HeLa cells.

Martinez et al. (2002) followed the accumulation of the RISC complex in HeLa cells. They detected single-stranded siRNAs as components of RISC and concluded that single-strand antisense siRNAs guide the cleavage of target RNA in the RNAi system in these cells. From *in vitro* data, Schwarz et al. (2002) concluded that in *D. melanogaster* and in humans, each siRNA guides the endonucleolytic cleavage of the target RNA at a single site, and that RNA-dependent RNA polymerases are not required for RNAi in man and fly.

An updated overview of the RNAi system in animals, with experimental evidence for the steps of this system in various animals as well as in man, was provided by Caplen et al. (2002) and by Chiu and Rana (2002).

1.4. Selected Reviews on RNAi and PTGS

In the following, selected reviews will be listed in approximately chronological order. Fire (1999); Bass (2000); Sharp and Zamore (2000); Catalanotto et al. (2000); Plasterk and Ketting (2000); Baulcombe (2001); Vionnet (2001); Matzke et al. (2001); Ringrose and Paro (2001); Tuschl (2001); Moazed (2001); Wang and Waterhouse (2001); Okamoto and Hirochika (2001); Hutvagner and Zamore (2002); Carmichael (2002); Allshire (2002); Plasterk (2002); Zamore (2002); Ahlquist (2002); Hannon (2002). Many of these reviews have exotic titles but two titles are noteworthy: "Making noise about silence: repression of repeated genes in animals" (Birchler et al., 2000) and "Listening to silence and understanding nonsense" (Cartegni et al., 2002).

There is a group of RNA species, termed *micro RNAs*. These will not be handled in this appendix, although they are related to RNA silencing. These RNA species were reviewed recently by Reinhart et al. (2002).

2. APPENDIX II: CHROMATIN REMODELING

The cells of maize contain 10^{10} bp of DNA, or, if this DNA were stretched into one double-strand, its length would attain 10 m. Nevertheless, all this DNA fits into a nucleus that has a diameter of a few μm (several million times less than the length of the dsDNA). This is made possible, in maize as well as in other eukaryotes, by several levels of coiling. In this coiling and supercoiling the histone proteins are complexed with the dsDNA in a very specific manner. Obviously, before either DNA replication or DNA transcription can take place, the "regular", tight DNA/histone complex has to be released. The details of this complex are presented

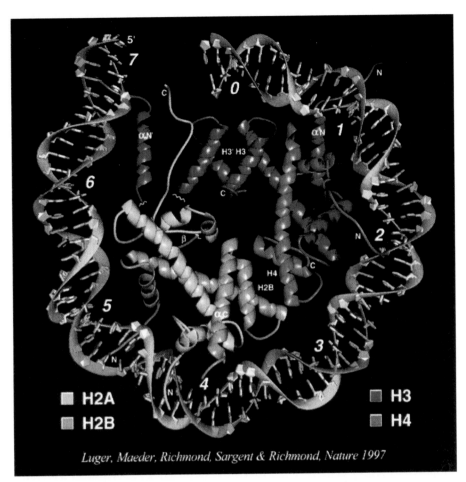

Figure 61. Nucleosome core particle: 73-bp half. The view is down the superhelix axis with the pseudodyad axis aligned vertically. The central base pair through which the dyad passes is above the SHLO label, O (SHL, superhelix axis location). Each SHL label represents one further DNA double helix turn from SHLO. The complete histone proteins primarily associated with the 73-bp superhelix half are shown (interparticle tail regions are not shown). The two copies of each histone pair are distinguished as unprimed and primed copies, where the histone fold of the unprimed copy is primarily associated with the 73-bp DNA half and the primed copy with the 72-bp half. The 4-helix bundles are labeled as H3' H3 and H2B H4; histone-fold extensions of H3 and H2B are labeled as αN', αN, and αC, respectively; the interface between the H2A docking domain and the H4C terminus as β; and N- and C-terminal tail regions as N or C. (From Luger et al., 1997).

in text books on molecular genetics and the complex is termed chromatin. Chromatin structure and modification have been the subjects of intensive study for many years and shall not be detailed in this appendix. Just one example from plants: Prof James Bonner and his associates at the California Institute of

Technology, started to investigate the chromatin of pea (e.g. Huang et al., 1960; Bonner et al., 1961; Huang and Bonner, 1962) in order to reveal the causal relationship between histone modifications and gene activity. Kitchen utensils (a blender and a modified washing machine) were recruited to obtain 500 mg of crude chromatin from about 12 kg of pea seeds.

These, and other studies in plants as well as in other eukaryotes, provided information on the basic structure of chromatin. An important component of the chromatin is the *nucleosome*. The latter consists of dsDNA wrapped two full turns (or about 146 bp) around a globular octamer of histone proteins. The octamer is made up of two tetramers, each of these consists of four different histone proteins: H2A, H2B, H3 and H4. Another histone protein, H1, binds outside the nucleosome core. It probably functions to stabilize the higher order of chromatin (supercoiling). After the two turns around the histone octamer the dsDNA is connected to another octamer (Figure 61). There is a length of 20 to 200 bp of DNA (spaces) between the nucleosomes. Thus, the chromatin attains a structure like beads (nucleosomes) on a string (of DNA). This is the structure that is actually visible by electron microscopy. The structure of the nucleosomes was described by Kornberg (1974) and updated 25 years later (Kornberg and Lorch, 1999). The spacer DNA can be cleaved by the bacterial enzyme micrococcal nuclease, leading to the separation of the nucleosomes from the string.

It was also found, many years ago, that when a specific region of the chromatin goes into the transcription phase, it undergoes a series of modifications. At the "resting" stage, between mitotic divisions and when the region is neither replicating nor being transcribed, the "beads on a string" is actually further coiled to form a "solenoid" of packed chromatin. There can be decondensation of the chromatin, which renders the region accessible to the transcription machinery. The decondensation is accompanied by specific acetylation of histone amino acids.

2.1. Chromatin Acetylation

The acetylation of histone and its association with transcription activity was already proposed by Alfrey et al. (1964). From the understanding of acetylation and other changes in histones that are correlated with gene activation, it became very clear that chromatin structure and modification cannot be regarded as being independent of gene activation and inactivation. Moreover, it became evident that chromatin is not merely a structure that keeps the DNA in its coiled and supercoiled configuration.

The acetylation and deacetylation of histones appeared to be a rather elaborate process that involved several enzymes (e.g. Vogelauer et al., 2000). The link between histone-acetyltransferases (HATs), chromatin modifications and gene activation was more recently reviewed by Marmorstein and Roth (2001). A cross-talk was revealed between acetylation and other histone modifications. Thus, developmental aberrations in mice and certain human cancers are associated with HAT mutations, stressing the importance of the HAT enzymes in normal cell growth and differentiation.

2.2. Phosphorylation, Methylation and Ubiquitination

In recent years, understanding of additional histone modifications, such as phosphorylation, methylation and ubiquitination, was increased by various studies that were reviewed by Berger (2002). These studies led to a novel concept: *the histone code.* This concept claims that the totality of modifications dictates a particular biological outcome. In other words, this concept/hypothesis says that covalent histone modifications that include acetylation, phosphorylation, methylation and ubiquitination are all involved in gene-specific regulation. These modifications can affect histone structure directly but they can also render histones accessible to other modifiers. HATs are frequently components of large, multi-subunit complexes that are recruited to the chromatin by interaction with other effectors. The interactions involved are further elaborated. For example, there is *bromodomain* (BrD) that interacts with acetylated lysine of the histone and the *chromodomain* (ChrD) that interacts with methylated lysine (see Berger, 2002, for details).

Histone phosphorylation (HK) at the serine-10 of H3 was found to be an important modification involved in the activation of transcription and in chromosome condensation during mitosis (Cheung et al., 2000) and heat-shock gene activation is accompanied, in *Drosophila*, with a dramatic increase in H3 serine-10 phosphorylation (Nowak and Corces, 2000).

We mentioned above, in dealing with silencing of transposition of transposable elements, that methylation of DNA is correlated with this silencing. DNA methylation and the methylation of histones are probably linked. Studies with the fungus *Neurospora crassa* (Tamaru and Selker, 2001) provided evidence for the dependence of DNA methylation on the methylation of lysine-9 in H3. The use of *N. crassa* for such studies was based on the fact that in this filamentous fungus the methylation takes place but it is *not* essential for the fungal survival. Methylation is essential in plants, mammals and yeasts.

It should be noted that although there is abundant information on DNA methylation in eukaryotic organisms, the process by which DMTase (DNA-methyltransferase) exerts its activity is not yet clear; it was assumed that DMTase has to interact with other proteins (e.g. Jeddeloch et al. 1999) – but "which and how" was not clarified; moreover, is sequence-specificity of the DNA required? Tamaru and Selker (2001) thus suggested that in eukaryotes, DMTases take clues, primarily from chromatin, as to when and where to exersise their activity. Their hypothesis is that chromatin remodeling factors may assist histone-modification enzymes (such as DMTase) to reach their target and/or to facilitate modification in the nucleosomes. A further, more recent study, with the plant *Arabidopsis* (Jackson et al., 2002) indicated that methylation of DNA in the cytosine of CpNpG is controlled by histone H3 lysine-9 methylation, through interaction of a plant-methylation enzyme (CMT3) with methylated chromatin.

Bird (2002) who recently reviewed, in detail, the subject of DNA methylation and epigenetic memory (especially in mammals), made an interesting suggestion. According to him, the accumulating information points to the possibility that *first* a region of DNA is silenced and only after it is silenced, does it attract the process of methylation. Bird noted that not all genes can be activated in an animal nucleus when it is put into a new cytoplasmatic environment that includes the appropriate protein factors, which have the potential to activate the promoters of these genes. An additional change is required to resurrect some silenced genes (regions of DNA). The nuclear DNA seems to have, at least in some genes, a very stable "epigenetic memory". Methylation of DNA varies considerably among animals. Briefly, m^5C

methylation does not exist in *C. elegans* (nematode), is very low in *Drosophila* (fruit fly) and is higher in *Ciona* (sea squirt).

In vertebrates, methylation is dispersed over the whole genome ("global methylation"). In man, methylation can affect about 1% of the DNA bases, meaning that most (70–80%) of the CpG can be methylated. In vertebrates, there is a drastic change in the level of methylation; it is very low at early development and then increases. Moreover, in mammals "islands" of unmethylated CpG were revealed in several genes. These "islands" may be methylated during further development. While how this happens is not yet clear, it is obvious that the methylation is highly programmed.

Experimental evidence for the impact of histone modification on DNA methylation is still limited, but as noted above, it was reported by Tamaru and Selker (2001) for *Neurospora*. It seemed plausible to Bird (2002) that a similar interaction between histone methylation and DNA methylation also exists in mammals. The methylation of DNA in mammalian genes seems to reassure the maintenance of a silenced state, rather than being the primary mechanism of gene silencing. Nevertheless, maintenance of methylation in mammalian DNA is important for the normal development of these animals. When methylation is impaired, several features of the organism's development and activities may go wrong; such as bursts of transpositions of transposable elements and the transcription – out of phase – of vital genes. The silencing of genes was also reviewed by Kouzarides (2002). Methylation of histone lysine is a major step in this silencing, but the silencing is a multistep process that includes DNA methylation.

Although DNA methylation is not covered in this appendix, it is related to chromatin remodeling. Readers should therefore be aware of the abundant information on methylation and demethylation of DNA in animals that has been published in recent years. A good start for this topic is the review of Hsieh (2000). The details of one complex, the methyl-CpG binding domains in humans, were provided by Ohki et al. (2001). The chemistry of this complex clarifies the question of recognition of methylation sites, on the DNA, by the proteins involved.

We shall return to histones. Recent studies have gradually indicated that all the histone proteins may be modified and in a range of locations. Moreover, the modifications are correlated (Briggs et al., 2002). For example, the conjugation of histone H2B with ubiquitin controls the outcome of methylation of a specific lysine residue (Lys4) in histone H3. The methylation of the latter lysine regulates silencing in the yeast *S. cerevisiae*. There are also other interactions of this kind. Ubiquitination of H2B is required for the methylation of Lys79 in H3. The latter is an exceptional H3 lysine because it is located away from the tail of H3, where other lysines may undergo methylation. On the other hand, the methylation of Lys36 of histone H3 is not related to ubiquitination in yeast (Briggs et al., 2002). If and when additional interactions of this kind are found, in yeast and in other organisms, we shall be faced with an even more intricate situation than known at present.

One could wonder: for such correlative interaction to happen, especially away from the histone tails, the specific amino acid residue should become accessible to the modifying machinery. How can this happen under conditions of the nucleosome being tightly-wrapped in DNA and when the whole chromatin is in the form of a tightly-coiled solenoid (this question was elaborated by Gregory, 2001, for acetylation of chromatin). Is the answer for these questions related to phases in which the chromatin is loose (mitosis, meiosis and transcription). The role of cell

phosphorylation (P). For example, histone H3 can be modified (numbers are distances from the NH^+_3 terminal) as follows: Lys4-M, Lys9-M & Ac, Ser10-P, Lys14-Ac, Lys18-Ac, Lys23-Ac, Lys27-M, Ser28-P. In histone H4 the modifications are: Ser1-P, Lys5-Ac, Lys8-Ac, Lys12-Ac, Lys16-Ac, Lys20-M. It is not clear whether there are additional possible modifications at the amino acids at the tails of H3 and H4. In addition to modification in Lys and Ser, there are also modifications in Arg, but the role of these latter modifications is not yet clear.

Bannister et al. (2002) handled the difference between acetylation and phosphorylation on the one hand and methylation on the other. While acetylation and phosphorylation are considered reversible, and can thus be changed by outside effectors, the methylation is considered to be biochemically more stable. The authors leave this question open and indicate that reversal of methylation cannot yet be ruled out.

The centromeric chromatin has also been handled in recent publications. Sullivan (2001) focused on the histone H3-like protein CENP-A that plays a role in the kinetochore assembly and may be a central element in the epigenetic maintenance of centromere identity. CENP-A was found in very diverse eukaryotic organisms. A review on several features of the centromere was provided by Sullivan et al. (2001). The centromeres of the plant genus *Beta* (beets) and its comparison with centromeres in other plants as well as in other organisms, was handled by Gindullis et al. (2001). Although these authors focused on the DNA component of the centromeres, rather than on the chromatin. It is noteworthy to mention one specific feature of the DNA in centromeres: this DNA includes Ty3/*gypsy*-like retrotransposons. With a smile we may consider this situation as another case of mummified transposable elements. The *Beta* publication handled the "body" of the mummy, while the mode of embalming (heterochromatin) was left unsolved.

2.5. Recent Reviews: The Overall Picture of Chromatin Remodeling

During the years 1999–2002 there was a vast improvement in the understanding of chromatin structure and remodeling of chromatin. This improvement owes much to the structural analysis of chromatin and especially of the fine-structure of the nucleosomes that preceded these years. The basic structure of the nucleosome was already provided many years ago (e.g. Kornberg, 1974) and a more detailed structural study, by X-ray crystal analysis at a 7 Å resolution, was reported by Richmond et al. (1984). Additional efforts in this line culminated in a very detailed image of the nucleosome, at 2.8 Å resolution by Luger et al. (1997). The latter publication showed, in atomic detail, how the histone protein octamer is assembled and how the 146 bp DNA double helix are organized into a superhelix around the octamer. The histone/histone and the histone/DNA interactions depend on the histone fold domains. The image (Figure 61) also showed that well-ordered structural elements extend from this motif. It especially showed that the amino terminal tails of the histone proteins pass over and between the gyred DNA superhelix, to contact neighboring particles. These terminal tails were found in later studies to contain the modifications (e.g. acetylation, methylation) on the respective lysine and serine amino acids, as indicated above.

Several reviews that handled chromatin remodeling appeared during the years 1999, 2000, 2001 and 2002. Here, only a few of them will be recorded. Those interested will be able to expand their knowledge on this subject by referring to the following reviews. The review of Struhl (1999) handles the fundamental difference between gene regulation in prokaryotic organisms (bacteria) and eukaryotes. Struhl summarized the information on the differences, concerning the regulatory logic of gene activation and silencing in these two types of organisms, and noted important differences in transcriptional regulatory mechanisms. The most important of these differences results from the fact the eukaryotic DNA is packaged into chromatin templates. The DNA of eukaryotes is complexed into an elaborate and dynamic chromatin structure that constitutes the chromosome. Not so in prokaryotes. Consequently, the ground state for transcription in prokaryotes is *non-restrictive*. The activators and repressors interact with the respective DNA motifs and affect the transcription. Once the RNA polymerase starts its transcription, the latter proceeds without further inhibitions imposed by the template. In eukaryotes, the transcriptional machinery is much more elaborate. Before the transcription complex (including PolII holoenzyme, TFIID, TATA box, etc.) can be established, there is a need for several chromatin modifying activities. Only thereafter, the promoter region attains an active state. The "ground state" for transcription, in eukaryotes is thus considered *restrictive*. In a subsequent review, Strahl and Allis (2000) elaborated the histone modifications that can take place in the amino acids at "tails" of H3, H4 and CENP-A. They proposed the term *histone code* for the result of distinct histone modifications and wrote about: " The language of covalent histone modifications".

The review of Zhang and Reinberg (2001) focused on histone methylation and the interplay between this methylation and other modifications in the histone "tails". The interactions can be between methylation (M) and acetylation (Ac) at the same amino acid (e.g. H3 Lys9) or between different amino acids (e.g. between M in H3 Lys9 and Ac at H3 Lys14).

Two additional reviews updated the information on histone methylation: Jenuwein (2001) and Jenuwein and Allis (2001). Jenuwein (2001) emphasized the differences between the acetylation and phosphorylation modifications and the methylation (of lysine). The latter is a relatively more stable modification and makes an important contribution to the establishment of the heterochromatic structure of chromosomal regions. The second review (Jenuwein and Allis, 2001) looks at several remodelings of the chromatin, and the role of these remodelings in regulating access to the "underlying" DNA. These changes in access dictate a dynamic transition between transcriptionally active and transcriptionally silent states. These authors suggested that the combinational nature of histone modifications (in their NH_3 "tails"), reveal a "*histone code*". This "code" extends the information-potential of the genetic code (in the DNA). Another review (Bannister et al., 2002) handles histone methylation as well as demethylation. This review focused on the question whether methylation is a "permanent" mark. Hence, this review surveyed the evidence for demethylation enzymes that could reverse the methylated state of lysines in histone "tails".

The review of Dhillon and Kamakaka (2002) highlights other aspects of the silencing mechanism imposed by chromatin modification. They noted that peaks of acetylated histone demarcate the boundaries between chromatin regions that are either active or inactive in transcription. This review also provided arguments that (in contrast to previous claims) DNA replication is *not* a prerequisite to transcription silencing. The review of Narlikar et al. (2002) analyzed the accumulated information on chromatin remodeling (e.g. methylation, acetylation and phosphorylation). They thus came up with models that describe how the activities of chromatin-modifying complexes can be coordinated to create a DNA template that is accessible to the transcription apparatus. The review of Li (2002) focused on the impact of chromatin modifications on the programming of development. Li handled mammal developmental issues, such as early embryogenesis and gametogenesis, as well as tissue-specific gene silencing and X chromosome inactivation. This review also analyzed the known links between DNA methylation, histone-modification and general chromatin remodeling.

Histone modification, chromatin remodeling and their impact on programmed gene activity are subjects of investigations in several laboratories and in diverse organisms. Interesting results are thus expected and a better understanding of the consequences of chromatin remodeling should be forthcoming. Nevertheless, what we already know about this subject is amazingly "wise" and complex. These elaborate control mechanisms can bring us into spiritual contemplation. The sentence of Psalms (Chapter 104) comes to mind: "O Lord, how manifold are thy works, in wisdom hast thou made them all; the earth is full of thy riches".

3. APPENDIX III: GENE TAGGING

Geneticists have their own language. Thus, there is *Forward Genetics* and *Reverse Genetics*. In *Forward Genetics* you have a mutation that was revealed either in a population of organisms, or by induced mutagenesis. The mutation could be an alteration in pigmentation, structure, metabolism, development, etc. Now, you would like to know what the DNA sequence is that encodes this mutation, and where in the genome this sequence is located. In *Reverse Genetics* you have a DNA sequence (that has the characteristics of a gene). Now, you would like to know which trait is encoded by this DNA sequence.

We shall see that techniques for gene tagging have improved vastly in recent years, although the fundamental approached did not change much. Here is an example for reverse genetics. A knot tied by King Gordius of Phrygia was supposed to be undone only by the future Master of Asia. Alexander (356–323 BC), son of Olympias and Philip, and King of Macedonia undid the knot with one stroke of his sword and marched into a successful conquest of Asia. As noted above, in reverse genetics we start with a segment of DNA that has a known nucleotide sequence but we do not know which gene is encoded by this sequence. The traditional procedure was to cause a specific alteration in the nucleotide sequence and trace the phenotypic change that results from this change in nucleotide sequence. The procedures for aiming into a specific sequence of genomic DNA could be rather elaborate. Now, we have the "sword" of Alexander the Great. This sword is the RNAi in animals and similar processes in plants and fungi. One can now use RNAi procedures to silence (by the destruction of specific mRNAs) only the gene that has a known sequence (or even when only a part of the sequence is known). The impact of this silencing can then be studied.

3.1. General Remarks on Gene Tagging

The localization of genes on chromosomes has been practised since the early years of genetic investigations. In the chapter on Historical Background, it was noted that due to intensive studies at Cornell and other places, during the early years of the

20th century, many genes were located on the 10 chromosomes of maize. In fact, several genes were located on each chromosome (Figure 1). The mapping of the genes was based on genetic crossings and selfings with the aid of cytological techniques. For example, whenever trisomics were available, cytogenetic techniques were used to locate genes on such chromosomes. Unfortunately, trisomic chromosomes exist in nature, even in man, where (in chromosome 21) they are correlated with Down's syndrome. Also, in those animals where males have only one X chromosome the respective traits can be assigned to the sex-linked X chromosome. After one or more genes are located on a specific chromosome or chromosomal arm, additional genes can be assigned to the same arm.

Mutations in natural populations were the first to be located but then investigators wished to "fill" the genetic maps with as many mutations as possible. Induced mutations were thus produced by chemical mutagenesis or by irradiation. A kind of break-through, in plants, occurred when *Agrobacterium*-mediated genetic transformation became a feasible practice (see Galun and Breiman, 1997). The insertion of T-DNA could then be used to disrupt genes. A relatively early example of the use of T-DNA-mediated gene tagging was reported by Koncz et al. (1989). This research was a "spiced" Hungarian-German-Belgian work. The T-DNA did not contain a promoter but it did contain a reporter gene. The reporter gene was thus activated when the T-DNA was inserted in a gene, downstream of the promoter of this gene. The investigators found that their T-DNA could serve as a mutator element and cause insertional inactivation of a gene and, at the same time, it heralded its presence by the reporter gene. In her comprehensive review on gene tagging V. Walbot (1992) devoted a chapter to gene tagging by T-DNA and discussed the advantages and disadvantages of this tagging in comparison with gene tagging by transposable elements.

As long as only very little data on DNA sequences in genomes of animals and plants were available, most of the gene tagging work was *Forward Genetics*, meaning that it was intended that once genes were mutated, the investigator was interested to know the DNA sequence that was affected by this (insertional) mutation. A DNA location that is *tagged* can then be isolated, cloned and sequenced. Various procedures were devised to isolate the flanking DNA on both sides of an inserted transposable element (or a T-DNA). One elegant one is the *Inverse PCR* (IPCR). An example of this procedure was provided by Koes et al. (1995) in a study in which 15 (!) Belgian and Dutch investigators tagged *Petunia* genes.

Here is a short description of IPCR. For regular PCR amplification of a DNA sequence, two primers are required, one for the upstream end of this sequence and another (in reverse) for the downstream end. With a template and these two primers, sufficient DNA can be amplified for sequencing analysis. However, how is this done when only the middle of this DNA sequence is known (the transposable element) while the flanking sequences, that are of interest, are unknown? The solution is that the whole fragment of DNA is first circularized. The circles are then cut specifically (by an appropriate endonuclease – this can be refined by introducing a rare restriction site into the transposable element). By this cut, the circles are linearized, but now the flanks are of the transposable element (that has a known DNA sequence) and the middle is the gene with an unknown sequence. Primers are synthesized for matching the flanking regions and then the linearized fragment is amplified by PCR, and sequenced.

Another procedure is to amplify a DNA sequence by holding its "tail". For this procedure, one PCR primer will be homologous to the transposable element. For the flanking region (the gene with the unknown sequence), several short primers are used. Experience showed that there is a fair chance that one of these short primers will serve the PCR and a fragment between it and the primer for the transposable element will be amplified. When 25 or more nucleotides of a gene are sequenced, the gene can be further analyzed by using these nucleotides to construct an appropriate (fluorescent or radioactive) primer in order to isolate the whole gene from a genomic library. DNA fragments of a genome, into which a T.E. has been inserted, can be isolated by *plasmid rescue* (e.g. Rommens et al., 1992). In this technique, the T.E. is engineered to contain a gene for resistance to an antibiotic compound and the fragments are inserted into bacterial plasmids. The latter are introduced into appropriate bacteria and the bacteria are then cultured in the presence of the antibiotic compound. The required plasmids are consequently multiplied and can be isolated. When the plasmid contain, in addition to the T.E., fragments of an inserted gene, the latter becomes available to the investigator.

In recent years, the genomes of several organisms have been fully sequenced. The complete genomic sequences of many prokaryotes (bacteria), including parasitic bacteria, are now available. In animals, the genomes of the nematode *C. elegans*, the fruit fly *D. melanogaster*, mosquitos, the zebra fish, the mouse as well as man, are available in databases. In plants, the genomes of rice and the model plant *Arabidopsis thaliana* are also available in databases. Thus, the trend of tagging was reversed – *Reverse Genetics* became a more prevalent direction of study. Although in many crop plants, especially those with extensive genomes such as maize and wheat, the percentage of sequenced DNA among the total genome is still very low, a high percentage of the coding regions have already been sequenced. The ingenuity of investigators provided an elegant solution for the study of complicated genomes. Ogh et al. (2002) devised a procedure, termed *microsynteny*, to study the tomato genome by the use of the full-information that exists for the *Arabidopsis thaliana* genome. The maize genome is surely enormous (about 2×10^9 bp) but the ambition of maize investigators is even greater. Consequently, Walbot (2000) reported on the launching of the huge project to "define all the genes (of maize) and find the phenotype of every individual gene". This project will be executed by saturation mutagenesis using transposable elements. Nature assisted the maize geneticists: albeit the total size of the genome is huge, the genes of maize are separated by many DNA repeats that are not genes and T.E. tend to insert preferentially into genes. Thus, gene tagging in maize is more efficient than appears from the size of this plant's genome.

Transposable elements (T.E.) are a relevant choice for insertional mutagenesis and gene tagging. The question is thus which of the T.E. should be used for each purpose. The descriptions of the various T.E. in the text of this book should assist the choice of favorable elements. Although endogeneous T.E. may be used, there is advantage in utilizing a T.E., which either does not exist in the tagged genome or is rare in it. Genetic transformation, or other methods, can be used to transfer an alien T.E. into a genome that is subjected to gene tagging. As has been frequently noted in this book, many T.E. systems consist of two types of elements (e.g. *Ac/Ds*). One of these is autonomous (e.g. *Ac*) while the other (e.g. *Ds*) can transpose only in the presence of the autonomous element or in the presence of the transposase required for the transposition. Such dual T.E. systems can be manipulated in one or more ways. For example, the mobility of the non-autonomous element can be stopped if the autonomous element (or the gene that supplies the transposase) is removed by appropriate crossings. The autonomous element itself can be immobilized by specific changes (i.e. removal of its flanking repeats). The elements can also be engineered to contain reporter genes. A very prevalent means of indicating the mobility of a T.E. is to insert it into a reporter or selectable gene. The excision of the T.E. from such a gene can thus signal the mobility of the T.E. Indeed, the examples of gene tagging provided below will describe such manipulations.

In some cases, the investigator is interested to obtain a "chain-reaction" of insertions. In these cases, the mobile elements, after insertion into a primary location, continue to "jump" into additional locations until the process is stopped (such as, by the removal of the autonomous element). A few T.E. have a tendency to transpose to a nearby location. These elements are useful for analyzing a gene in great detail: the T.E. may insert into the promoter, exons, introns or downstream of the coding region of the gene. The consequences of these various insertions can then be evaluated. Moreover, insertion into the promoter can be into different specific sequences and in each case the insertion may cause a different change in the regulation of gene expression. A family of T.E. may be restricted to a very small range of organisms and will not be mobile outside the genus where the T.E. was initially detected (e.g. the *P* element of *Drosophila*). In other cases, a T.E. found naturally in one genus (e.g. the *Ac/Ds* in *Zea*) is mobile even when transferred to very different genera (from monocots to dicots). In general terms, the *Tc1/Mariner* superfamily seems to have no phylogenetic limitations and the respective T.E. (or synthetic T.E. based on them) can be used for insertional mutagenesis and gene tagging in phylogenetically diverse organisms. Some T.E. have a low mutability (e.g. a forward mutation frequency of about 10^{-5} for unlinked genes in *Ac/Ds*), while other T.E. have about a hundred-fold higher mutation frequency (such as in *MuDR*). These features and additional characteristics of the various T.E. are detailed in the main text of this book. Investigators wishing to utilize T.E. for gene tagging should become familiar with these features and characteristics before this utilization.

For additional information on transposon-mediated mutagenesis and gene tagging, the reviews of Dooner (1989), Haring et al. (1991), Osborne and Baker (1995), Hamer et al. (2001) and Ramachandran and Sundaresan (2001) are recommended.

3.2. Examples of Gene Tagging Investigations

This appendix does not provide a comprehensive review on the topic of gene tagging and its relation to transposable elements, but is intended to lead those interested to the pertinent literature. In the following examples, especially those performed in recent years, the respective authors usually furnished background literature that is useful for those interested to study further the gene tagging subject.

The first two examples of gene tagging by T.E. will be from the nematode *Caenorhabditis elegans*. The research of Greenwald (1985) was performed in Cambridge, England, where S. Brenner brought this nematode to fame. The gene that was isolated, *lin-12*, controls a developmental process in this worm. The worm inhabited the actively transposing *Tc1* elements and seven mutations of *lin-12* were found to be associated with *Tc1* insertions in this locus. Briefly, Greenwald "cleaned" the *C. elegans* worms, that had *lin-12* mutations, from superfluous *Tc1* elements and found a DNA fragment (of 3.0 kbp) in the genome, which after crossings, segregated with the mutation and the *Tc1* element. The part of this fragment that was not the *Tc1* element (ca 1.4 kbp), was then used as a probe to isolate the *lin-12* gene. Several such experiments indicated that a specific 2.9 kbp restriction fragment of the worm's genome contained the gene or part of it. By DNA sequencing, it was found that although this restriction fragment did not contain the whole gene, it contained three exons of it. The exons encoded 11 peptide units and they were homologous to peptides in certain mammalian proteins, such as epidermal growth factors.

Moerman et al. (1986) also used *Tc1* to tag a *C. elegans* gene. The authors focused on the *unc-22* gene that is important in the assembly and function of body-wall musculature. It was found that *Tc1* is responsible for mutating the *unc-22* gene. Here, as in the previous example, the mutation region was first "cleaned" from superfluous *Tc1* by transfer of the region from the "Bergerac" background (that has a lot of *Tc1* elements) to a "Bristol" strain background (that contains only a few *Tc1* elements). By digestions of DNA from mutants and wild-type worms with specific endonucleases, running the fragments on gels and size comparisons, the authors revealed fragments that apparently had the additional *Tc1* sequence that caused the insertional-inoculation of *unc-22*. Such fragments (of ca 3.8 kbp) were cloned. One was indeed the fragment that contained *Tc1* and apparently was flanked by *unc-22*.

In *Drosophila*, many years before its genome was fully sequenced, Forwards Genetics was assisted by a transposable element. Searles et al. (1982) utilized the *P* element for this purpose. These authors exploited the P-M hybrid dysgenesis syndrome caused by the *P* element and intended to clone a gene that caused a mutation of RNA polymerase II (RPII). This gene was already previously mapped on the X chromosome, but nothing on the DNA level of this gene was known. The authors had a lambda-phage genomic library of hybrid dysgenesis flies that contained the mutation. They also had a probe for the *P* element. They thus found a 10 kbp fragment that contained 1.3 kbp of the *P* element. They cloned the non-*P* part of the 10 kbp fragment and used it in *in situ* hybridization on polytene chromosomes. The non-*P* fragment hybridized on the X chromosome at the site where this mutation was mapped. Several additional manipulations and tests verified that the authors had isolated a clone that contained the RPII gene.

In another example of the use of a transposable element to detect *Drosophila* genes, O'Kane and Gehring (1987) used the "Trojan Horse" manipulation (enhancer trapping). Their Trojan Horse was a truncated *P* element that retained its weak natural promoter. Inside this "horse" the investigators engineered the *lacZ* gene, without its promoter. The *lacZ* will not be expressed unless the Trojan Horse is inserted closely downstream of a strong promoter (enhancer) in the fly's genome. The expression of the *lacZ* was revealed by β-galactosidase staining of *in situ* embryos obtained from flies that were transformed with the *P-lacZ* construct. This *in situ* staining procedure detected many cell-type and tissue-specific patterns of *lacZ* expression. The procedure could thus detect elements that regulate gene expression in *Drosophila* and generate cell-type markers. Subsequently, this could lead to the isolation and characterization of fly genes.

The *P* elements, or components of these elements, were useful in subsequent gene tagging studies, but this usage is limited to *Drosophila* flies, since the range of *P* activity is restricted to these flies. The *Drosophila melanogaster* genome is now fully sequenced, thus further Forward Genetics is no longer required. For

Reverse Genetics, there are now appropriate methods such as those based on RNAi, by which a specific gene (sequence) can be silenced.

We shall now "jump" on the phylogenetic ladder of animals and reach vertebrates. Our next example will be the *Sleeping Beauty* (*SB*). This is an artificial T.E. of the *Tc1/Mariner* superfamily that was mentioned in the text of this book. *Sleeping Beauty* was constructed (Ivics et al., 1997) by a combined effort of an international research team from St. Paul, Minnesota, Amsterdam, The Netherlands and Szeged, Hungary. Although vertebrates (including man) harbor many Class II transposable elements, these elements are "fossils" that do not transpose and thus are not useful for gene tagging. The "origin" of *SB* was an element of the *Tc1/Mariner* type, that is (as all other T.E. of this kind in vertebrates) transpositionally inactive. The investigators focused on an element in a salmonid fish. They "repaired" all the defects of this transposable element by genetic engineering and came up with *SB*. *SB* has a remarkable ability to transpose in a wide range of vertebrate animals: fish, mouse and man. Izsvak et al. (2000) stated that *Sleeping Beauty* (*SB*) "is the only active DNA-based transposon system of vertebrate origin that is available for experimental manipulations". The activity of *SB* in various vertebrates varies considerably. It is very high in cell lines from human, hamster, *Xenopus* and Fathead minnow fish, but low in other cell lines of human, and in cell lines from rabbit, dog, sheep, quail and carp. The authors characterized the requirements for *SB* transposition. One observation indicated that the level of transposase expression that is induced by the *SB* (but is also regulated by the host) is a decisive factor in the level of *SB* transposition. Obviously, an element that is transposing in a given organism (e.g. man) is useful for a variety of purposes, such as serving as a gene vector (including human gene therapy) and in gene tagging. Indeed, recently the *SB* integration machinery was incorporated into adenovirus vectors for use in gene therapy (Yant et al., 2002). Interestingly, the latter authors revealed a limitation in the *SB* system: in order to transpose, the *SB* transposons need to circularize. From *in vivo* experiments with mouse liver it appears that the integration of the *SB* is random and that a "payload" of a DNA fragments can be carried into the genome by this system. Hence, the *SB* system appears to be a suitable tool for gene tagging.

Gene tagging in plants started with the study of Fedoroff et al. (1984) who isolated the bronze locus of maize by using *Ac*. This study of Fedoroff and associates was followed by numerous other investigations as reviewed by Gierl and Saedler (1992). These authors reviewed the main characteristics of plant transposable elements (e.g. *Ac/Ds*, *En/Spm*, *Mu*) and already then listed 12 cases of maize genes and 7 cases of snapdragon genes that were cloned or identified by T.E. A later review of Martienssen (1998) added numerous cases of gene tagging in plants and also updated the innovations in molecular genetics that are useful for this tagging.

A good example of an early study on gene tagging in *Arabidopsis* is the study of Bancroft et al. (1992). These authors (as others before them) used the "two-element transposon tagging" system. In this system, one element, the autonomous *Ac*, supplies the transposase while the other element, *Ds*, is the one that is aimed to be inserted into genes. The *Ds* was engineered to facilitate its mobility and to carry tags to identify it. The *Ac* can be removed by outcrossing, preventing further transposition. Several later studies on gene tagging in *Arabidopsis* were performed, some with advanced techniques, such as "enhancer trap" (e.g. Springer et al., 1995; Sundaresan et al., 1995; Wilson et al., 1996; Bhatt et al., 1996; Tissier et al., 1999; Speulman et al., 1999; Parinov et al., 1999). It should be noted that these studies were performed before the full genomic nucleotide sequence of *Arabidopsis* was available to the public.

The change in the trend of gene tagging research that usually takes place after the sequence of nucleotides in an organisms genome become available, is exemplified by the work of N. Fedoroff and associates (now at Pennsylvania State University, University Park, PA). This work (Raina et al., 2002) took into account that the data on the genomic sequence of *Arabidopsis thaliana* were already accessible to the public. The investigators thus "filled" the 5 chromosomes of this plant with *Ds* insertions and reported on 260 such independent single-transposon insertion lines. The information was made available to the public. By the nucleotide sequences that flank the *Ds* in each line and by reference to the *A. thaliana* database, each insertion can be precisely located. The impact of the specific insertions on the plants can thus be evaluated and serve functional genomic studies.

The genus *Petunia* is not an ultimate model genus for plant geneticists, but it does have features that have rendered it a frequent object for genetic studies. Thus, several studies on gene tagging were performed with *Petunia*. We shall mention only a few of these. Gerats et al. (1989) of the Free University of Amsterdam (The Netherlands) reviewed previous information on the inheritance of flower pigmentation and the mutations that were induced by the mobility of endogeneous T.E. elements. They also discussed the impact of introducing alien T.E. into *Petunia* and the possible interaction between endogeneous and alien T.E. Due to similarities between the two types of T.E., cross-activation is possible; thus, *Ac* from maize may activate non-autonomous *Petunia* T.E. and cause pigment mutations. With the characterization of some *Petunia* T.E., the authors suggest that *Petunia* gene tagging can start. A similar study was reported by investigators (Renckens et al., 1996) in a neighboring country (Belgium). Renckens and associates used endogeneous T.E. (*dTph1, dTph3* and *dTph4*) and utilized haploid plants. Their study led to their conclusion that haploid petunia is an excellent system for gene tagging. Another team of investigators also from the Free University of Amsterdam, Souer et al. (1995) also studied gene tagging in *Petunia*. The latter investigators used one specific T.E. *dTph1*. They utilized the inverse PCR procedure to identify the DNA sequence into which the T.E. was inserted.

Tomato (*Lycopersicon esculentum*) is also not a model plant for genetic studies, but due to its economic importance, it attracted numerous studies on gene tagging. Osborne et al. (1991) briefly reviewed previous reports on insertional gene inactivations by the *Ac/Ds* system that led to the cloning of plant genes. The basic strategy of these investigators was to introduce *Ac* into the T-DNA (of *Agrobacterium*) together with selectable genes (Hyg^R and *NptII* for resistances to hygromycin and kanamycin, respectively). This strategy also allowed the evaluation of how far could the *Ac* "jump" after it was inserted with the T-DNA. The tomato sequences that flanked the inserted elements were cloned and sequenced.

A very different approach was taken by Jones et al. (1994). These investigators aimed to isolate a tomato gene (*Cf-9*) for resistance to the fungal pathogen *Cladosporium fulvum*. They performed gene tagging by the use of *Ds*. To do so the authors used a tomato line that already had a *Ds* element that was close (3 centrimorgans) to the *Cf-9* locus. By appropriate crossings they introduced, into the same tomato genome, an *Ac* that can mobilize *Ds* but is itself non-mobile. Additional genetic manipulations led to tomato progeny that will be killed by *C. fulvum* infection unless the *Ds* inserted *Cf-9*. From 160,000 such seedlings 118 survived, but only a small fraction of the latter were of the type hoped for. The DNA sequence was consequently determined. A different strategy to tag a tomato gene (actually derived from *L. hirsutum*) for resistance to *C. fulvum* (race 5), by the *Ac/Ds*, T.E. was reported by Takken et al. (1998).

A comprehensive approach was undertaken by A. Levy and associates (the Weizmann Institute of Science, Rehovot, Israel) to bridge between tomato genes and their function (see the recent review, Emmanuel and Levy, 2002). Levy and associates (Meissner et al., 2000) started their project by developing a comprehensive system. For that they chose a tiny cultivar of tomato, the Micro-Tom. This cultivar differs from standard tomato cultivars only by two genes, but it can be planted very densely (the authors claimed that up to 1,357 plants/m^2 can be grown; this primary number of plants is a bit enigmatic). Micro-Tom plants flower very early and produce (small) mature fruits at two to three months after sowing. Numerous mutations were induced in Micro-Tom by EMS and by insertional mutagenesis. For the latter mutagenesis the authors built suitable constructs, based on *Ac/Ds* with selectable genes and reporter genes. Constructs could be moved into the tomato plants by *Agrobacterium*-mediated genetic transformation since Micro-Tom can be efficiently transformed by this technique.

This system was further applied for large scale transposon tagging of tomato (Meissner et al., 2000). The manipulated *Ds* was stabilized after it was inserted into numerous Micro-Tom plants. The manipulation of *Ds* included the engineering of a *Ds* construct into the borders of a T-DNA element. The *Ds* itself contained reporter genes (e.g. a gene, *GUS* for the expression of β-glucuronidase and *LUC*, a gene for the expression of luciferase). This enabled one to trace the mobility of *Ds* and the location of its insertion. A different T-DNA construct caused the production of transposase. The authors used LUC, without a promoter for promoter trapping and GUS, with a weak promoter, for enhancer trapping. The study indicated that the Micro-Tom *Ac/Ds* system developed by Levy and associates could lead to a high-throughput insertional mutagenesis of tomato, and that both Forward and Reverse Genetics can be achieved by this system. The authors detected 108 families with LUC expression, of these 55% were detected in flowers, 11% in fruits and 4% in seedlings. This indicated that *Ds* was frequently inserted into genes of tomato. The different insertion frequencies in fruits, flowers and seedlings could have several reasons. One of these is that more genes are required to structure a flower than a fruit. However, there are other possible reasons. The enhancer and promoter trapping achieved by the study of Meissner et al. (2000) was not reported previously in tomato. The Micro-Tom *Ac/Ds* system (e.g. Meissner et al., 1997) could also be useful for the isolation of transcriptional regulatory regions. As all the phases of fruit development in Micro-Tom plants are identical to those in regular tomato cultivars, this system also has value for tomato breeders. This system was further evaluated in a recent review (Emmanuel and Levy, 2002). The review and the article of Meissner et al. (2000) contain references to numerous gene tagging studies in *Arabidopsis*, tomato and in other plants, that have not been mentioned in this appendix.

CHAPTER 9

REFERENCES

Abe, H., Ohbayashi, F., Sugasaki, T., Kanchara, M., Teradam T., Shimada, T., Kawai, S., Mita, K., Kanamori, Y., Tamamoto, M.-T., and Oshiki, T. (2001) Two novel *Pao*-like retrotransposons (*Kamikaze* and *Yamato*) from the silkworm species *Bombyx mori* and *B. mandarina*: common structural features of *Pao*-like elements. *Mol. Genet. Genomics* 265: 375–385.

Adams, J.W., Kaufman, R.E., Kretschmer, P.J., Harrison, M., and Nienhuis, A.W. (1980) A family of long reiterated DNA sequences, one copy of which is next to the human beta globin gene. *Nucl. Acid Res.* 8: 6113–6128.

Adams, M.D., Celniker, S.E., Holt, R.A., and over 190 additional authors. (2000) The genome sequence of *Drosophila melanogaster*. *Science* 287: 2185–2197.

Agrawal, G.K., Yamazaki, M., Kobayashi, M., Hirochika, R., Miyao, A., and Hirochika, H. (2001) Screening of the rice viviparous mutants generated by endogenous retrotransposon Tos17 insertion. Tagging of a Zeaxanthin epoxidase gene and a novel OsTATC gene. *Plant Physiol.* 125: 1248–1257.

Ahlquist, P. (2002) RNA-dependent RNA polymerases, viruses, and RNA silencing. *Science* 296: 1270–1273.

Allfrey, V., Faulkner, R.M., and Mirsky, A.E. (1964) Acetylation and methylation of histones and their possible role in the regulation of RNA synthesis. *Proc. Natl. Acad. Sci. (USA)* 51: 786–794.

Allshire, R. (2002) RNAi and heterochromatin – a hushed-up affair. *Science* 297: 1818–1819.

Anaya, N. and Roncero, M.I.G. (1995) Skippy, a retrotransposon from the fungal plant pathogen *Fusarium oxysporum*. *Mol. Gen. Genet.* 249: 637–647.

Aravin, A.A., Naumova, N.M., Tulin, A.V., Vagin, V.V., Rozovsky, Y.M., and Gvozdev, V.A. (2001) Double-stranded RNA-mediated silencing of genomic tandem repeats and transposable elements in the *D. melanogaster* germline. *Curr. Biol.* 11: 1017–1027.

Arkhipova, I. and Meselson, M. (2000) Transposable elements in sexual and ancient asexual taxa. *Proc. Natl. Acad. Sci. (USA)* 97: 14473–14477.

Athma, P. and Peterson, T. (1991) *Ac* induces homologous recombination at the maize P locus. *Genetics* 128: 163–173.

Athma, P., Grotewold, E., and Peterson, T. (1992) Insertional mutagenesis of the maize P gene by intragenic transposition of *Ac*. *Genetics* 131: 199–209.

Atkinson, P.W., Warren, W.D., and O'Brochta, D.A. (1993) The *hobo* transposable element of *Drosophila* can be cross-mobilized in houseflies and excises like the *Ac* element of maize. *Proc. Natl. Acad. Sci. (USA)* 90: 9693–9697.

Auge-Gouillou, C., Bigot, Y., Pollet, N., Hamelin, M.H., Meunier-Rotival, M., and Periquet, G. (1995) Human and other mammalian genomes contain transposons of the *Mariner* family. *FEBS Lett.* 368: 541–546.

Avramova, Z. (2002) Heterochromatin in animals and plants, similarities and differences. *Plant Physiol.* 129: 40–49.

Avramova, Z., SanMiguel, P., Georgiera, E., and Bennetzen, J.L. (1995) Matrix attachment regions and transcribed sequences within a long chromosomal continuum containing maize *adhl. Plant Cell* 7: 1667–1680.

Avramova, Z., Tikhonov, A., Chen, M., and Bennetzen, J.L. (1998) Matrix attachment regions and structural colinearity in the genomes of two grass species. *Nucl. Acid Res.* 26: 761–767.

Bae, Y.-A., Moon, S.-Y., Kong, Y., Cho, S.-Y. and Rhyu, M.-G. (2001) *CsRN1* a novel active retrotransposon in a parasitic trematode, *Clonorchis sinensis*, discloses a new phylogenetic clade of Ty3/*gypsy*-like LTR retrotransposons. *Mol. Biol. Evol.* 18: 1474–1483.

Baker, B., Schell, J., Lörz, H., and Fedoroff, N. (1986) Transposition of the maize controlling element "Activator" in tobacco. *Proc. Natl. Acad. Sci. (USA)* 83: 4844–4848.

Baltimore, D. (1970) Viral RNA-dependent DNA polymerase. *Nature* 226: 1209–1211.

Bancroft, I., Bhatt, A.M., Sjodin, C., Scofield, S., Jones, J.D.G., and Dean, C. (1992) Development of an efficient two-element transposon tagging system in *Arabidopsis thaliana. Mol. Gen. Genet.* 233: 449–461.

Banks, J., Kingsbury, J., Rabay, V., Schiefelbeing, J.W., Nelson, O., and Fedoroff, N. (1985). The *Ac* and *Spm* controlling element families in maize. *Cold Spring Harbor Symp. Quant. Biol.* 50: 307–311.

Bannister, A.J., Schneider, R., and Kouzarides, T. (2002) Histone methylation: dynamic or static? *Cell* 109: 801–806.

Barth, P.T. and Datta, N. (1977) Two naturally occurring transposons indistinguishable from Tn7. *J. Gen. Bacteriol.* 102: 129–134.

Barth, P.T., Datta, N., Hedges, R.W., and Grinter, N.J. (1976) Transposition of a deoxyribonucleotide sequence encoding trimethoprim and streptomycin resistances from R483 to other replicons. *J. Bacteriol.* 125: 800–810.

Bass, B.L. (2000) Double-stranded RNA as a template for gene silencing. *Cell* 101: 235–238.

Baulcombe, D. (2001) Diced defence. *Nature* 409: 295–296.

Bayev, A.A. Jr., Krayev, A.S., Lyubomirskaya, N.V., Ilyin, Y.V., Skryabin, K.G., and Georgiev, G.P. (1980) The transposable element *Mdg3* in *Drosophila melanogaster* is flanked with the perfect direct and mismatched inverted repeats. *Nucl. Acid Res.* 8: 3263–3273.

Bayev, A.A. Jr., Lyubomirskaya, N.V., Dzhumagaliev, E.B., Ananiev, E.V., Amiantova, I.G., and Ilyin Y.V. (1984) Structural organization of transposable element *mdg4* from *Drosophila melanogaster* and a nucleotide sequence of its long terminal repeats. *Nucl. Acid Res.* 12: 3707–3723.

Beall, E.L. and Rio, D.C. (1998) Transposase makes critical contacts with, and is stimualted by, single-stranded DNA at the P element termini *in vitro. EMBO J.* 17: 2122–2136.

Becker, D., Lutticke, R., Li, M., and Starlinger, P. (1992) Control of excision frequency of maize transposable element *Ds* in *Petunia* protoplasts. *Proc. Natl. Acad. Sci. (USA)* 89: 5552–5556.

Becker, H.-A. and Kunze, R. (1997) Maize *Activator* transposase has a bipartite DNA binding domain that recognizes subterminal sequences and the terminal inverted repeats. *Mol. Gen. Genet.* 254: 219–230.

Beguiristain, T., Grandbastien, M.-A., Puigdomenech, P., and Casacuberta, J.M. (2001) Three Tnt1 subfamilies show different stress-associated patterns of expression in tobacco. Consequences for retrotransposon control and evolution in plants. *Plant Physiol.* 127: 212–221.

Behrens, U., Fedoroff, N., Laird, A., Müller-Neumann, M., Starlinger, P., and Yoder, J. (1984) Cloning of the *Zea mays* controlling element *Ac* from the *wx*-m7 allele. *Mol. Gen. Genet.* 194: 346–347.

Bellfort, M., Derbyshire, V., Parker, M.M., Cousinean, B., and Lambowitz, A.M. (2002) Mobile introns: pathways and proteins. In: *Mobile DNA II*. Craig, N., Craigie, R., Gellert, M., and Lambowitz, A.M. (eds). ASM Press, Washington, DC. pp. 761–783.

Bender, J. (2001) A vicious cycle: RNA silencing and DNA methylation in plants. *Cell* 106: 129–132.

Benito, M.I. and Walbot, V. (1994) The terminal inverted repeat sequences of *MuDR* are functionally active promoters in maize cells. *Maydica* 39: 255–264.

Bennetzen, J.I. (2000) Transposable element contributions in plant gene and genome evolution. *Plant Mol. Biol.* 42: 251–269.

Bennetzen, J.L. (1984) Transposable element *Mu1* is found in multiple copies only in Robertson's mutator maize lines. *J. Mol. Appl. Genet.* 2: 519–524.

Bennetzen, J.L. (1996) The contributions of retroelements to plant genome organization, function and evolution. *Trends Microbiol.* 4: 347–353.

Bennetzen, J.L., Springer, P.S., Cresse, A.D., and Hendrick, M. (1993) Specificity and regulation of the mutator transposable element system in maize. *Crit. Rev. Plant Sciences* 12: 57–95.

Bennetzen, J.L., Swanson, J., Taylor, W.C., and Freeling, M. (1984) DNA insertion in the first intron of maize *Adh1* affects message levels: Cloning of pogenitor and mutant *Adh1* alleles. *Proc. Natl. Acad. Sci. (USA)* 81: 4125–4128.

Berg, D.E., Davis, J., Allet, B., and Rochaix, J.-D. (1975) Transposition of R factor genes to bacteriophage λ. *Proc. Natl. Acad. Sci. (USA)* 72: 3628–3632.

Berger, B. and Haas, D. (2001) Transposase and cointegrase: specialized transposition proteins of the bacterial insertion sequence IS21 and related elements. *Cell. Mol. Life Sci.* 58: 403–419.

Berger, S.L. (2002) Histone modifications in transcriptional regulation. *Curr. Opin. Genet. and Develop.* 12: 142–148.

Bernstein, E., Caudy, A.A., Hammond, S.M., and Hannon, G.J. (2001) Role for a bidentate ribonuclease in the initiation step of RNA interference. *Nature* 409: 363–366.

Besansky, N.J. (1990) A retrotransposable element from the mosquito *Anopheles gambiae*. *Mol. Cell. Biol.* 10: 863–871.

Bhasin, A., Goryshin, I.Y., and Reznikoff, W.S. (1999) Hairpin formation in Tn5 transposition. *J. Biol. Chem.* 274: 37021–37029.

Bhatt, A.M., Page, T., Lawson, E.J.R., Lister, C., and Dean, C. (1996) Use of *Ac* as an insertional mutagen in *Arabidopsis*. *Plant J.* 9: 935–945.

Bi, Y.-N. and Laten, H.M. (1996) Sequence analysis of a cDNA containing the *gag* and *prot* regions of the soybean retrovirus-like element, *SIRE-1*. *Plant Mol. Biol.* 30: 1315–1319.

Biery, M.C., Steward, F., Stellwagen, A.E., Raleigh, E.A., and Craig, N.L. (2000) A simple *in vitro* Tn7-based transposition system with low target site selectivity for genome and gene analysis. *Nucl. Acid Res.* 28: 1067–1077.

Bingham, P.M. and Zachar, S. (1989) Retrotransposons and the FB transposon from *Drosophila melanogaster*. In: *Mobile DNA*. Berg, D.E. and Howe M.M. (eds.) ASM Press, Washington, DC. pp. 485–502.

Bingham, P.M., Kidwell, M.G., and Rubin, G.M. (1982) The molecular basis of P-M hybrid dysgenesis: The role of the P element, a P strain-specific transposon family. *Cell* 29: 995–1004.

Birchler, J.A., Bhadra, M.P., and Bhadra, U. (2000) Making noise about silence: repression of repeated genes in animals. *Curr. Opin. Genet. Develop.* 10: 211–216.

Bird, A. (2002) Methylation patterns and epigenetic memory. *Genes Develop.* 16: 6–21.

Blackman, R.K. and Gelbart, W.M. (1989) The transposable element *hobo* of *Drosophila melanogaster*. In: *Mobile DNA*. Berg, D.E. and Howe, M.M. (eds.) ASM Press, Washington DC. pp. 523–529.

Boehm, U., Heinlein, M., Behrens, U., and Kunze, R. (1995) One of three nuclear localization signals of maize *Activator* (*Ac*) transposase overlaps the DNA-binding domain. *Plant J.* 7: 441–451.

Boeke, J.D. (2002) Putting mobile DNA to work: the toolbox. In: *Mobile DNA II*. Craig, N.L., Craigie, R., Gellert, M., and Lambowitz, A.M. (eds.) ASM Press, Washington, DC. pp. 24–37.

Boeke, J.D. and Sandmeyer, S.B. (1991) Yeast transposable elements. In: *The Molecular and Cellular Biology of the Yeast Saccharomyces*. Vol. 1. Broach, J.R., Pringle, J.R., and Jones, E.W. (eds.) Cold Spring Harbor Lab. Press. pp. 193–261.

Boeke, J.D., Garfinker, D.J., Styles, C.A., and Fink, G.R. (1985) Ty elements transpose through an RNA intermediate. *Cell* 40: 491–500.

Bonas, U., Sommer, H., and Saedler, H. (1984a) The 17-kb *Tam1* element of *Antirrhinum majus* induces a 3-bp duplication upon integration into the chalcone synthase gene. *EMBO J.* 3: 1015–1019.

Bonas, U., Sommer, H., Harrison, B.J., and Saedler, H. (1984b) The transposable element *Tam1* of *Antirrhinum majus* is 17 kb long. *Mol. Gen. Genet.* 194: 138–143.

Bonner, J., Huang, R.-C., and Maheshwari, N. (1961) The physical state of newly synthesized RNA. *Proc. Nat. Acad. Sci. (USA)* 47: 1548–1554.

Bottinger, P., Steinmetz, A., Schieder, O., and Pickardt, T. (2001) *Agrobacterium*-mediated transformation of *Vicia faba*. *Mol. Breeding* 8: 243–254.

Bowen, N.J. and McDonald, J.F. (2001) *Drosophila* euchromatic LTR retrotransposons are much younger than the host species in which they reside. *Genome Res.* 11: 1527–1540.

Branciforte, D. and Martin, S.L. (1994) Developmental and cell type specificity of *LINE-1* expression in mouse testis: implications for transposition. *Mol. Cell. Biol.* 14: 2584–2592.

Bregliano, J.-C. and Kidwell, M.G. (1983) Hybrid dysgenesis determinants. In: *Mobile Genetic Elements*. Shapiro, J.A. (ed.) Academic Press, NY. pp. 363–410.

Breiman, A., Felsenberg, T., and Galun, E. (1989) Is *No* region variability in wheat invariably caused by tissue culture? *Theor. Appl. Genet.* 77: 809–814.

Brenner, S. (1974) The genetics of *Caenorhabditis elegans*. *Genetics* 77: 71–94.

Brenner, S., Elgar, G., Sanford, R., Macrae, A., Venkatesh, B., and Aparicio, S. (1993) Characterization of the pufferfish (*Fugu*) genome as a compact vertebrate genome. *Nature* 366: 265–268.

Briggs, S.D., Xiao, T., Sun, Z.-W, Caldwell, J.A., Shahanowiz, J., Hunt, D.E., Allis, C.D., and Strahl, B.D. (2002) Trans-histone regulatory pathway in chromatin. *Nature* 418: 498.

Brink, R.A. and Nilan, R.A. (1952) The relation between light variegated and medium variegated pericarp in maize. *Genetics* 37: 519–544.

Britten, R.J., McCormack, T.J., Mears, T.L., and Davidson, E.H. (1995) *Gypsy*/Ty3-class retrotransposons integrated in DNA of herning, tunicate and echinogerms. *J. Mol. Evol.* 40: 13–24.

Briza, J., Niedermeierova, H., Pavingerova, D., Thomas, C.M., Klimyuk, V.I., and Jones, J.D.G. (2002) Transposition patterns of unlinked transposed *Ds* elements from two T-DNA location in tomato chromosomes 7 and 8. *Mol. Genet. Genomics* 266: 882–890.

Bucheton, A. (1995) The relationship betwee the *flamenco* gene and *gypsy* in *Drosophila*: how to tame a retrovirus. *Trends Genet.* 11: 349–353.

Bureau, T.E. and Wessler, S.R. (1992) *Tourist*: A large family of small inverted repeat elements frequently associated with maize genes. *Plant Cell* 4: 1283–1294.

Bureau, T.E. and Wessler, S.R. (1994a) *Stowaway*: A new family of inverted repeat elements associated with the genes of both monocotyledonous and dicotyledonous plants. *Plant Cell* 6: 907–916.

Bureau, T.E. and Wessler, S.R. (1994b) Mobile inverted-repeat elements of the *Tourist* family are associated with the genes of many cereal grasses. *Proc. Natl. Acad. Sci. (USA)* 91: 1411–1415.

Bureau, T.E., Ronald, P.C., and Wessler, S.R. (1996). A computer-based systematic survey reveals the predominance of small inverted-repeat elements in wild-type rice genes. *Proc. Natl. Acad. Sci. (USA)* 93: 8524–8529.

Burke, W.D., Eickbush, D.G., Xiong, Y., Jakubczak, J., and Eickbush, T.H. (1993) Sequence relationship of retrotransposable elements R1 and R2 within and between divergent insect species. *Mol. Biol. Evol.* 10: 163–185.

Burr, B. and Burr, F.A. (1980) Detection of changes in maize DNA at the *Shrunken* locus due to the intervention of *Ds* elements. *Cold Spring Harbor Symp. Quant. Biol.* 45: 463–465.

Burr, B. and Burr, F.A. (1981) Controlling-element events at the *Shrunken* locus in maize. *Genetics* 98: 143–156.

Burr, B. and Burr, F.A. (1982) *Ds* controlling elements of maize at the *Shrunken* locus are large and dissimilar insertions. *Cell* 29: 977–986.

Burwinkel, B. and Kilimann, M.W. (1998) Unequal homologous recombination between *LINE-1* elements as a mutational mechanism in human genetic disease. *J. Mol. Biol.* 277: 513–517.

Cambareri, E.B., Helber, J., and Kinsey, J.A. (1994) *Tad1-1*, an active *LINE*-like element of *Neurospora crassa. Mol. Gen. Genet.* 242: 658–665.

Cameron, J.R., Loh, E.Y., and Davis, R.W. (1979) Evidence for tranposition of dispersed repetitive DNA families in yeast. *Cell* 16: 739–751.

Campbell, A. (1980) Some general questions about movable elements and their implications. *Cold Spring Harbor Symp. Quant. Biol.* 45: 1–9.

Campbell, A. (1983) Bacteriophage λ. In: *Mobile Genetic Elements*. Shapiro, J.A. (ed.) Academic Press, NY. pp. 65–103.

Campuzano, S., Balcells, L., Villares, R., Carramolino, L., Garcia-Alonso, L., and Modollell, J. (1986) Excess function *Hairy-wing* mutations caused by *gypsy* and *copia* insertions within structural genes of the achaete-scute locus of *Drosophila. Cell* 44: 303–312.

Cann, R.L., Stoneking, M., and Wilson (1987) Mitochondrial DNA and human evolution. *Nature* 325: 31–36.

Caplen, N.J., Parrish, S., Imani, F., Fire, A., and Morgan, R.A. (2002) Specific inhibition of gene expression by small double-stranded RNAs in invertebrate and vertebrate systems. *Proc. Natl. Acad. Sci. (USA)* 98: 9742–9747.

Cardon, G.H., Frey, M., Seadler, H., and Gierl, A. (1993). Mobility of the maize transposable element *En/Spm* in *Arabidopsis thaliana. Plant J.* 3: 773–784.

Carmichael, G.G. (2002) Silencing viruses wth RNA. *Nature* 418: 379–380.

Carmirand, A., St.-Pierre, B., Marineau, C., and Brisson, N. (1990) Occurrence of a *copia*-like transposable element in one of the introns of the potato starch phosphorylase gene. *Mol. Gen. Genet.* 224: 33–39.

Carpenter, R. and Coen, E.S. (1990) Floral homeotic mutations produced by transposon-mutagenesis in *Antirrhinum majus. Genes Develop.* 4: 1483–1493.

Carroll, B.J., Klimyuk, V.I., Thomas, C.M., Bishop, G.J., Harrison, K., Scofield, S.R., and Jones, J.D.G. (1995) Germinal transpositions of the maize element *Dissociation* from T-DNA loci in tomato. *Genetics* 139: 407–420.

Cartegni, L., Chew, S.L., and Krainer, A.R. (2002) Listening to silence and understanding nonsense: exonic mutations that affect splicing. *Nature Rev.* 3: 285–298.

Casacuberta, E., Casacuberta, J.M., Puigdomenech, P., and Monfort, A. (1998) Presence of miniature inverted-repeat transposable elements (*MITEs*) in the genome of *Arabidopsis thaliana*: characterisation of the emigrant family of elements. *Plant J.* 16: 79–85.

Casacuberta, J.M. and Granbastien, M.-A. (1993) Characterisation of LTR sequences involved in the protoplast specific expression of the tobacco *Tnt1* retrotransposon. *Nucl. Acid Res.* 21: 2087–2093.

Casacuberta, J.M., Vernhettes, S., and Grandbastien, M.-A. (1995) Sequence variability within the tobacco retrotransposon *Tnt1* population. *EMBO J.* 14: 2670–2678.

Casavant, N.C., Sherman, A.M., and Wichman, H.A. (1996) Two persistent *LINE*-1 lineages in *Peromyuscus* have unequal rates of evolution. *Genetics* 142: 1289–1298.

Catalanotto, C., Azzsalin, G., Macino, G., and Cogoni, C. (2000) Gene silencing in worms and fungi. *Nature* 404: 245.

Chaleff, D.T. and Fink, G.R. (1980) Genetic events associated with an insertion mutation in yeast. *Cell* 21: 227–237.

Chalmers, R., Sewitz, S., Lipkow, K., and Crellin, P. (2000) Complete nucleotide sequence of Tn10. *J. Bacteriol.* 182: 2970–2972.

Chalvet, F., Teysset, L., Terzian, C., Prud'homme, N., Santamaria, P., Bucheton, A., and Pelisson, A. (1999) Proviral amplification of the *gypsy* endogenous retrovirus of *Drosophila melanogaster* involves *env*-independent invasion of the female germline. *EMBO J.* 18: 2659–2669.

Chandler, M. and Mahillon, J. (2002) Insertion sequences revisited. In: *Mobile DNA II.* Craig, N.L., Craigie, R., Gellert, M., and Lambowitz, A.M. (eds) AMS Press, Washington, DC. pp. 305–366.

Chandler, V.L. and Hardeman, K.J. (1992) The *Mu* elements of *Zea mays*. *Adv. Genet.* 30: 77–122.

Chandler, V.L. and Walbot, V. (1986) DNA modification of a maize transposable element correlates with loss of activity. *Proc. Natl. Acad. Sci. (USA)* 83: 1767–1771.

Chen, J.-Y. and Fonzi, W.A. (1992) A temperature-regulated, retrotransposon-like element from *Candida albicans*. *J. Bacteriol.* 174: 5624–5632.

Cheung, A., Allis, C.D., and Sassone-Corsi, P. (2000) Signaling to chromatin through histone modifications. *Cell* 103: 263–271.

Chiu, Y.-L, and Rana, T.M. (2002) RNAi in human cells: basic structural and functional features of small interfering RNA. *Mol. Cell* 10: 549–561.

Chomet, P., Lisch, D., Hardeman, K.J., Chandler, V.L., and Freeling, M. (1991) Identification of a regulatory transposon that controls the *Mutator* transposable element system in maize. *Genetics* 129: 261–270.

Chomet, P.S., Wessler, S., and Dellaporta, S.L. (1987) Inactivation of the maize transposable element *Activator (Ac)* is associated with its DNA modification. *EMBO J.* 6: 295–302.

Christensen, S., Pont-Kingdon, G., and Carroll, D. (2000) Target specificity of the endonuclease from the *Xenopus laevis* non-long terminal repeat retrotransposon, *Tx1L. Mol. Cell. Biol.* 20: 119–1226.

Chuang, C.-F. and Meyerowitz, E.M. (2000) Specific and heritable genetic interference by double-stranded RNA in *Arabidopsis thaliana. Proc. Nat. Acad. Sci. (USA)* 97: 4985–4990.

Coen, E.S. and Carpenter, R. (1986) Transposable elements in *Antirrhinum majus*: generators of genetic diversity. *Trends Genet.* 2: 292–296.

Coen, E.S. and Meyerowitz, E.M. (1991) The war of the whorls: genetic interactions controlling flower development. *Nature* 353: 31–37.

Coen, E.S., Carpenter, R., and Martin, C. (1986) Transposable elements generate novel spatial patterns of gene expression in *Antirrhinum majus. Cell* 47: 285–296.

Coen, E.S., Robbins, T.P., Almeida, J., Hudson, A., and Carpenter, R. (1989) Consequences and mechanisms of transposition in *Antirrhinum majus*. In: *Mobile DNA.* Berg, D.E. and Howe, M.M. (eds). ASM Press, Washington, DC. pp. 413–436.

Cogoni, C. and Macino, G. (1999) Gene silencing in *Neurospora crassa* requires a protein homologous to RNA-dependent RNA polymerase. *Nature* 399: 166–169.

Cohen, D.E. and Lee, J.T. (2002) X-chromsome inactivation and the search for chromosome-wide silencers. *Curr. Opin. Genet. Develop.* 12: 219–224.

Comfort, N.C. (2001) *The Tangled Field: Barbara McClintock's Search for the Patterns of Genetic Control.* Harvard University Press, Cambridge, MA, 357 p.

Corces, V.G. and Geyer, P.K. (1991) Interactions of retrotransposons with the host genome. *Trends Genet.* 7: 86–90.

Coupland, G., Baker, B., Schell, J., and Starlinger, P. (1988) Characterization of the maize transposable element *Ac* by internal deletions. *EMBO J.* 7: 3653–3659.

Courage, U., Döring, H.-P., Frommer, W.-B., Kunze, R., Laird, A., Merckelbach, A., Müller-Neumann, M., Riegel, J., Starlinger, P., Tillmann, E., Weck, E., Werr, W., and Yoder, J. (1984) Transposable elements *Ac* and *Ds* at the shrunken, waxy, and alcohol dehydrogenase 1 loci in *Zea mays* L. *Cold Spring Harbor Symp. Quant. Biol.* 49: 329–338.

Courage-Tebbe, U., Döring, H.-P., Fedoroff, N., and Starlinger, P. (1983) The controlling element *Ds* at the shrunken locus in *Zea mays*: structure of the unstable *shm5933* allele and several revertants. *Cell* 34: 383–393.

Courtial, B., Feuerbach, F., Eberhard, S., Rohmer, L., Chiapello, H., Camilleri, C., and Lucas, H. (2001) *Tnt1* transposition events are induced by *in vitro* transformation of *Arabidopsis thaliana*, and transposed copies integrate into genes. *Mol. Genet. Genomics* 265: 32–42.

Craig, N.L. (1989) Transposon Tn7. In: *Mobile DNA.* Berg, D.E. and Howe, M.M. (eds) ASM Press, Washington, DC. pp. 211–225.

Craig, N.L. (1996) Transposon Tn7. *Curr. Topics Microbiol. Immunol.* 204: 27–48.

Craig, N.L. (2002) Tn7. In: *Mobile DNA II.* Craig, N.L., Craigie, R., Gellert, M., and Lambowitz, A.M. (eds) ASM Press, Washington, DC. pp. 423–456.

D'Ambrosio, E., Waitzkin, S.D., Witney, F.R., Salemme, A., and Furano, A.V. (1986) Structure of the highly repeated, long interspersed DNA family (*LINE* or *L1rn*) of the rat. *Mol. Cell. Biol.* 6: 411–424.

Daboussi, M.J. (1996) Fungal transposable elements: generators of diversity and genetic tools. *J. Genet.* 75: 325–339.

Daboussi, M.J. and Langin, T. (1994) Transposable elements in the fungal plant pathogen *Fusarium oxysporum. Genetica* 93: 49–59.

Daboussi, M.J., Daviere, J.-M, Graziani, S., and Langin, AT. (2002) Evolution of the *Fot1* transposons in the genus *Fusarium*: Discontinuous distribution and epigenetic inactivation. *Mol. Biol. Evol.* 19: 510–520.

Daboussi, M.J., Langin, T., and Brygoo, Y. (1992) *Fot1*, a new family of fungal transposable elements. *Mol. Gen. Genet.* 232: 12–16.

Dalle Nogare, D.E., Clark, M.S., Elgar, G., Frame, I.G., and Poulter, R.T.M. (2002) Xena, a full-length basal retroelement from tetradontid fish. *Mol. Biol. Evol.* 19: 247–255.

Dalmay, T., Hamilton, A., Rudd, S., Angell, S., and Baulcombe, D.C. (2000) An RNA-dependent RNA polymerase gene in *Aabidopsis* is required for posttranscriptional gene silencing mediated by a transgene but not by a virus. *Cell* 101: 543–553.

Dalmay, T., Horsefield, R., Braunstein, T.H., and Baulcombe, D.C. (2001) SDE3 encodes an RNA helicase required for post-transcriptional gene silencing in *Arabidopsis. EMBO J.* 20: 2069–2077.

Dangl, J.L. and Jones, J.D.G. (2001) Plant pathogens and integrated defence responses to infection. *Nature* 411: 826–833.

Dash, S. and Peterson, P.A. (1994) Frequent loss of the *En* transposable element after excision and its relation to chromosome replication in maize. *Genetics* 136: 653–671.

Davies, D.R., Braam, L.M., Reznikoff, W.S., and Rayment, I. (1999) The three-dimensional structure of a Tn5 transposase-related protein determined to 2.9-A resolution. *J. Biol. Chem.* 274: 11904–11913.

Davies, D.R., Goryshin, I.Y., Reznikoff, W.S., and Rayment, I. (2000) Three-dimensional structure of the Tn5 synaptic complex transposition intermediate. *Science* 289: 77–85.

Day, A., Schirmer-Rahire, M., Kuchka, M.R., Mayfield, S., and Rochaix, J.-D. (1998) A transposon with unusual arrangement of long terminal repeats in the green alga *Chlamydomonas reinhardtii*. *EMBO J.* 7: 1917–1927.

De Keukeleire, P., Maes, T., Sauer, M., Zethof, J., Van Montagu, M., and Gerats, T. (2001) Analysis by transposon display of the behavior of the *dTph1* element family during ontogeny and inbreeding of *Petunia*-hybrida. *Mol. Genet. Genomics* 265: 72–81.

Dean, C., Sjodin, C., Page, T., Jones, J., and Lister, C. (1992) Behaviour of the maize transposable element *Ac* in *Arabidopsis thaliana*. *Plant J.* 2: 69–81.

DeBerardinis, R.J., Goodier, J.L., Ostertag, E.M., and Kazazian, Jr. H.H. (1998) Rapid amplification of a retrotransposon subfamily is evolving the mosue genome. *Nature Genet.* 20: 288–290.

Deininger, P.L. (1989) *SINEs*: Short interspersed repeated DNA elements in higher eucaryotes In: *Mobile DNA*. Berg, D.E. and Howe, M.M. (eds.) ASM Press, Washington, DC. pp. 619–637.

Deininger, P.L. and Batzer, M.A. (1999) Alu repeats and human disease. Mol. Genet. Met. 67: 183–193.

Dej, K.J., Gerasimova, T., Corces, V.G., and Boeke, J.D. (1998) A hotspot for the *Drosophila gypsy* retroelement in the *ovo* locus. *Nucl. Acid Res.* 26: 4019–4024.

Dhillon, N. and Kamakaka, R.T. (2002) Breaking through to the other side: silencers and barriers. *Curr. Opin. Genet. Develop.* 12: 188–192.

Di Nocera, P.P. (1988) Close relationship between non-viral retroposons in *Drosophila melanogster*. *Nucl. Acid Res.* 16: 4041–4052.

Di Nocera, P.P. and Casari, G. (1987) Related polypeptides are encoded by *Drosophila* F elements, I factors, and mammalian L1 sequences. *Proc. Natl. Acad. Sci. (USA)* 84: 5843–5847.

Dillon, N. and Festenstein, R. (2002) Unravelling heterochromatin: competition between positive and negative factors regulates accessibility. *Trends Genet.* 18: 252–258.

Doak, T.G., Doerder, F.P., Jahn, C.L., and Herrick, G. (1994) A proposed superfamily of transposase genes: transposon-like elements in ciliated protozoa and a common "D35E" motif. *Proc. Natl. Acad. Sci. (USA)* 91: 942–946.

Dobinson, K.F., Harris, R.E., and Hamer, J.E. (1993) *Grasshopper*, a long terminal repeat (LTR) retroelement in the phytopathogenic fungus *Magnaporthe grisea*. *Mol. Plant Microbe Inter.* 6: 114–126.

Dombroski, B.A., Mathia, S.L., Nanthakumar, E., Scott, A.F., and Kazazian, Jr. H.H. (1991) Isolation of an active human transposable element. *Science* 254: 1805–1808.

Dooner, H.P. (1989) Tagging genes with maize transposable elements. An overview. *Maydica* 34: 73–88.

Döring, H.-P., Geiser, M., and Starlinger, P. (1981) Transposable element *Ds* at the shrunken locus in *Zea mays*. *Mol. Gen. Genet.* 184: 377–380.

Döring, H.-P., Tillmann, E., and Starlinger, P. (1984) DNA sequence of the maize transposable element *Dissociation*. *Nature* 307: 127–130.

Döring, H.P., Freeling, M., Hake, S., Johns, M.A., Kunze, R., Merckelbach, A., Salamini, F., and Starlinger, P. (1984) A *Ds*-mutation of the *Adh1* gene in *Zea mays* L. *Mol. Gen. Genet.* 193: 199–204.

Duvernell, D.D. and Turner, B.J. (1998) *Swimmer 1*, a new low-copy-number *LINE* family in teleost genomes with sequence similarity to mammalian L1. *Mol. Biol. Evol.* 15: 1791–1793.

Eichinger, D.J. and Boeke, J.D. (1988) The DNA intermediate in yeast Ty1 element transposition copurifies with virus-like particles: cell-free Ty1 transposition. *Cell* 54: 955–966.

Eickbush, T.H. and Malik, H.S. (2002) Origins and evolution of retrotransposons. In: *Mobile DNA II*. Craig, N.L., Craigie, R., Gellert, M., and Lambowitz, A.M. (eds) ASM Press, Washington, DC. pp. 1111–1146.

El Amrani, A., Marie, L., Ainouche, A., Nicolas, J., and Couee, I. (2002) Genome-wide distribution and potential regulatory functions of AtATE, a novel family of miniature inverted-repeat transposable elements in *Arabidopsis thaliana*. *Mol. Genet. Genomics* 267: 459–471.

Elbashir, S.M., Harborth, J., Lendeckel, W., Yalcin, A., Weber, K., and Tuschl, T. (2001b) Duplexes of 21-nucleotide RNAs mediate RNA interference in cultured mammalian cells. *Nature* 411: 494–498.

Elbashir, S.M., Lendeckel, W., and Tuschl, T. (2001a) RNA interference is mediated by 21- and 22-nucleotide RNAS. *Genes Develop.* 15: 188–200.

Elder, R.T., St. John, T.P., Stinchcomb, D.T., and Davis, R.W. (1980) Studies on the transposable element Ty1 of yeast. I. RNA homologous to Ty1. *Cold Spring Harbor Symp. Quant. Biol.* 45: 581–584.

Elgar, G., Sanford, R., Aparicio, S., Macrae, A., Venkatesh, B., and Brenner, S. (1996) Small is beautiful: comparative genomics with pufferfish (*Fugu rubripes*). *Trends Genet.* 12: 145–149.

Emerson, R.A. (1914) The inheritance of a recurring somatic variation in variegated ears of maize. *Amer. Naturalist* 48: 87–115.

Emerson, R.A., Beadle, G.W., and Fraser, A.C. (1935) A summary of linkage studies in maize. *Memoir* 180: 1–83, Cornell University, Agr. Exp. Station, Ithaca, NY.

Emmanuel, E. and Levy, A.A. (2002) Tomato mutants as tools for functional genomics. *Curr. Opin. Plant Biol.* 5: 112–117.

Emmons, S.W. and Yesner, L. (1984) High-frequency excision of transposable element Tc1 in the nematode *Caenorhabditis elegans* is limited to somatic cells. *Cell* 36: 599–605.

Emmons, S.W., Rosenzweig, B., and Hirsh, D. (1980) Arrangement of repeated sequences in the DNA of the nematode *Caenorhabditis elegans*. *J. Mol. Biol.* 144: 481–500.

Emmons, S.W., Yesner, L., Ruan, K., and Katzenberg, D. (1983) Evidence for a transposon in *Caenorhabditis elegans*. *Cell* 32: 55–65.

Engels, W.R. (1983) The P family of transposable elements in *Drosophila*. *Ann. Rev. Genet.* 17: 315–344.

Engels, W.R. (1989) P elements in *Drosophila melanogaster*. In: *Mobile DNA*. Berg, D.E. and Howe, M.M. (eds) ASM Press, Washington, DC. pp. 437–484.

Engels, W.R. (1996) P elements in *Drosophila*. *Curr. Topics Microbiol. Immunol.* 204: 104–123.

Errede, B., Cardillo, T.S., Sherman, E., Dubois, E., Deschamps, J., and Uliame, J.M. (1980a) Mating signals control expression of mutations resulting from insertion of a transposable repetitive element adjacent to diverse yeast genes. *Cell* 22: 427–436.

Errede, B., Cardillo, T.S., Wever, G., and Sherman, F. (1980b) Studies on transposable elements of yeast. I. ROAM mutations causing increased expression of yeast genes: their activation by signals directed toward conjugation functions and their formation by insertion of Ty1 repetitive elements. *Cold Spring Harbor Symp. Quant. Biol.* 45: 593–602.

Esnault, C., Maestre, J., and Heidmann, T. (2000) Human line retrotransposons generate processed pseudogenes. *Nature Genet.* 24: 363–367.

Essers, L., Adolphs, R.H., and Kunze, R. (2000) A highly conserved domain of the maize activator transposase is involved in dimerization. *Plant Cell* 12: 211–223.

Evgen'ev, M.B., Zelentsova, H., Poluectova, H., Lyosin, G.T., Veleikodonskaja, V., Pyatkov, K.I., Zhirotovsky, L.A., and Kidwell, M.G. (2000) Mobile elements and chromosomal evolution in the group *virilis* of *Drosophila*. *Proc. Natl. Acad. Sci. (USA)* 97: 11337–11342.

Evgen'ev, M.B., Zelentsova, H., Shostak, N., Kozitsina, M., Barskyi, V., Lankenau, D.-H., and Croces, V.G. (1997). *Penelope*, a new family of transposable elements and its possible role in hybrid dysgenesis in *Drosophila viridis*. *Proc. Natl. Acad. Sci. (USA)* 94: 196–201.

Fanning, T. and Singer, M. (1987b) The LINE-1 DNA sequences in four mammalian orders predict proteins that conserve homologies to retrovirus proteins. *Nucl. Acid Res.* 15: 2251–2260.

Fanning, T.G. (1983) Size and structure of the highly repetitive BAM H1 element in mice. *Nucl. Acid Res.* 11: 5073–5091.

Fanning, T.G. and Singer, M.F. (1987a) LINE-1: a mammalian transposable element. *Biochim. Biophys. Acta* 910: 203–212.

Farabaugh, F.J. and Fink, G.R. (1980) Insertion of the eukaryotic transposable element Ty1 creates a 5-base pair duplication. *Nature* 286: 352–356.

Farman, M.L., Tosa, Y., Nitta, N., and Leong, S.A. (1996) *Maggy*, a retrotransposon in the genome of the rice blast fungus *Magnaporthe grisea*. *Mol. Gen. Genet.* 251: 665–674.

Fawcett, D.H., Lister, C.K., Kellett, E., and Finnegan, D.J. (1986) Transposable elements controlling I-R hybid dysgenesis in *D. melanogaster* are similar to mammalian LINEs. *Cell* 47: 1007–1015.

Fedoroff, N. (1989a) Maize transposable elements. In: *Mobile DNA*. Berg, D.E. and Howe, M.M. (eds.) ASM Press, Washington DC, pp. 375–411.

Fedoroff, N. (1989b) The heritable activation of cryptic suppressor-mutator elements by an active element. *Genetics* 121: 591–608.

Fedoroff, N. (2000) Transposons and genome evolution in plants. *Proc. Natl. Acad. Sci. (USA)* 97: 7002–7007.

Fedoroff, N. (2002) Control of mobile DNA. In: *Mobile DNA II*. Craig, N., Craigie, R., Gellert, M., and Lambowitz, A.M. (eds) ASM Press, Washington, DC. pp. 997–1007.

Fedoroff, N. and Botstein, D. (1992) *The Dynamic Genome – Barbara McClintock's Ideas in the Century of Genetics*. Cold Spring Harbor Laboratory Press, NY. 442 p.

Fedoroff, N., Shure, M., Kelly, S., Johns, M., Furtek, D., Schiefelbein, J., and Nelson, O. (1984) Isolation of *Spm* controlling elements from maize. *Cold Spring Harbor Symp. Quant. Biol.* 49: 339–345.

Fedoroff, N., Wessler, S., and Shure, M. (1983) Isolation of the transposable maize controlling elements *Ac* and *Ds*. *Cell* 35: 235–242.

Fedoroff, N.V. (1983) Controlling elements in maize. In: *Mobile Genetic Elements*. Shapiro, J.A. (ed.) Academic Press, NY. pp. 1–35.

Fedoroff, N.V. (1989c) About maize transposable elements and development. *Cell* 56: 181–191.

Fedoroff, N.V. (1995) DNA methylation and activity of the maize *Spm* transposable element. *Curr. Topics. Microbiol. Immun.* 197: 144–164.

Fedoroff, N.V., Furtek, D.B., and Nelson, O.E. Jr. (1984) Cloning of the bronze locus in maize by a simple and generalizable procedure using the transposable controlling element *Activator* (*Ac*). *Proc. Natl. Acad. Sci. (USA)* 81: 3825–3829.

Felder, H., Herzceg, A., de Chastonay, Y., Aeby, P., and Tobler, H. (1994) *Tas*, a retrotransposon from the parasitic nematode *Ascaris lumbricoides*. *Gene* 149: 219–225.

Feng, Q., Schumann, G., and Boeke, J.D. (1998) Retrotransposon R1Bm endonuclease cleaves the target sequence. *Proc. Natl. Acad. Sci. (USA)* 95: 2083–2088.

Feschotte, C. and Mouches, C. (2000) Evidence that a family of miniature inverted-repeat transposable elements (MITEs) roam the *Arabidopsis thaliana* genome has arisen from a *pogo*-like DNA transposon. *Mol. Biol. Evol.* 17: 730–737.

Feschotte, C. and Wessler, S.R. (2002) *Mariner*-like transposases are widespread and diverse in flowering plants. *Proc. Natl. Acad. Sci. (USA)* 99: 280–285.

Feschotte, C., Jiang, N., and Wessler, S.R. (2002) Plant transposable elements: where genetics meets genomics. *Nature Rev. Genetics* 3: 329–341.

Fiandt, M., Szybalski, W., and Malamy, M.H. (1972) Polar mutations in *lac*, *gal* and Phage λ consist of a few IS-DNA sequences inserted with either orientation. *Mol. Gen. Genet.* 119: 223–231.

Fincham, J.R.S. and Sastry, G.R.K. (1974) Controlling elements in maize. *Ann. Rev. Genet.* 8: 15–50.

Fink, G.R., Farabaugh, P.J., Roeder, G.S., and Chaleff, D. (1980) Transposable elements (Ty) in yeast. *Cold Spring Harbor Symp. Quant. Biol.* 45: 575–580.

Finnegan, D.J., Rubin, G.M., Young, M.W., and Hogness, D.S. (1977) Repeated gene families in *Drosophila melanogaster. Cold Spring Harbor Symp. Quant. Biol.* 42: 1053–1063.

Finnegan, E.J., Taylor, B.H., Craig, S., and Dennis, E.S. (1989) Transposable elements can be used to study cell lineages in transgenic plants. *Plant Cell* 1: 757–764.

Fire, A. (1999) RNA triggered gene silencing. *Trends Genet.* 15: 358–363.

Fire, A., Xu, S., Montgomery, M.K., Kostas, S.A., Driver, S.E., and Mello, C.C. (1998) Potent and specific genetic interference by double-stranded RNA in *Caenorhabditis elegans. Nature* 391: 806–811.

Fladung, M. and Ahuja, M.R. (1997) Excision of the maize transposable element *Ac* in periclinal chimeric leaves of *35S-Ac-rolC* transgenic aspen-*Populus. Plant Mol. Biol.* 33: 1097–1103.

Flavell, A.J., Dunbar, E., Anderson, R., Pearce, S.R., Hartley, R., and Kumar, A. (1992a) Ty1-*copia* group retrotransposons are ubiquitous and heterogeneous in higher plants. *Nucl. Acid Res.* 20: 3639–3644.

Flavell, A.J., Pearce, S.R., and Kumar, A. (1994) Plant transposable elements and the genome. *Curr. Opin. Genet. Develop.* 4: 838–844.

Flavell, A.J., Smith, D.B., and Kumar, A. (1992b) Extreme heterogeneity of Ty1-*copia* group retrotansposons in plants. *Mol. Gen. Genet.* 231: 233–242.

Fling, M. and Richards, C. (1983) The nucleotide sequence of the trimethoprim-resistant dihydro-date reductase gene harbored by Tn7. *Nucl. Acid Res.* 11: 5147–5158.

Foster, T.J., Howe, T.G.B., and Richmond, K.M.V. (1975) Translocation of tetracycline resistance determination from R100-1 to the *Escherichia coli* K-12 chromosome. *J. Bacteriol.* 124: 1153–1158.

Foster, T.M., Lough, T.J., Emerson, S.J., Lee, R.H., Bowman, J.L., Forster, R.L.S., and Lucas, W.J. (2002) A surveillance system regulates selective entry of RNA into the shoot apex. *Plant Cell* 14: 1497–1508.

Frank, M.J., Liu, D., Tsay, Y.-F., Ustach, C., and Crawford, N.M. (1997) *Tag1* is an autonomous transposable element that shows somatic excision in both *Arabidopsis* and tobacco. *Plant Cell* 9: 1745–1756.

Frankel, R. (1956) Graft-induced transmission to progeny of cytoplasmic male sterility in *Petunia. Science* 124: 684–685.

Franz, G., Loukeris, T.G., Dialektaki, G., Thompson, C.R.L., Savakis, C. (1994) Mobile *Minos* elements from *Drosophila hydei* encode a two-exon transposase with similarity to the paired DNA-binding domain. *Proc. Natl. Acad. Sci. (USA)* 91: 4746–4750.

Freeling, M. (1984) Plant transposable elements and insertion sequences. *Ann. Rev. Plant Physiol.* 35: 277–298.

Freund, R. and Meselson, M. (1984) Long terminal repeat nucleotide sequence and specific insertion of the *gypsy* transposon. *Proc. Natl. Acad. Sci. (USA)* 81: 4462–4464.

Frey, M., Reinecke, J., Grant, S., Saedler, H., and Gierl, A. (1990) Excision of the *En/Spm* transposable element of *Zea mays* requires two element-encoded proteins. *EMBO J.* 9: 4037–4044.

Froschauer, A., Korting, C., Bernhardt, W., Nanda, I., Schmid, M., Schartl, M., and Volff, J-N. (2001) Genomic plasticity and melanoma formation in the fish *Xiphophorus. Mar. Biotechnol.* 3: S72–S80.

Fukuchi, A., Kikuchi, F., and Hirochika, H. (1993) DNA fingerprinting of cultivated rice with rice retrotransposon probes. *Jpn. J. Genet.* 68: 195–204.

Fusswinkel, H., Schein, S., Courage, U., Starlinger, P., and Kunze, R. (1991) Detection and abundance of mRNA and protein encoded by transposable elements *Activator* (*Ac*) in maize. *Mol. Gen. Genet.* 225: 186–192.

Galas, D.J. and Chandler, M. (1981) On the molecular mechanism of transposition. *Proc. Natl. Acad. Sci. (USA)* 78: 4858–4862.

Galas, D.J. and Chandler, M. (1989) Bacterial insertion sequences. In: *Mobile DNA*. Berg, D.E. and Howe, M.M. (eds) ASM Press, Washington, DC. pp. 109–162.

Galindo, M.I., Bigot, Y., Sanchez, M.D., Periquet, G., and Pascual, L. (2001) Sequences homologous to the *hobo* transposable element in E strains of *Drosophila melanogaster*. *Mol. Biol. Evol.* 18: 1532–1539.

Galun, E. (1988) Application of molecular methods to modern Citrus taxonomy. In: *Proc. Sixth Intern. Citrus Congress*. Goren, R. and Mendel, K. (eds.) Balaban Publishers, Philadelphia/Rehovot, pp. 295–311.

Galun, E. and Breiman, A. (1997) *Transgenic Plants*. Imperial College Press, London, 376 p.

Galun, E. and Galun, E. (2001) *The Manufacture of Medical and Health Products by Transgenic Plants*. Imperial College Press, London, 332 p.

Garber, K., Bilic, I., Pusch, O., Tohme, J., Bachmair, A., Schweizer, D., and Jantsch, V. (1999) The Tpv2 family of retrotransposons of *Phaseolus vulgaris*: structure, integration characteristics and use for genotype classification. *Plant Mol. Biol.* 39: 797–807.

Garrett, J.E. and Carroll, D. (1986) *Tx1*: a transposable element from *Xenopus laevis* with some unusual properties. *Mol. Cell. Biol.* 6: 933–941.

Garrett, J.E., Knutzon, D.S., and Carroll, D. (1989) Composite transposable elements in the *Xenopus laevis* genome. *Mol. Cell. Biol.* 9: 3018–3027.

Geiser, M., Weck, E., Döring, H.P., Werr, W., Courage-Tebbe, U., Tillmann, E., and Starlinger, P. (1982) Genomic clones of a wild-type allele and a transposable element-induced mutant allele of the sucrose synthase gene of *Zea mays* L. *EMBO J.* 1: 1455–1460.

George, M. Jr. and Ryder, O.A. (1986) Mitochondrial DNA evolution in the genus *Equus*. *Mol. Biol. Evol.* 3: 535–546.

Gerasimova, T., Ilyin, Y., Mizrokhi, L.J., Semjonova, L., and Georgiev, G.P. (1984a) Mobilization of the transposable element moly4 by hybrid dysgenesis generates a family of unstable cut mutations in *Drosophila melanogaster*. *Mol. Gen. Genet.* 193: 488–492.

Gerasimova, T.I., Matyunina, L.V., Ilyin, Y.V., and Georgiev, G.P. (1984b) Simultaneous transposition of different mobile elements: relation to multiple mutagenesis in *Drosophila melanogaster*. *Mol. Gen. Genet.* 194: 517–522.

Gerasimova, T.I., Mizrokhi, L.J., and Georgiev, G.P. (1984c) Tansposition bursts in genetically unstable *Drosophila melanogaster*. *Nature* 309: 714–716.

Gerats, A.G.M., Beld, M., Huits, H., and Prescott, A. (1989) Gene tagging in *Petunia hybrida* using homologous and heterologous transposable elements. *Develop. Genet.* 10: 561–568.

Gerats, A.G.M., Huits, H., Vrijlandt, E., Marana, C., Souer, E., and Beld, M. (1990) Molecular characterization of a nonautonomous transposable element (dTph1) of petunia. *Plant Cell* 2: 1121–1128.

Gierl, A. (1996) The *En/Spm* transposable element of maize. *Curr. Topics Microbiol. Immunol.* 204: 145–159.

Gierl, A. and Saedler, H. (1992) Plant transposable elements and gene tagging. *Plant Mol. Biol.* 19: 39–49.

Gierl, A., Lutticke, S., and Saedler, H. (1988) TnpA product encoded by the transposable element En-1 of *Zea mays* is a DNA binding protein. *EMBO J.* 7: 4045–4053.

Gierl, A., Saedler, H., and Peterson, P.A. (1989) Maize transposable elements. *Ann. Rev. Genet.* 23: 71–85.

Gindullis, F., Desel, C., Galasso, I., and Schmidt, T. (2001) The large-scale organization of the centromeric region in Beta species. *Genome Res.* 11: 253–265.

Goff, S.A., Ricke, D., Lan, T.-H. et al. (2002) A draft sequence of the rice genome (*Oryza sativa* L. ssp. *japonica*) *Science* 296: 92–100.

Gonzalez, P. and Lessios, H.A. (1999) Evolution of sea urchin retroviral-like (SURL) elements: evidence from 40 echinoid species. *Mol. Biol. Evol.* 16: 938–952.

Goodier, J.L. and Davidson, W.S. (1994) Tc1 transposon-like sequences are widely distributed in salmonids. *J. Mol. Biol.* 241: 26–34.

Goodier, J.L., Ostertag, E.M., Du, K., and Kazazian, Jr. H.H. (2001) A novel active L1 retrotransposon subfamily in the mouse. *Genome Res.* 11: 1677–1685.

Goodwin, T.J.D. and Poulter, R.T.M. (2001) The DIRS1 group of retrotransposons. *Mol. Biol. Evol.* 18: 2067–2082.

Gorbunova, V. and Levy, A.A. (1997a) Non-homologous DNA end joining in plant cells is associated with deletions and filler DNA insertions. *Nucl. Acid Res.* 25: 4650–4657.

Gorbunova, V. and Levy, A.A. (1997b) Circularized *Ac/Ds* transposons: formation, structure and fate. *Genetics* 145: 1161–1169.

Gottgens, B., Barton, L.M., Grafham, D., Vaudin, M., and Green, A.R. (1999) Tdr2, a new zebrafish transposon of the Tc1 family. *Gene* 239: 373–379.

Grandbastien, M.A. (1998) Activation of plant retrotransposons under stress conditions. *Trends Plant Sci.* 3: 181–187.

Grandbastien, M.A., Spielmann, A., and Coboche, M. (1989) Tnt1, a mobile retoviral-like transposable element of tobacco isolated by plant cell genetics. *Nature* 337: 376–380.

Grandbastien, M.A., Spielmann, A., Pouteau, S., Huttner, E., Lonquest, M., Kunert, K., Meyer, C., Rouze, P., and Caboche, M. (1991) Characterization of mobile endogenous *copia*-like transposable elements in the geome of Solanaceae. In: *Plant Molecular Biology 2*. Hermann, R.G. and Larkins, B. (eds.) Plenum Press, New York, pp. 333–343.

Green, M.M. (1977) The case for DNA insert mutations in *Drosophila*. In: *DNA Insertion Elements, Plasmids and Episomes*. Bukhari, A.I., Shapiro, J.A., and Adhya, A.B. (eds) Cold Spring Harbor Lab. Press, pp. 437–445.

Greenblatt, I.M. (1974) Movement of modulator in maize: a test of an hypothesis. *Genetics* 77: 671–678.

Greenwald, I. (1985) *lin-12*, a nematode homeoticagene, is homologans to a set of mammalian proteins that includes epidermal growth factor. *Cell* 43: 583–590.

Gregory, P.D. (2001) Transcription and chromatin converge: lessons from yeast genetics. *Curr. Opin. Genet. Develop.* 11: 142–147.

Grewal, S.I.S. and Elgin, S.C.R. (2002) Heterochromatin: new possibilities for the inheritance of structure. *Curr. Opin. Genet. Develop.* 12: 178–187.

Grimaldi, G. and Singer, M.F. (1983) Members of the KpnI family of long interspersed repeated sequences join and interrupt α-satellite in the monkey genome. *Nucl. Acid Res.* 11: 321–339.

Grimaldi, G., Queen, C., and Singer, M.F. (1981) Interspersed repeated sequences in the African green monkey genome that are homologous to the human Alu family. *Nucl. Acid Res.* 9: 5553–5569.

Grimaldi, G., Skowronski, J., and Singer, M.F. (1984) Defining the beginning and end of KpnI family segments. *EMBO J.* 3: 1753–1759.

Grindley, N.D.F. (2002) The movement of Tn3-like elements: transposition and cointegrate resolution. In: *Mobile DNA II*. Craig, N.L., Craigie, R., Gellert, M., and Lambowitz, A.M. (eds.) ASM Press, Washington, DC. pp. 272–303.

Grishok, A., Tabara, H., and Mello, C.C. (2000) Genetic requirements for inheritance of RNAi in *C. elegans*. *Science* 287: 2494–2497.

Grotewold, E., Athma, P., and Peterson, T. (1991) A possible hot spot for *Ac* insertion in the maize P gene. *Mol. Gen. Genet.* 230: 329–331.

Gueiros-Filho, F.J. and Beverley, S.M. (1997) Trans-kingdom transposition of the *Drosophila* element *Mariner* within the protozoan leishmania. *Science* 276: 1716–1719.

Gupta, M., Bertram, I., Shepherd, N.S., and Saedler, H. (1983) *Cin1*, a family of dispersed repetitive elements in *Zea mays*. *Mol. Gen. Genet.* 192: 373–377.

Gupta, R., He, Z., and Luan, S. (2002a) Functional relationship of cytochrome c(6) and plastocyanin in *Arabidopsis*. *Nature* 417: 567–571.

Gupta, R., Ting, J.T.L., Sokolov, L.N., Johnson, S.A., and Luan, S. (2002b). A tumor suppressor homolog, AtPTEN1, is essential for pollen development in *Arabidopsis*. *Plant Cell* 14: 2496–2507.

Hamer, L., DeZwaan, T.M., Montenegro-Chamorro, M.V., Frank, S.A., and Hamer, J.E. (2001) Recent advances in large-scale transposon mutagenesis. *Curr. Opin. Chem. Biol.* 5: 67–73.

Hamilton, A.J. and Baulcombe, D.C. (1999) A species of small antisense RNA in posttranscriptional gene silencing in plants. *Science* 286: 950–952.

Hammond, S.M., Bernstein, E., Beach, D., and Hannon, G.J. (2000) An RNA-directed nuclese mediates post-transcriptional gene silencing in *Drosophila* cells. *Nature* 404: 293–296.

Han, Y. and Grierson, D. (2002) The influence of inverted repeats on the production of small antisense RNAs involved in gene silencing. *Mol. Genet. Genomics* 267: 629–635.

Haniford, D.B. (2002) Transposon Tn10. In: *Mobile DNA II*. Craig, N.L., Craigie, R., Gellert, M., and Lambowitz, A.M. (eds.) ASM Press, Washington, DC. pp. 457–483.

Hannon, G.J. (2002) RNA interference. *Nature* 418: 244–251.

Haren, L., Polard, P., Ton-Hoang, B., and Chandler, M. (1998) Multiple oligomerisation domains in the IS911 transposase: A leucine zipper motif is essential for activity. *J. Mol. Biol.* 283: 29–41.

Haring, M.A., Rommens, C.M.T., Nijkamp, J.J., and Hille, J. (1991) The use of transgenic plants to understand transposition mechanisms and to develop transposon tagging strategies. *Plant Mol. Biol.* 16: 449–461.

Harris, L.J., Baillie, D.L., and Rose, A.M. (1988) Sequence identity between an inverted repeat family of transposable elements in *Drosophila* and *Caenorhabditis*. *Nucl. Acid Res.* 16: 5991–5998.

Harrison, B.J. and Carpenter, R. (1973) A comparison of the instability at the *nivea* and *pallida* loci of *Antirrhinum majus*. *Heredity* 31: 309–323.

Harrison, B.J. and Fincham, J.R.S. (1964) Instability at the *Pal* locus in *Antirrhinum majus*. *Heredity* 19: 237–258.

Hartl, D.L. (1989) Transposable element *Mariner* in *Drosophila* species. In: *Mobile DNA*. Berg, D.E. and Howe, M.M. (eds.) ASM Press, Washington, DC. pp. 531–536.

Hartl, D.L. (2001) Discovery of the transposable element *Mariner*. *Genetics* 157: 471–476.

Hartl, D.L., Lohe, A.R., and Lozovskaya, E.R. (1997) Modern thoughts on an ancient marinere: Function, evolution, regulation. *Annu. Rev. Genet.* 31: 337–358.

Hauser, C., Fusswinkel, H., Li, J., Oellig, C., Kunze, R., Muller-Neumann, M., Heinlein, M., Starlinger, P., and Doerfler, W. (1988) Overproduction of the protein encoded by the maize transposable element *Ac* in insect cells by a baculovirus vector. *Mol. Gen. Genet.* 214: 373–378.

Hedges, R.W. and Jacob, A.E. (1974) Transposition of ampicillin resistance from RP4 to other replicons. *Mol. Gen. Genet.* 132: 31–40.

Heffron, F., McCarthy, B.J., Ohtsubo, H., and Ohtsubo, E. (1979b) DNA sequence analysis of the transposon Tn3: three genes and three sites involved in transposition. *Cell* 18: 1153–1163.

Heffron, F., Rubens, C., and Falkow, S. (1975b). Translocation of a plasmid DNA sequence which mediates ampicillin resistance: molecular nature and specificity of insertion. *Proc. Nat. Acad. Sci. (USA)* 72: 3623–3627.

Heffron, F., So, M., and McCarthy, B.J. (1979a). Insertion mutations affecting transposition of Tn3 and replication of a ColE1 derivative. *Cold Spring Harbor Symp. Quant. Biol.* 43: 1279–1285.

Heffron, F., Sublett, R., Hedges, R.W., Jacob, A., and Falkow, S. (1975a) Origin of the TEM beta-lactamase gene found on plasmids. *J. Bacteriol.* 122: 250–256.

Heierhorst, J., Lederis, K., and Richter, D. (1992) Presence of a member of the Tc1-like transposon family from nematodes and *Drosophila* within the vasotoc in gene of a primitive vertebrate, the Pacific hagfish *Eptatretus stouti*. *Proc. Natl. Acad. Sci. (USA)* 89: 6798–6802.

Heinlein, M. (1996) Excision patterns of *activator* (*Ac*) and *dissociation* (*Ds*) elements in *Zea mays* L.: Implications for the regulation of transposition. *Genetics* 144: 1851–1869.

Heinlein, M., Brattig, T., and Kunze, R. (1994) *In vivo* aggregation of maize *Activator* (*Ac*) transposase in nuclei of maize endosperm and *Petunia* protoplasts. *Plant J.* 5: 705–714.

Henig, R.M. (2000) *The Monk in the Garden*. Houghton Mifflin Co. Boston, 292 p.

Henikoff, S. (1992) Detection of *Caenorhabditis* transposon homologs in diverse organisms. *New Biol.* 4: 382–388.

Henikoff, S. and Plasterk, R.H.A. (1988) Related transposons in *C. elegans* and *D. melanogaster*. *Nucl. Acid Res.* 16: 6234.

Hershberger, R.J., Benito, M.-I., Hardeman, K.J., Warren, C., Chandler, V.L., and Walbot, V. (1995) Characterization of the major transcripts encoded by the regulatory *MuDR* transposable element of maize. *Genetics* 140: 1087–1098.

Hershberger, R.J., Warren, C.A., and Walbot, V. (1991) Mutator activity in maize correlates with the presence and expression of the *Mu* transposable element *Mu9*. *Proc. Natl. Acad. Sci. (USA)* 88: 10198–10202.

Heslop-Harrison, J.S., Schwartzacher, T., Anamthawat-Jonsson, K., Leitch, A.R., Shi, M., and Leitch, I.J. (1991) *In situ* hybridization with automated chromosome denaturation. *J. Met. Cell. Mol. Biol.* 3: 109–116.

Higashiyama, T., Noutoshi, Y., Fujie, M., and Yamada, T. (1997) *Zepp*, a line-like retrotransposon accumulated in the *Chlorella* telomeric region. *EMBO J.* 16: 3715–3723.

Hiraizumi, Y. (1971) Spontaneous recombination in *Drosophila melanogaster* males. *Proc. Natl. Acad. Sci. (USA)* 68: 268–270.

Hirochika, H. (1993) Activation of tobacco retrotransposons during tissue culture. *EMBO J.* 12: 2521–2528.

Hirochika, H. (2001) Contribution of the Tos17 retrotransposon to rice functional genomics. *Curr. Opin. Plant Biol.* 4: 118–122.

Hirochika, H., Fukuchi, A., and Kikuchi, F. (1992) Retrotransposon families in rice. *Mol. Gen. Genet.* 233: 209–216.

Hirochika, H., Okamoto, H., and Kakutami, T. (2000) Silencing of retrotransposons in *Arabidopsis* and reactivation by the *ddm1* mutation. *Plant Cell* 12: 357–368.

Hirochika, H., Otsuki, H., Yoshikawa, M., Otsuki, Y., Sugimoto, K., and Takeda, S. (1996b) Autonomous transposition of the tobacco retrotransposon Tto1 in rice. *Plant Cell* 8: 725–734.

Hirochika, H., Sugimoto, K., Otsuki, Y., Tsugawa, H., and Kanda, M. (1996a) Retrotransposons of rice involved in mutations induced by tissue culture. *Proc. Natl. Acad. Sci. (USA)* 93: 7783–7788.

Hirsch, H.-J., Starlinger, P., and Brachet, P. (1972) Two kinds of insertions in bacterial genes. *Mol. Gen. Genet.* 119: 191–206.

Holmes, S.E., Xombroski, B.A., Krebs, C.M., Boehm, C.D., and Kazazian, Jr. H.H. (1994) A new retrotransposable human L1 element from the LRE2 locus on chromosome 1q produces a chimaeric insertion. *Nature Genet.* 7: 143–148.

Houck, C.M., Rinehart, F.P., and Schmid, C.W. (1979) A ubiquitous family of repeated DNA sequences in the human genome. *J. Mol. Biol.* 132: 289–306.

Hsieh, C.-L. (2000) Dynamics of DNA methylation pattern. *Curr. Opin. Genet. Devel.* 10: 224–228.

Hu, W., Das, O.P., and Messing, J. (1995) Zeon-1 a member of a new maize retrotransposon family. *Mol. Gen. Genet.* 248: 471–480.

Hua-Van, A., Langin, T., and Daboussi, M.-J. (2001a) Evolutionary history of the *impala* transposon in *Fusarium oxysporum. Mol. Biol. Evol.* 19: 1959–1969.

Hua-Van, A., Pamphile, J.A., Langin, T., and Daboussi, M.-J. (2001b) Transposition of autonomous and engineered *impala* transposons in *Fusarium oxysporum* and a related species. *Mol. Gen. Genomics* 264: 724–731.

Huang, R.-C. and Bonner, J. (1962) Histone, a repressor of chromosomal RNA synthesis. *Proc. Natl. Acad. Sci. (USA)* 48: 1216–1222.

Huang, R.-C., Maheshwari, N., and Bonner, J. (1960) Enzymatic synthesis of RNA. *Biochem. Biophys. Res. Comm.* 3: 689–694.

Hull, M.W., Erickson, J., Johnston, M., and Engelke, D.R. (1994) tRNA genes as transcriptional repressor elements. *Mol. Cell Biol.* 14: 1266–1277.

Hutvagner, G. and Zamore, P.D. (2002) RNAi: nature abhors a double-strand. *Curr. Opin. Genet. Devel.* 12: 225–232.

Ikeda, K., Nakayashiki, H., Takagi, M., Tosa, Y., and Mayama, S. (2001) Heat shock, copper sulfate and oxidative stress activate the retrotransposon MAGGY resident in the plant pathogenic fungus *Magnaporthe grisea. Mol. Genet. Genomics* 266: 318–325.

Ilyin, Y.V., Chmeliauskaite, V.G., Ananiev, E.V., Lyubomirskaya, N.V., Kulguskin, V.V., Bayev, A.A. Jr., and Georgiev, G.P. (1980b) Mobile dispersed genetic element MDG1 of *Drosophila melanogaster*: structural organization. *Nucl. Acid Res.* 8: 5333–5346.

Ilyin, Y.V., Chmeliauskaite, V.G., Kulguskin, V.V., and Georgiev, G.P. (1980c) Mobile dispersed genetic element MDG1 of *Drosophila melanogaster*: transcription pattern. *Nucl. Acid Res.* 8: 5347–5361

Ilyin, Y.V., Chmeliauskaite,V.G., Ananiev, E.V., and Georgiev, G.P. (1980a) Isolation and characterization of a new family of mobile dispersed genetic elements, mdg3, in *Drosophila melanogaster. Chromosoma* 81: 27–53.

Inouye, S., Yuki, S., and Saigo, K. (1986) Complete nucleotide sequence and genome organization of a *Drosophila* transposable genetic element, 297. *Eur. J. Biochem.* 154: 417–425.

Ising, G. and Ramel, C. (1973) The behavior of a transposing element in *Drosophila melanogaster. Genetics* 73: s123.

Ivics, Z., Hackett, P.B., Plasterk, R.H., and Izsvak, Z. (1997) Molecular reconstruction of *Sleeping Beauty*, a Tc1-like transposon from fish, and its transposition in human cells. *Cell* 91: 501–510.

Ivics, Z., Izsvak, Z., Minter, A., and Hackett, P.B. (1996) Identification of functional domains and evolution of Tc1-like transposable elements. *Proc. Natl. Acad. Sci. (USA)* 93: 5008–5013.

Izsvak, Z., Ivics, Z., and Hackett, P.B. (1995) Characterization of a Tc1-like transposable element in zebrafish (*Danio rerio*). *Mol. Gen. Genet.* 247: 312–322.

Izsvak, Z., Ivics, Z., and Plasterk, R.H. (2000) *Sleeping Beauty*, a wide host-range transposon vector for genetic transformation in vertebrates. *J. Mol. Biol.* 302: 93–102.

Jackson, J.P., Lindroth, A.M., Caom X., and Jacobsen, S.E. (2002) Control of CpNpG DNA methylation by KRYPTONITE histone 43 methyltransferase. *Nature* 416: 556–560.

Jacob, F. and Monod, J. (1959) Genes de structure et genes de regulation dans la biosynthese de proteins. *Comp. Rend.* 249: 1282–1284.

Jacobson, J.W., Medhora, M.M., and Hartl, D.L. (1986) Molecular structure of a somatically unstable transposable element in *Drosophila. Proc. Natl. Acad. Sci. (USA)* 83: 8684–8688.

Jagadeeswaran, P., Forget, B.G., and Weissman, S.M. (1981) Short interspersed repetitive DNA elements in eucaryotes: transposable DNA elements generated by reverse transcription of RNA pol III transcripts? *Cell* 26: 141–142.

Jahn, C.L., Doktor, S.Z., Frels, J.S., Jaraczewski, J.W., and Krikau, M.F. (1993) Structures of the *Euplotes crassus Tec1* and *Tex2* elements: identification of putative transposase coding regions. *Gene* 133: 71–78.

Jakubczak, J.L., Burke, W.D., and Eickbush, T.H. (1991) Retrotransposable elements R1 and R2 interrupt the rRNA genes of most insects. *Proc. Natl. Acad. Sci. (USA)* 88: 3295–3299.

Jakubczak, J.L., Xiong, Y., and Eickbush, T.H. (1990) Type I (R1) and Type II (R2) ribosomal DNA insertions of *Drosophila melanogaster* are retrotransposable elements closely related to those of *Bombyx mori. J. Mol. Biol.* 212: 37–52.

Jeddeloch, J.A., Stokes, T.L., and Richards, E.J. (1999) Maintenance of genomic methylation requires a SW12/SNF2 like protein. *Nature Genet.* 22: 94–97.

Jelinek, W.R., Toomey, T.P., Leinwand, L., Duncan, C.H., Biro, P.A., Choudary, P.V., Weissman, S.M., Rubin, C.M., Houck, C.M., Deininger, P.L., and Schmid, C.W. (1980) Ubiquitous, interspersed repeated sequences in mammalian genomes. *Proc. Natl. Acad. Sci. (USA)* 77: 1308–1402.

Jensen, S., Gassama, M.-P., and Heidmann, T. (1999) Taming of transposable elements by homology-dependent gene silencing. *Nature Genet.* 21: 209–212.

Jenuwein, T. (2001) Re-SET-ting heterochromatin by histone methyltransferases. *Trends Cell Biol.* 11: 266–273.

Jenuwein, T. and Allis, C.D. (2001) Translating the histone code. *Science* 293: 1074–1080.

Johns, M.A., Mottinger, J., and Freeling, M. (1985) A low copy number *copia*-like transposon in maize. *EMBO J.* 4: 1093–1102.

Jones, D.A., Thomas, C.M., Hammond-Kosack, K.E., Balint-Kurti, P.J., and Jones, J.D.G. (1994) Isolation of the tomato *Cf-9* gene for resistance to *Cladosporium fulvum* by transposon tagging. *Science* 266: 789–793.

Jones, J.D.G., Carland, F., Lim, E., Ralston, E., and Dooner, H.K. (1990) Preferential transposition of the maize element *Activator* to linked chromosomal locations in tobacco. *Plant Cell* 2: 701–707.

Jones, J.D.G., Carland, F.M., Maliga, P., and Dooner, H.K. (1989) Visual detection of transposition of the maize element activator (*Ac*) in tobacco seedlings. *Science* 244: 204–207.

Jordan, E., Saedler, H., and Starlinger, P. (1968) O⁰ and strong-polar mutations in the *gal* operon are insertions. *Mol. Gen. Genet.* 102: 353–363.

Judelson, H.S. (2002) Sequence variation and genomic amplification of a family of *Gypsy*-like elements in the oomycete genus *Phytophthora. Mol. Biol. Evol.* 19: 1313–1322.

Kapitonov, C. and Jurka, J. (1996) The age of alu subfamilies. *J. Mol. Evol.* 42: 59–65.

Karran, P. (2000) DNA double strand break repair in mammalian cells. *Curr. Opin. Genet. Develop.* 10: 144–150.

Katziotis, A., Schmidt, T., and Heslop-Harrison, S. (1996) Chromosomal and genomic organization of Ty1-*copia*-like retrotransposon sequences in the genus *Avena. Genome* 39: 410–417.

Katzir, N., Rechavi, G., Cohen, J.B., Unger, T., Simoni, F., Segal, S., Cohen, D., and Givol, D. (1985) Retroposon insertion into the cellular oncogene *c-myc* in canine transmissible venereal tumor. *Proc. Natl. Acad. Sci. (USA)* 82: 1054–1058.

Kaufman, P.D. and Rio, D.C. (1992) P element tranposition *in vitro* proceeds by a cut-and-paste mechanism and uses GTP as a cofactor. *Cell* 69: 27–39.

Kazazian, H.H., Wong, C., Youssoufian, H., Scott, A.F., Phillips, D.G., and Antonarakis, S.E. (1988) Haemophilia A resulting from *de novo* insertion of L1 sequences represents a novel mechanism for mutation in man. *Nature* 332: 164–166.

Kazazian, Jr. H.H. (2000) L1 retrotransposons shape the mammalian genome. *Science* 289: 1152–1153.

Keller, E.F. (1983) *A Feeling for the Organism – The Life and Work of Barbara McClintock.* W.H. Freeman and Co., NY. 235 p.

Keller, J., Jones, J.D.G., Harper, E., Lim, E., Carland, F., Ralston, E.J., and Dooner, H.K. (1993) Effects of gene dosage and sequence modification on the frequency and timing of transposition maize element *Activator (Ac)* in tobacco. *Plant Mol. Biol.* 21: 157–170.

Kempken, F. and Kuck, U. (1998) Transposons in filamentous fungi – facts and perspectives. *BioEssays* 20: 652–659.

Kennedy, A.K., Haniford, D.B., and Mizuuchi, K. (2000) Single active site catalysis of the successive phosphoryl transfer steps by DNA transposases: insights from phosphorothioate stereoselectivity. *Cell* 101: 295–305.

Ketting, R.F. and Plasterk, R.H.A. (2000) A genetic link between co-suppression and RNA interference in *C. elegans. Nature* 404: 296–298.

Ketting, R.F., Haverkamp, T.H.A., Van Luenen, H.G.A.M., and Plasterk, R.H.A. (1999) *mut-7* of *C. elegans*, required for transposon silencing and RNA interference, is a homolog of Werner syndrome helicase and RNaseD. *Cell* 99: 133–144.

Kidwell, M.G., Kidwell, J.F., and Sved, J.A. (1977) Hybrid dysgenesis in *Drosophila melanogaster*: A syndrome of aberrant traits including mutations, sterility and male recombination. *Genetics* 86: 813–833.

Kim, A., Terzian, C., Santamaria, P., Pelisson, A., Prud'homme, N., and Bucheton, A. (1994) Retroviruses in invertebrates: The *gypsy* retrotransposon is apparently an infectious retrovirus of *Drosophila melanogaster. Proc. Natl. Acad. Sci. (USA)* 91: 1285–1289

Kim, J.M., Vanguri, S., Boeke, J.D., Gabriel, A., and Voytas, D.F. (1998) Transposable elements and genome organization: A comprehensive survey of retrotransposons revealed by the complete *Saccharomyces cerevisiae* genome sequence. *Genome Res.* 8: 464–478.

Kimberland, M.L., Divoky, V., Prchal, J., Schwahn, U., Berger, W., and Kazazian, Jr. H.H. (1999) Full-length human L1 insertions retain the capacity for high frequency retrotransposition in cultured cells. *Human Mol. Genet.* 8: 1557–1560.

Kimmel, B.E., Ole-Moiyoi, O.K., and Young, J.R. (1987) Ingi, a 5.2 kb dispersed sequence element from *Trypanosoma brucei* that carries half of a smaller mobile element at either end and has homology with mammalian LINEs. *Mol. Cell. Biol.* 7: 1465–1475.

Kinsey, J.A. and Helber, J. (1989) Isolation of a transposable element from *Neurospora crassa. Proc. Natl. Acad. Sci. (USA)* 86: 1929–1933.

Kishima, Y., Yamashita, S., and Mikami, T. (1997) Immobilized copies with a nearly intact structure of the transposon *Tam3* in *Antirrhinum majus*: implications for the *cis*-element related to the transposition. *Theor. Appl. Genet.* 95: 1246–1251.

Kishima, Y., Yamashita, S., Martin, C., and Mikami, T. (1999) Structural conservation of the transposon *Tam3* family in *Antirrhinum majus* and estimation of the number of copies able to transpose. *Plant Mol. Biol.* 39: 299–308.

Kleckner, N. (1989) Tn 10. In: *Mobile DNA.* Berg, D.H. and Howe, M.M. (eds) ASM Press, Washington, DC. pp. 227–268.

Kleckner, N., Chalmers, R.M., Dwon, D., Sakai, J., and Bolland, S. (1996) Tn10 and IS10 transposition and chromosome rearrangements: mechanism and regulation *in vivo* and *in vitro. Curr. Topics Microbiol. Immunol.* 204: 50–82.

Kleckner, N., Reichardt, K., and Botstein, D. (1975) Mutagenesis by insertion of a drag-resistance element carrying an inverted repetition. *J. Mol. Biol.* 97: 561–575.

Klobutcher, L.A. and Jahn, C.L. (1991). Developmentally controlled genomic rearrangements in ciliated protozoa. *Curr. Opin. Genet. Devel.* 1: 397–403.

Knapp, S., Coupland, G., Uhrig, H., Starlinger, P., and Salamini, F. (1988) Transposition of the maize transposable element *Ac* in *Solanum tuberosum. Mol. Gen. Genet.* 213: 285–290.

Knoop, V., Unseld, M., Marienfeld, J., Brandt, P., Sunkel, S., Ullrich, H., and Brennicke, A. (1996) *Copia, gypsy-* and line-like retrotransposon fragments in the mitochondrial genome of *Arabidopsis thaliana. Genetics* 142: 579–585.

Koes, R., Souer, E., Van Houwelingen, A. Mur, L., Spelt, C., Quattrocchio, F., Wing, J., Oppedijk, Ahmed, S., Maes, T., Gerats, T., Hoogeveen, P., Meesters, M., Kloos, D., and Mol, J.N.M. (1995) Targeted gene inactivation in petunia by PCR-based selection of transposon insertion mutants. *Proc. Natl. Acad. Sci. (USA)* 92: 8149–8153.

Koga, A., Suzuki, M., Inagaki, H., Bessho, Y., and Hori, H. (1996) Transposable element in fish. *Nature* 383: 30.

Koncz, C., Martini, N., Mayer, Hoffer, R.D., Koncz-Kalman, Z., Korber, H., Redei, G.P., and Schell, J. (1989) High frequency T-DNA mediated gene tagging in plants. *Proc. Natl. Acad. Sci. (USA)* 86: 8467–8471.

Konieczny, A., Voytas, D.F., Cummings, M.P., and Ausubel, F.M. (1991) A superfamily of *Arabidopsis thaliana* retrotranspons. *Genetics* 127: 801–809.

Kornberg, R.D. (1974) Chromatin structure: a repeating unit of histones and DNA. *Science* 184: 868–871.

Kornberg, R.D. and Lorch, Y. (1999) Twenty-five years of the nucleosome, fundamental particle of the eukaryote chromosome. *Cell* 98: 285–294.

Kouzarides, T. (2002) Histone methylation in transcriptional control. *Curr. Opin. Genet. Devel.* 12: 198–209.

Kulguskin, V.V., Ilyin, Y.V., and Georgiev, G.P. (1981) Mobile dispersed genetic element MDG1 of *Drosophila melanogaster*: nucleotide sequence of long terminal repeats. *Nucl. Acid Res.* 9: 3451–3465.

Kumar, A. and Bennetzen, J.L. (1999) Plant retrotransposons. *Ann. Rev. Genet.* 33: 479–532.

Kunze, R. (1996) The maize transposable element *activator* (*Ac*). Curr. *Topics Microbiol. Immun.* 204: 162–194.

Kunze, R. and Starlinger, P. (1989) The putative transposase of transposable element *Ac* from *Zea mays* L. interacts with subterminal sequences of *Ac. EMBO J.* 8: 3177–3185.

Kunze, R., Behrens, U.Y., Courage-Franzkowiak, U., Feldmar, S., Kuhn, S., and Lutticke, R. (1993) Dominant transposition-deficient mutants of maize *Activator* (*Ac*) transposase. *Proc. Natl. Acad. Sci. (USA)* 90: 7094–7098.

Kunze, R., Stochaj, U., Laufs, J., and Starlinger, P. (1987) Transcription of transposable element *Activator* (*Ac*) of *Zea mays* L. *EMBO J.* 6: 1555–1563.

Laha, T., Loukas, A., Verity, C.K., McManus, D.P., and Brindley, P.J. (2001) *Gulliver*, a long terminal repeat retrotransposon from the genome of the oriental blood fluke *Schistosoma japonicum. Gene* 264: 59–68.

Lam, W.L., Lee, T.-Sh., and Gilbert, W. (1996b) Active transposition in zebrafish. *Proc. Natl. Acad. Sci. (USA)* 93: 10870–10875.

Lam, W.L., Seo, P., Robison, K., Samant, V., and Gilbert, W. (1996a) Discovery of amphibian Tc1-like transposon families. *J. Mol. Biol.* 257: 359–366.

Lampe, D.J., Grant, T.E., and Robertson, H.M. (1998) Factors affecting transposition of the *Himar1 Mariner* transposon *in vitro. Genetics* 149: 179–187.

Lampe, D.J., Walden, K.K.O., and Robertson, H.M. (2001) Loss of transposase-DNA interaction may underlie the divergence of *Mariner* family transposable elements and the ability of more than one *Mariner* to occupy the same genome. *Mol. Biol. Evol.* 18: 954–961.

Langin, T., Capy, P., and Daboussi, M.-J. (1995) The transposable element *impala*, a fungal member of the *Tc1-mariner* superfamily. *Mol. Gen. Genet.* 246: 19–28.

Laten, H.M. and Morris, R.O. (1993) *SIRE-1*, a long interspaced repetitive DNA element from soybean with weak sequence similarity to retrotransposons: initial characterization and partial sequence. *Gene* 134: 153–159.

Laten, H.M., Majumdar, A., and Gaucher, E.A. (1998) *SIRE-1*, a *copia*/Ty1-like retroelement from soybean, encodes a retroviral envelope-like protein. *Proc. Natl. Acad. Sci. (USA)* 95: 6897–6902.

Le, Q.-H., Wright, S., Yu, Z., and Bureau, T. (2000) Transposon diversity in *Arabidopsis thaliana*. *Proc. Natl. Acad. Sci. (USA)* 97: 7376–7381.

Leeton, P.R.J. and Smyth, D.R. (1993) An abundant LINE-like element amplified in the genome of *Lilium speciosum*. *Mol. Gen. Genet.* 237: 97–104.

Lenoir, A., Cournoyer, B., Warwick, S., Picard, G., and Deragon, J.-M. (1997) Evolution of SINE S1 retroposons in Cruciferae plant species. *Mol. Biol. Evol.* 14: 934–941.

Lenoir, A., Lavie, L., Prieto, J.-L., Goubely, C., Cote, J.-C., Pelissier, T., and Deragou, J.-M. (2001) The evolutionary origin and genomic organization of SINEs in *Arabidopsis*. *Mol. Biol. Evol.* 18: 2315–2322.

Lerman, M.I., Thayer, R.E., and Singer, M.F. (1983) Kpn I family of long interspersed repeated DNA sequences in primates: polymorphism of family members and evidence for transcription. *Proc. Natl. Acad. Sci. (USA)* 80: 3966–3970.

Levin, H.L. (2002) Newly identified retrotransposons of the *gypsy*/Ty3 class in fungi, plants and vertebrates. In: *Mobile DNA II*. Craig, N.L., Craigie, R., Gellert, M., and Lambowitz, A.M. (eds.) ASM Press, Washington, DC. pp. 684–704.

Levin, H.L., Weaver, D.C., and Boeke, J.D. (1990) Two related families of retrotransposons from *Schizosaccharomyces pombe*. *Mol. Cell Biol.* 10: 6791–6798.

Levis, R.W., Ganesan, R., Houtchens, K., Tolar, L.A., and Shee, F.-M. (1993) Transposons in place of telomeric repeats at a *Drosophila* telomere. *Cell* 75: 1083–1093.

Levy, A.A. and Walbot, V. (1990) Regulation of the timing of transposable element excision during maize development. *Science* 248: 1534–1537.

Levy, A.A., Britt, A.B., Luehrsen, K.R., Chandler, V.L., Warren, C., and Walbot, V. (1989) Developmental and genetic aspects of mutator excision in maize. *Devel. Genet.* 10: 520–531.

Li Destri Nicosia, M.G., Brocard-Masson, C., Demals, S., Hua Van, A., Daboussi, M.J., and Scazzocchio, C. (2001) Heterologous transposition in *Aspergillus nidulans*. *Mol. Microbiol.* 39: 1330–1344.

Li, E. (2002) Chromatin modification and epigenetic programming in mammalian development. *Nature Rev. Genet.* 3: 662–673.

Li, H.-W., Lucy, A.P., Guo, H.-S., Li, W.-X., Ji, L.-H., Wong, S.-M., and Ding, S.-W. (1999) Strong host resistance targeted against a viral suppressor of the plant gene silencing defence mechanism. *EMBO J.* 18: 2683–2691.

Liao, L.W., Rosenzweig, B., and Hirsh, D. (1983) Analysis of a transposable element in *Caenorhabditis elegans*. *Proc. Natl. Acad. Sci. (USA)* 80: 3585–3589.

Lichten, M. (2001) Meiotic recombination: Breaking the genome to save it. *Curr. Biol.* 11: R253–R256.

Lichtenstein, C. and Brenner, S. (1981) Site-specific properties of Tn7 transpostion into the *E. coli* chromosome. *Mol. Gen. Genet.* 183: 380–387.

Lichtenstein, C. and Brenner, S. (1982) Unique insertion site of Tn7 in the *E. coli* chromosome. *Nature* 297: 601–603.

Lifschytz, E. and Falk, R. (1968) Fine structure analysis of a chromosome segment in *Drosophila melanogaster* analysis of X-ray induced lethals. *Mutation Res.* 6: 235 244.

Lim, J.K. (1988) Intrachromosomal rearrangements mediated by *hobo* transposons in *Drosophila melanogaster*. *Proc. Natl. Acad. Sci. (USA)* 85: 9153–9157.

Linares, C., Irigoyen, M.L., and Fominaya, A. (2000) Identification of C-genome chromosomes involved in intergenomic translocations in *Avena sativa* L., using cloned repetitive DNA sequences. *Theor. Appl. Genet.* 100: 353–360.

Linares, C., Loarce, Y., Serna, A., and Fominaya, A. (2001) Isolation and characterization of two novel retrotransposons of the Ty1 *copia* group in oat genomes. *Chromosoma* 110: 115–123.

Linares, C., Serna, A., and Fominaya, A. (1999) Chromosomal organization of a sequence related to LTR-like elements of Ty1-*copia* retrotransposons in *Avena* species. *Genome* 42: 706–713.

Lindauer, A., Fraser, D., Brüderlein, M., and Schmitt, R. (1993) Reverse transcriptase families and a *copia*-like retrotransposon, *Osser*, in the green alga *Volvox carteri*. *FEBS Lett.* 319: 261–266.

Lindbo, J.A., Silva-Rosales, L., Proebsting, W.M., and Gougherty, W.G. (1993) Induction of a highly specific antiviral state in transgenic plants: implications for regulation of gene expression and virus resistance. *Plant Cell* 5: 1749–1759.

Linder-Basso, D., Foglia, R., Zhu, P., and Hillman, B.I. (2001) *Crypt1*, an active *Ac*-like transposon from the chestnut blight fungus, *Cryphonectria parasitica*. *Mol. Genet. Genomics* 265: 730–738.

Lingner, J., Hughes, T.R., Shevchenko, A., Mann, M., Lundblad, V., and Cech, T.R. (1997) Reverse transcriptase motifs in the catalytic subunit of telomerase. *Science* 276: 561–567.

Lisch, D., Carey, C.C., Dorweiler, J.E., and Chandler, V.L. (2002) A mutation that prevents paramutation in maize also reverses *Mutator* transposon methylation and silencing. *Proc. Natl. Acad. Sci. (USA)* 99: 6130–6135.

Liu, D. and Crawford, N.M. (1997) Characterization of the germinal and somatic activity of the *Arabidopsis* transposable element *Tag1*. *Genetics* 148: 445–456.

Liu, D. and Crawford, N.M. (1998) Characterization of the putative transposase mRNA of *Tag1*, which is ubiquitously expressed in *Arabidopsis* and can be induced by *Agrobacterium*-mediated transformation with *dTag1* DNA. *Genetics* 149: 693–701.

Liu, D., Wang, R., Galli, M., and Crawford, N.M. (2001) Somatic and germinal excision activities of the *Arabidopsis* transposon *Tag1* are controlled by distinct regulatory sequences within *Tag1*. *Plant Cell* 13: 1851–1863.

Liu, D., Zhang, S., Fauquet, C., and Crawford, N.M. (1999) The *Arabidopsis* transposon *Tag1* is active in rice, undergoing germinal transposition and restricted, late somatic excision. *Mol. Gen. Genet.* 262: 413–420.

Long, D., Swinburne, J., Martin, M., Wilson, K., Sundberg, E., Lee, K., and Coupland, G. (1993) Analysis of the frequency of inheritance of transposed *Ds* elements in *Arabidopsis* after activation by a CaMV 35S promoter fusion to the *Ac* transposase gene. *Mol. Gen. Genet.* 241: 627–636.

Lönning, W.-E. and Saedler, H. (2002) Chromosome rearrangements and transposable elements. *Ann. Rev. Genet.* 36: 389–410.

Luan, D.D., Korman, M.H., Jakubczak, J.L., and Eickbush, T.H. (1993) Reverse transcription of R2Bm RNA is primed by a nick at the chromosomal target site: A mechanism for non-LTR retrotransposition. *Cell* 72: 595–605.

Lucas, H., Moore, G., Murphy, G., and Flavell, R.B. (1992) Inverted repeats in the long terminal repeats of the wheat transposon *Wis2*-1A. *Mol. Biol. Evol.* 9: 716–728.

Luger, K., Mäder, A.W., Richmond, R.K., Sargent, D.F., and Richmond, T.J. (1997) Crystal structure of the nucleosome core particle at 2.8A resolution. *Nature* 389: 251–260.

Luo, G., Ivics, Z., Izsvak, Z., and Bradley, A. (1998) Chromosomal transposition of a *Tc1/mariner*-like element in mouse embryonic stem cells. *Proc. Natl. Acad. Sci. (USA)* 95: 10769–10773.

Lyubomirskaya, N.V., Smirnova, J.B., Razorenova, O.V., Karpova, N.N., Surkob, S.A., Avedisov, S.N., Kim, A.I., and Ilyin, Y.V. (2001) Two variants of the *Drosophila melanogaster* retrotransposon *gypsy* (mdg4): structural and functional differences, and distribution in fly stocks. *Mol. Genet. Genomics* 265: 367–374.

Mack, A.M. and Crawford, N.M. (2001) The *Arabidopsis* TAG1 transposase has an N-terminal zinc finger DNA binding domain that recognizes distinct subterminal motifs. *Plant Cell* 13: 2319–2331.

Mahillon, J. and Chandler, M. (1998) Insertion Sequences. *Microbiol. Mol. Biol. Rev.* 62: 725–774.

Malamy, M.H. (1966) Frameshift mutations in the lactose operon of *E. coli. Cold Spring Harbor Symp. Quant. Biol.* 31: 189–201.

Malamy, M.H., Fiandt, M., and Szybalski, W. (1972) Electron microscopy of polar insertions in the *lac* operon of *Escherichia coli. Mol. Gen. Genet.* 119: 207–222.

Malik, H.G.S. and Eickbush, T.H. (1998) The RTE class of non-LTR retrotransposons is widely distributed in animals and is the origin of many SINEs. *Mol. Biol. Evol.* 15: 1123–1134.

Malik, H.S. and Eickbush, T.H. (1999) Modular evolution of the integrase domain in the Ty3/*Gypsy* class of LTR retrotransposons. *J. Virology* 73: 5186–5190.

Mallory, A.C., Ely, L., Smith, T.H., Marathe, R., Anandalakshmi, R., Fagard, M., Vaucheret, H., Pruss, G., Bowman, L., and Vance, V.B. (2001) HC-Pro suppression of transgene silencing eliminates the small RNAs but not transgene methylation or the mobile signal. *Plant Cell* 13: 571–583.

Manuelidis, L. (1982) Nucleotide sequence definition of a major human repeated DNA, the Hind III 1.9 kb family. *Nucl. Acid Res.* 10: 3211–3219.

Marion-Poll, A., Marin, E., Bonnefoy, N., and Pautot, V. (1993) Transposition of the maize autonomous element *Activator* in transgenic *Nicotiana plumbaginifolia* plants. *Mol. Gen. Genet.* 238: 209–217.

Marlor, R.L., Parkhurst, S.M., and Corces, V.G. (1986) The *Drosophila melanogaster gypsy* transposable element encodes putative gene products homologous to retroviral proteins. *Mol. Cell. Biol.* 6: 1129–1134.

Marmorstein, R. and Roth, S.Y. (2001) Histone acetyltransferases: function, structure and catalysis. *Curr. Opin. Genet. Devel.* 11: 155–161.

Marracci, S., Batistoni, R., Pesole, G., Citti, L., and Nardi, I. (1996) *Gypsy*/Ty3-like elements in the genome of the terrestrial salamander hydromantes (Amphibia, Urodela). *J. Mol. Evol.* 43: 584–593.

Martienssen, R.A. (1998) Functional genomics: probing plant gene function and expression with transposons. *Proc. Natl. Acad. Sci. (USA)* 95: 2021–2026.

Martin, C.R., Carpenter, R., Sommer, H., Saedler, H., and Goen, E.S. (1985) Molecular analysis of instability in the flower pigmentation of *Antirrhinum majus* following isolation of the *pallida* locus by transposon tagging. *EMBO J.* 4: 1625–1630.

Martin, F., Maranon, C., Olivares, M., Alonso, C., and Lopez, M.C. (1995) Characterization of a non-long terminal repeat retrotransposon cDNA (L1Tc) from *Trypanosoma cruzi*: Homology of the first ORF with the ape family of DNA repair enzymes. *J. Mol. Biol.* 247: 49–59.

Martinez, J., Patkaniowska, A., Urlaub, H., Luhrmann, R., and Tuschl, T. (2002) Single-stranded antisense siRNAs guide target RNA cleavage in RNAi. *Cell* 110: 563–574.

Masson, P., Strem, M., and Fedoroff, N. (1991) The *tnpA* and *tnpD* gene products of the *Spm* element are required for transposition in tobacco. *Plant Cell* 3: 73–85.

Masson, P., Surosky, R., Kingsbury, J.A., and Fedoroff, N.V. (1987) Genetic and molecular analysis of the *Spm*-dependent *a-m2* alleles of the maize a locus. *Genetics* 117: 117–137.

Mathias, S.L., Scott, A.F., Kazazian, Jr. H.H., Boeke, J.D., and Gabriel, A. (1991) Reverse transriptase encoded by a human transposable element. *Science* 254: 1808–1810.

Matsuoka, Y. and Tsunewaki, K. (1996) Wheat retrotransposon families identified by reverse transcriptase domain analysis. *Mol. Biol. Evol.* 13: 1384–1392.

Matsuoka, Y. and Tsunewaki, K. (1999) Evolutionary dynamics of Ty1-*copia* group retrotransposons in grass shown by reverse transcriptase domain analysis. *Mol. Biol. Evol.* 16: 208–217.

Matthews, G.D., Goodwin, T.J.D., Butler, M.I., Berryman, T.A., and Poulter, R.T.M. (1997) PCal, a highly unusual Ty1/*copia* retrotransposon from the pathogenic yeast *Candida albicans. J. Bacteriol.* 179: 7118–7128.

Matzke, M.A., Matzke, A.J.M., Pruss, G.J., and Vance, V.B. (2001) RNA-based silencing strategies in plants. *Curr. Opin. Genet. Devel.* 11: 221–227.

Maxam, A.M. and Gilbert, W. (1977) A new method for sequencing DNA. *Proc. Natl. Acad. Sci. (USA)* 74: 560–564.

May, E.W. and Craig, N.L. (1996) Switching from cut-and-paste to replicative Tn7 transposition. *Science* 272: 401–404.

McClintock, B. (1938) The fusion of broken ends of sister half chromatids following chromatid breakage at meiotic anaphases. *Miss. Agric. Exp. Stn. Res. Bull.* 190: 1–48.

McClintock, B. (1941) The stability of broken ends of chromosomes in *Zea mays. Genetics* 26: 234–282.

McClintock, B. (1942a) Maize genetics. *Carnegie Institute of Washington, Year Book* 41: 181–186.

McClintock, B. (1942b) The fusion of broken ends of chromosomes following nuclear fusion. *Proc. Natl. Acad. Sci. (USA)* 28: 458–463.

McClintock, B. (1943) Maize genetics. *Carnegie Institute of Washington, Year Book* 42: 148–152.

McClintock, B. (1945a) Cytogenetic studies of maize and neurospora. *Carnegie Institute of Washington, Year Book* 44: 108–112.

McClintock, B. (1945b) Presidents Report. *Carnegie Institute of Washington, Year Book* 44: 60–61.

McClintock, B. (1946) Maize genetics. *Carnegie Institute of Washington, Year Book* 45: 176–188.

McClintock, B. (1947) Cytogenetic studies of maize and neurospora. *Carnegie Institute of Washington, Year Book* 46: 146–152.

McClintock, B. (1948) Mutable loci in maize. *Carnegie Institute of Washington, Year Book* 47: 155–169.

McClintock, B. (1951) Chromosome organization and genic expression. *Cold Spring Harbor Symp. Quant. Biol.* 16: 13–47.

McClintock, B. (1952) Mutable loci in maize. *Carnegie Institute of Washington, Year Book* 51: 212–219.

McClintock, B. (1953a) Induction of instability at selected loci in maize. *Genetics* 38: 579–599.

McClintock, B. (1953b) Mutation in maize. *Carnegie Institute of Washington, Year Book* 52: 227–237.

McClintock, B. (1954) Mutations in maize and chromosomal aberrations in *Neurospora. Carnegie Institute of Washington, Year Book* 53: 254–260.

McClintock, B. (1956) Controlling elements and the gene. *Cold Spring Harbor Symp. Quan. Biol.* 21: 197–216.

McClintock, B. (1961a) Some parallels between gene control systems in maize and in bacteria. *Amer. Naturalist.* 95: 265–277.

McClintock, B. (1961b) Further studies of the suppressor-mutator system of control of gene action in maize. *Carnegie Institute of Washington, Year Book* 60: 469–476.

McClintock, B. (1968) The states of a gene locus in maize. *Carnegie Institute of Washington, Year Book* 66: 664–672.

McClintock, B. (1978) Mechanisms that rapidly reorganize the genome. *Stadler Genetic Symp.* 10: 25–48.

McClintock, B. (1984) The significance of responses of the genome to challenge. *Science* 226: 792–801.

McElroy, D., Louwerse, J.D., McElroy, S.M., and Lemaux, P.G. (1997) Development of a simple transient assay for *Ac/Ds* activity in cells of intact barley tissue. *Plant J.* 11: 157–165.

Meissner, R., Chague, V., Hu, Q., Emmanuel, E., Elkind, Y., and Levy, A.A. (2000) A high throughput system for transposon tagging and promoter trapping in tomato. *Plant J.* 22: 265–274.

Meissner, R., Jacobson, Y., Melamed, S., Levyatuv, S., Shalev, G., Ashri, A., Elkind, Y., and Levy, A. (1997) A new model system for tomato genetics. *Plant J.* 12: 1465–1472.

Melayah, D., Bonnivard, E., Chalhoub, B., Audeon, C., and Grandbastien, M.-A. (2001) The mobility of the tobacco Tnt1 retrotransposon correlates with its transcriptional activation by fungal factors. *Plant J.* 28: 159–168.

Mhiri, C., Morel, J.-B., Vernhettes, S., Casacuberta, J.M., Lucas, H., and Grandbastien, M.A. (1997) The promoter of the tobacco Tnt1 retrotransposon is induced by wounding and by abiotic stress. *Plant Mol. Biol.* 33: 257–266.

Miesfeld, R., Krystal, M., and Arnheim, N. (1981) A member of a new repeated sequence family which is conserved throughout eucaryotic evolution is found between the human δ and β globin genes. *Nucl. Acid Res.* 9: 5931–5947.

Miki, Y., Nishisho, I., Horii, A., Miyoshi, Y., Utsumomiya J., Kinzler, K.W., Vogelstein, B., and Nakamura, Y. (1992) Disruption of the APC gene by a retrotransposal insertion of L1 sequence in a colon cancer. *Cancer Res.* 52: 643–645.

Miura, A., Yonebayashi, S., Watanabe, K., Toyama, T., Shimada, H., and Kakutani, T. (2001) Mobilization of transposons by a mutation abolishing full DNA methylation in *Arabidopsis*. *Nature* 411: 212–214.

Mizrokhi, L.J., Obolenkova, L.A., Priimagi, A.F., Ilyin, Y.V., Gerasimova, T.I., and Georgiev, G.P. (1985) The nature of unstable insertion mutations and reversions in the locus cut of *Drosophila melanogaster*: molecular mechanism of transposition memory. *EMBO J.* 4: 3781–3787.

Mizuuchi, M. and Baker, T.A. (2002) Chemical mechanisms for mobilizing DNA. In: *Mobile DNA II.* Craig, N.L., Craigie, R., Gellert, M., and Lambowitz, A.M. (eds) ASM Press, Washington, DC. pp. 12–23.

Moazed, D. (2001) Common themes in mechanisms of gene silencing. *Mol. Cell* 8: 489–498.

Modolell, J., Bender, W., and Meselson, M. (1983) *Drosophila melanogaster* mutations suppressible by the suppressor of Hairy-wing are insertions of a 7.3-kilobase mobile element. *Proc. Natl. Acad. Sci. (USA)* 80: 1678–1682.

Moerman, D.G., Benian, G.M., and Waterston, R.H. (1986) Molecular cloning of the muscle gene *unc-22* in *Caenorhabditis elegans* by Tc1 transposon tagging. *Proc. Natl. Acad. Sci. (USA)* 83: 2579–2583.

Monte, J.V., Flavell, R.B., and Gustafson, J.P. (1995) WIS.2-1A: an ancient retrotransposon in the Triticeae tribe. *Theor. Appl. Genet.* 91: 367–372.

Moore, G., Cheung, W., Schwarzacher, T., and Flavell, R. (1991) BIS 1 a major component of the cereal genome and a tool for studying genomic organization. *Genomics* 10: 469–476.

Moore, G.A. (2001) Oranges and lemons: clues to the taxonomy of *Citrus* from molecular markers. *Trends Genet.* 17: 536–540.

Moran, J.V., Holmes, S.E., Naas, T.P., DeBerardinis, R.J., Boeke, J.D., and Kazazian, Jr. H.H. (1996) High frequency retrotransposition in cultured mammalian cells. *Cell* 87: 917–927.

Moreau-Mhiri, C., Morel, J.B., Audeon, C., Ferault, M., Grandbastien, M.A., and Lucas, H. (1996) Regulation of expression of the tobacco *Tnt1* retrotransposon in heterologous species, following pathogen related stress. *Plant J.* 9: 409–419.

Morgan, G.T. (1995) Identification in the human genome of mobile elements spread by DNA-mediated transposition. *J. Mol. Biol.* 254: 1–5.

Mouches, C., Bensaadi, N., and Salvado, J.-C. (1992) Characterization of a LINE retroposon dispersed in the genome of three non-sibling *Aedes* mosquito species. *Gene* 120: 183–190.

Mourrain, P., Beclin, C., Elmayan, T., Feuerbach, F., Godon, C., Morel, J.-B., Jouette, D., Lacombe, A.-M., Nikic, S., Picault, N., Remoue, K., Sanial, M., Vo, T.-A., and Vaucheret, H. (2000) *Arabidopsis* SGS2 and SGS3 genes are required for posttranscriptional gene silencing and natural virus resistance. *Cell* 101: 533–542.

Muller-Neumann, M., Yoder, J.I., and Starlinger, P. (1984) The DNA sequence of the transposable element *Ac* of *Zea mays* L. *Mol. Gen. Genet.* 198: 19–24.

Murphy, N.B., Pays, A., Tebabi, P., Coquelet, H., Guyaux, M., Steinert, G.M., and Pays, E. (1987) *Trypanosoma brucei* repeated element with unusual structural and transcriptional properties. *J. Mol. Biol.* 195: 855–871.

Muszynski, M.G., Gierl, A., and Peterson, P.A. (1993). Genetic and molecular analysis of a three-component transposable-element system in maize. *Mol. Gen. Genet.* 237: 105–112.

Myers, E.W., Sutton, G.G., Delcher, A.L., Dew, I.M. Fasulo, D.P., et al. (2000) A whole-genome assembly of *Drosophila*. *Science* 287: 2196–2204.

Nakagawa, Y., Machida, C., Machida, Y., and Toriyama, K. (2000) Frequency and pattern of transposition of the maize transposable element *Ds* in transgenic rice plants. *Plant Cell Physiol.* 4: 733–742.

Nakamura, T.M., Morin, G.B., Chapman, K.B., Weinrich, S.L., Andrews, W.H., Lingner, J., Harley, C.B., and Cech, T.R. (1997) Telomerase catalytic subunit homologs from fission yeast and human. *Science* 277: 955–959.

Nakaya, R., Nakamura, A., and Murata, T. (1960) Resistance transfer agents in Shigella. *Biochem. Biophys. Res. Comm.* 3: 654–659.

Narlikar, G.J., Fen, H.-Y., and Kingston, R.E., (2002) Cooperation between complexes that regulate chromatin structure and transcription. *Cell* 108: 475–487.

Ngo, H., Tschudi, C., Gull, K., and Ullu, E. (1998) Double-stranded RNA induces mRNA degradation in *Trypanosoma brucei*. *Proc. Natl. Acad. Sci. (USA)* 95: 14687–14692.

Nikaido, M., Rooney, A.P., and Okada, N. (1999) Phylogenetic relationships among cetartiodactyls based on insertions of short and long interspersed elements: Hippopotamuses are the closest extant relatives of whales. *Proc. Natl. Acad. Sci. (USA)* 96: 10261–10266.

Noma, K., Ohtsubo, E., and Ohtsubo, H. (1999) Non-LTR retrotransposons (LINEs) as ubiquitous components of plant genomes. *Mol. Gen. Genet.* 261: 71–79.

Noutoshi, Y., Ito, Y., Kanetani, S., Fujie, M., Usami, S., and Yamada, T. (1998) Molecular anatomy of a small chromosome in the green alga *Chlorella vulgaris*. *Nucl. Acid Res.* 26: 3900–3907.

Nowak, S.J. and Corces, V.G. (2000) Phosphorylation of histone H3 correlates with transcriptionally active loci. *Genes Develop.* 14: 3003–3013.

O'Hare, K. and Rubin, G.M. (1983) Structures of P transposable elements and their sites of insertion and excision in the *Drosophila melanogaster* genome. *Cell* 34: 25–35.

O'Kane, C.J. and Gehring, W.J. (1987) Detection *in situ* of genomic regulatory elements in *Drosophila*. *Proc. Natl. Acad. Sci. (USA)* 84: 9123–9127.

Ogh, K., Hardeman, K., Ivanchenko, M.G., Ellard-Ivey, M., Nebenfuhr, A., White, T.J., and Lomax, T.L. (2002) Fine mapping in tomato using microsynteny with the *Arabidopsis* genome: the *Diageotropica (Dgt) locus*. *Genome Biol.* 3: 1–11.

Ogiwara, I., Miya, M., Ohshima, K., and Okada, N. (2002) V-SINEs: A new superfamily of vertebrate SINEs that are widespread in vertebrate genomes and retain a strongly conserved segment within each repetitive unit. *Genome Res.* 12: 316–324.

Ohki, I., Shimotake, N., Fujita, N., Jee, J.-G., and Ikegami, T. (2001) Solution structure of the methyl-CpG binding domain of human MBD1 in complex with methylated DNA. *Cell* 105: 487–497.

Ohtsubo, H., Ohmori, H., and Ohtsubo, E. (1979) Nucleotide-sequence analysis of Tn3 (Ap): implications for insertion and deletion. *Cold Spring Harbor Symp. Quant. Biol.* 43: 1269–1277.

Ohtzubo, E. and Sekine, Y. (1996) Bacterial insertion sequences. *Curr. Topics Microbiol. Immunol.* 204: 1–26.

Okada, N. (1991) SINEs. *Curr. Opin. Genet. Devel.* 1: 498–504.

Okamoto, H. and Hirochika, H. (2000) Efficient insertion mutagenesis of *Arabidopsis* by tissue culture-induced activation of the tobacco retrotransposon Tto1. *Plant J.* 23: 291–304.

Okamoto, H. and Hirochika, H. (2001) Silencing of transposable elements in plants. *Trends Plant Sci.* 6: 527–534.

Oliveira, C., Chew, J.S.K., Porto-Foresti, F., Dobson, M.J., and Wright, J.M. (1999) A LINE2 repetitive DNA sequence from the cichlid fish, *Oreochromis niloticus*: sequence analysis and chromosomal distribution. *Chromosoma* 108: 457–468.

Oosumi, T., Belknap, W.R., and Garlick, B. (1995) Mariner transposons in humans. *Nature* 378: 672–672.

Osborne, B.I. and Baker, B. (1995) Movers and shakers: maize transposons as tools for analyzing other plant genomes. *Curr. Opin. Cell Biol.* 7: 406–413.

Osborne, B.I., Corr, C.A., Prince, J.P., Hehl, R., Tanksley, S.D., McCormick, S., and Baker, B. (1991) *Ac* transposition from a T-DNA can generate linked and unlinked clusters of insertions in the tomato genome. *Genetics* 129: 833–844.

Ostertag, E.M. and Kazazian, H.H. Jr (2001) Biology of mammalian L1 retrotransposons. *Ann. Rev. Genet.* 35: 501–538.

Palauqui, J.-C., Elmayan, T., Pollien, J.-M., and Vaucheret, H. (1997) Systemic acquired silencing: transgene-specific post-transcriptional silencing is transmitted by grafting from silenced stocks to non-silenced scions. *EMBO J.* 16: 4738–4745.

Pardue, M.-L. and DeBaryshe, P.G. (2002) Telomeres and transposable elements. In: *Mobile DNA II*. Craig, N.L., Craigie, R., Gellert, M., and Lambowitz, A.M. (eds). ASM Press, Washington, DC. pp. 870–890.

Parinov, S., Sevugan, M., Ye, D., Yang, W.-C., Kumaran, M., and Sunaresan, V. (1999) Analysis of flanking sequences from *Dissociation* insertion lines: a database for reverse genetics in *Arabidopsis. Plant Cell* 11: 2263–2270.

Parkhurs, S.M. and Corces, V.G. (1985) Forked, Gypsys, and suppressors in *Drosophila. Cell* 41: 429–437.

Pearce, S.R., Harrison, G., Heslop-Harrison, P.J.S., Flavell, A.J., and Kumar, A. (1997) Characterization and genomic organization of Ty1-*copia* group retrotransposons in rye (*Secale cereale*). *Genome* 40: 617–625.

Pearce, S.R., Harrison, G., Li, D., Heslop-Harrison, J.S., Kumar, A., and Flavell, A.J. (1996) The Ty1-*copia* group of retotransposons in *Vicia* species: copy number, seuqene heterogeneity and chromosomal location. *Mol. Gen. Genet.* 250: 305–315.

Peleman, J., Cottyn, B., Van Camp, W., Van Montagu, M., and Inze, D. (1991) Transient occurrence of extrachromosomal DNA of an *Arabidopsis thaliana* transposon-like element, Tat1. *Proc. Natl. Acad. Sci. (USA)* 88: 3618–3622.

Pelissier, T., Tutois, S., Deragon, J.M., Tourmente, S., Genestier, S., and Picard, G. (1995) *Athila*, a new retroelement from *Arabidopsis thaliana. Plant Mol. Biol.* 29: 441–452.

Pelissier, T., Tutois, S., Tourmente, S., Dergon, J.M., and Picard, G. (1996) DNA regions flanking the major *Arabidopsis thaliana* satellite are principally enriched in *Athila* retroelement sequences. *Genetics* 97: 141–151.

Pereira, A., Cuypers, H., Gierl, A., Schwarz-Sommer, Z., and Saedler, H. (1986) Molecular analysis of the *En/Spm* transposable element system of *Zea mays. EMBO J.* 5: 835–841.

Pereira, A., Schwarz-Sommer, Z., Gierl, A., Bertram, I., Peterson, P.A., and Saedler, H. (1985) Genetic and molecular analysis of the enhancer (*En*) transposable element system of *Zea mays. EMBO J.* 4: 17–23.

Peterson, P.A. (1953) A mutable pale green locus in maize. *Genetics* 38: 682–683.

Peterson, P.A. (1960) The pale green mutable system in maize. *Genetics* 45: 115–133.

Peterson, P.A. (1961) Mutable a1 of the *En* system in maize. *Genetics* 46: 759–771.

Peterson, P.A. (1965) A relationship between the *Spm* and *En* control systems in maize. *Amer. Naturalist* 99: 391–398.

Peterson, P.A. (1970) The *En* mutable system in maize. *Theor. Appl. Genet.* 40: 367–377.

Peterson, P.A. (1987) Mobile elements in plants. *Crit. Rev. Plant Sci.* 6: 105–208.

Peterson, P.A. and Bianchi, A. (1999) *Maize Genetics and Breeding in the 20th Century.* World Scientific Publ., Singapore, 379 p.

Peterson, T. (1990) Intragenic transposition of *Ac* generates a new allele of the maize P gene. *Genetics* 126: 469–476.

Picard, G.J., Lavige, M., Bucheton, A., and Bregliano, J.C. (1977) Non-mendelian female sterility in *Drosophila melanogaster*: physiological pattern of embryo/lethality. *Biol. Cell.* 29: 89–98.

Pickeral, O.K., Makalowski, W., Boguski, M.S., and Boeke, J.D. (2000) Frequent human genomic DNA transduction driven by LINE1 retrotransposition. *Genet. Res.* 10: 411–415.

Plasterk, R.H.A. (1996) The *Tc1/mariner* transposon family. *Curr. Topics Microbiol. Immunol.* 204:125-143.

Plasterk, R.H.A. (2002) RNA silencing: The genome's immune system. *Science* 296: 1263–1265.

Plasterk, R.H.A. and Ketting, R.F. (2000) The silence of the genes. *Curr. Opin. Genet. Devel.* 10: 562–567.

Plasterk, R.H.A. and van Luenen, H.G.A.M. (2002) The *Tc1/mariner* family of transposable elements. In: *Mobile DNA II.* Craig, N.L., Craigie, R., Gellert, M., and Lambowitz, A.M. (eds). ASM Press, Washington, DC. pp. 519–532.

Plasterk, R.H.A., Izsvak, Z., and Ivics, Z. (1999) Resident aliens. *Trends Genet.* 15: 326–332.

Pohlman, R.F., Fedoroff, N.V., and Messing, J. (1984) The nucleotide sequence of the maize controlling element activator. *Cell* 37: 635–643.

Polard, P. and Chandler, M. (1995a) An *in vivo* transposase-catalyzed, single-stranded DNA circularization reaction. *Genes Devel.* 9: 2846–2858.

Polard, P. and Chandler, M. (1995b) Bacterial transposases and retrovral integrases. *Mol. Microbiol.* 15: 13–23.

Polard, P., Prere, M.F., Fayet, O., and Chandler, M. (1992) Transposase-induced excision and circularization of the bacterial insertion sequence IS911. *EMBO J.* 11: 5079–5090.

Polard, P., Ton-Hoang, B., Haren, M.L., Betermier, M., Walczak, R., and Chandler, M. (1996) IS911-mediated transpositional recombination *in vitro. J. Mol. Biol.* 264: 68–81.

Pontecorvo, G., De Felice, B., and Carfagna, M. (2000) A novel repeated sequence DNA originated from a Tc1-like transposon in water green frog *Rana esculenta. Gene* 261: 205–210.

Pontecorvo, G., Roper, J.A., Hemmons, L.M., Macdonald, R.D., and Bulton, A.W. J. (1953) The genetics of *Aspergillus nidulans. Advan. Genet.* 5: 141–238.

Potter, S.S. (1982) DNA sequence of a foldback transposable element in *Drosophila. Nature* 297: 201–204.

Poulter, R.T.M. and Butler, M. (1998) A retrotransposon family from the pufferfish (fugu) *Fugu rubripes. Gene* 215: 241–249.

Poulter, R.T.M., Butler, M.I., and Ormandy, J. (1999) A LINE element from the pufferfish (fugu) *Fugu rubripes* which shows similarity to the CR1 family of non-LTR retrotransposons. *Gene* 227: 169–179.

Pouteau, S., Grandbastien, M.A., and Boccara, M. (1994) Microbial elicitors of plant defence responses activate transcription of a retrotransposon. *Plant J.* 5: 535–542.

Pouteau, S., Huttner, E., Grandbastien, M.A., and Caboche, M. (1991) Specific expression of the tobacco Tnt1 retrotransposon in protoplasts. *EMBO J.* 10: 1911–1918.

Pozueta-Romero, J., Houlne, G., and Schantz, R. (1996) Nonautonomous inverted repeat Alien transposable elements are associated with genes of both monocotyledonous and dicotyledonous plants. *Gene* 171: 147–153.

Pozueta-Romero, J., Klein, M., Houlne, G., Schantz, M.-L., Meyer, B., and Schantz, R. (1995) Characterizaton of a family of genes encoding a fruit-specific wound-stimulated protein of bell pepper (*Capsicum annuum*): identification of a new family of transposable elements. *Plant Mol. Biol.* 28: 1011–1025.

Presting, G.G., Malysheva, L., Fuchs, J., and Schubert, I. (1998) A Ty3/*gypsy* retrotransposon-like sequence localizes to the centromeric regions of cereal chromosomes. *Plant J.* 16: 721–728.

Priimagi, A.F., Mizrokhi, L.J., and Ilyin, Y.V. (1988) The *Drosophila* mobile element *jockey* belongs to LINEs and contains coding sequences homologous to some retroviral proteins. *Gene* 70: 253–262.

Prud'homme, N., Gans, M., Masson, M., Terziun, C., and Bucheton, A. (1995) *Flamenco*, a gene controlling the *gypsy* retrovirus of *Drosophila melanogaster. Genetics* 139: 697–711.

Purugganan, M.D. and Wessler, S.R. (1994) Molecular evolution of *magellan*, a maize Ty3/*gypsy*-like retrotransposon. *Proc. Natl. Acad. Sci. (USA)* 91: 11674–11678.

Qin, M., Robertson, D.S., and Ellingboe, A.H. (1991) Cloning of the *Mutator* transposable element *MuA2*, a putative regulator of somatic mutability of the *al-Mum2* allele in maize. *Genetics* 129: 845–854.

Radice, A.D., Bugaj, B., Fitch, D.H.A., and Emmons, S.W. (1994) Widespread occurrence of the *Tc1* transposon family: *Tc1*-like transposons from teleost fish. *Mol. Gen. Genet.* 244: 606–612.

Raina, R., Cook, D., and Fedoroff, N. (1993) Maize *Spm* transposable element has an enhancer-insensitive promoter. *Proc. Natl. Acad. Sci. (USA)* 90: 6355–6359.

Raina, S., Mahalingan, R., Chen, F., and Fedoroff, N. (2002) A collection of sequences and mapped *Ds* transposon insertion sites in *Arabidopsis thaliana. Plant Mol. Biol.* 50: 93–110.

Raizada, M.N. and Walbot, V. (2000) The late developmental pattern of *Mu* transposon excision is conferred by a cauliflower mosaic virus 35S-driven *MURA* cDNA in transgenic maize. *Plant Cell* 12: 5–21.

Raizada, M.N., Nan, G.-L., and Walbot, V. (2001) Somatic and germinal mobility of the rescueMu transposon in transgenic maize. *Plant Cell* 13: 1587–1608.

Ramachandran, S. and Sundaresan, V. (2001) Transposons as tools for functional genomics. *Plant Physiol. Biochem.* 39: 243–252.

Randolph, L.F. and McClintock, B. (1926) Polyploidy in *Zea mays. Amer. Naturalist* 60: 99–102.

Redei, G.P. (1998) *Genetic Manual.* World Scientific Publ., Singapore, 1142 p.

Reinhart, B.J., Weinstein, E.G., Rhodes, M.W., Bartel, B., and Bartel, D.P. (2002) MicroRNAs in plants. *Genes Devel.* 16: 1616–1626.

Renckens, S., De Greve, H., Beltran-Herrera, J., Toong, L.T., Deboeck, F., De Rycke, R., Van Montagu, M., and Hernalsteens, J.-P. (1996) Insertion mutagenesis and study of transposable elements using a new unstable virescent seedling allele for isolation of haploid petunia lines. *Plant J.* 10: 533–544.

Reznikoff, W.S. (2002) Tn5 transposition. In: *Mobile DNA II.* Craig, N.L., Craigie, R. Gellert, M., and Lambowitz, A.M. (eds.) ASM Press, Washington, DC. pp. 403–422.

Reznikoff, W.S., Bhasin, A., Davies, D.R., Goryshin, I.Y., Mahnke, L.A., Naumann, T., Rayment, I., Steiniger-White, M., and Twining, S.S. (1999) Tn5: A molecular window on transposition. *Biochem. Biophys. Res. Commun.* 266: 729–734.

Rhoades, M.M. (1984) The early years of maize genetics. *Ann. Rev. Genet.* 18: 1–29.

Rhoades, M.M. and McClintock, B. (1935) The cytogenetics of maize. *Bot. Rev.* 1: 292–325.

Rhodes, P.R. and Vodkin, L.O. (1988) Organization of the *Tgm* family of transposable elements in soybean. *Genetics* 120: 597–604.

Richards, E.J. (1997) DNA methylation and plant development. *Trends Genet.* 13: 319–323.

Richards, E.J. and Elgin, S.C.R. (2002) Epigmetic codes for heterochromatin formation and silencing: rounding up the usual suspects. *Cell* 108: 489–500.

Richmond, T.J., Finch, J.T., Rushton, B., Rhodes, D., and Klug, A. (1984) Structure of the nucleosome core particle at 7Å resolution. *Nature* 311: 532–537.

Ringrose, L. and Paro, R. (2001) Cycling silence. *Nature* 412: 493–494.

Rio, D.C. (1991) Regulation of *Drosophila* P element transposition. *Trends Genet.* 7: 282–287.

Rio, D.C. (2002) P transposable elements in *Drosophila melanogaster*. In: *Mobile DNA II.* Craig, N.L., Criagie, R., Gellert, M., and Lambowitz, A.M. (eds.) ASM Press, Washington, DC. pp. 484–518.

Robert, V., Prud'homme, N., Kim, A., Bucheton, A., and Pelisson, A. (2001) Characterization of the *flamenco* region of the *Drosophila melanogaster* genome. *Genetics* 158: 701–713.

Robertson, D.S. (1978) Characterization of a mutator system in maize. *Mutation Res.* 51: 21–28.

Robertson, D.S. (1980) The timing of *Mu* activity in maize. *Genetics* 94: 969–978.

Robertson, D.S. (1981) *Mutator* activity in maize: timing of its activation in ontogeny. *Science* 213: 1515–1517.

Robertson, D.S. (1983) A possible dose-dependent inactivation of *Mutator* (*Mu*) in maize. *Mol. Gen. Genet.* 191: 86–90.

Robertson, D.S. (1985) Differential activity of the maize mutator *Mu* at different loci and in different cell lineages. *Mol. Gen. Genet.* 200: 9–13.

Robertson, D.S. and Mascia, P.N. (1981) Tests of 4 controlling-element systems of maize for mutator activity and their interaction with *Mu* mutator. *Mutation Res.* 84: 283–289.

Robertson, D.S. and Stinard, P.S. (1987) Genetic evidence of *Mutator*-induced deletions in the short arm of chromosome 9 of maize. *Genetics* 115: 353–361.

Robertson, D.S. and Stinard, P.S. (1989) Genetic analyses of putative two-element systems regulating somatic mutability in *Mutator*-induced aleurone mutants of maize. *Dev. Genet.* 10: 482–506.

Robertson, H.M. (1993) The *mariner* transposable element is widespread in insects. *Nature* 362: 241–245.

Robertson, H.M. (1996) Members of the *pogo* superfamily of DNA-mediated transposons in the human genome. *Mol. Gen. Genet.* 252: 761–766.

Robertson, H.M. (1997) Multiple *Mariner* transposons in flatworms and hydras are related to those of insects. *J. Heredity* 88: 195–201.

Roeder, G.S. and Fink, G.R. (1980) DNA rearrangements associated with a transposable element in yeast. *Cell* 21: 239–249.

Roeder, G.S. and Fink, G.R. (1983) Transposable elements in yeast. In: *Mobile Genetic Elements.* Shapiro, J.A. (ed.). Academic Press, NY. pp. 299–328.

Roeder, G.S., Farabaugh, P.J., Chaleff, D.T., and Fink, G.R. (1980) The origin of gene instability in yeast. *Science* 209: 1375–1380.

Rohr, C.J.B., Ranson, H., Wang, X., and Besansky, N.J. (2002) Structure and evolution of *mtanga*, a retrotransposon actively expressed on the Y chromosome of the African malaria vector *Anopheles gambiae*. *Biol. Mol. Evol.* 19: 149–162.

Rommens, C.M.T., Rudenko, G.N., Dijkwel, P.P., Van Haaren, M.J.J., Ouwerkerk, B.P.F., Blok, K.M., Nijkamp, J.J., and Hille, J. (1992) Characterization of the *Ac/Ds* behaviour in transgenic tomato plants using plasmid rescue. *Plant Mol. Biol.* 20: 61–70.

Rommens, C.M.T., Van der Biezen, A., Ouserkerk, P.B.E., Nijkamp, H.J.J., and Hille, J. (1991) *Ac*-induced disruption of the double *Ds* structure in tomato. *Mol. Gen. Genet.* 228: 453–458.

Ros, F. and Kunze, R. (2001) Regulation of *Activator/Dissociation* transposition by replication and DNA methylation. *Genetics* 157: 1723–1733.

Rose, A.M. and Snutch, T.P. (1984) Isolation of the closed circular form of the transposable element Tc1 in *Caenorhabditis elegans. Nature* 311: 485–487.

Rosenzweig, B., Liao, L.W., and Hirsch, D. (1983) Sequence of the transposable element Tc1. *Nucl. Acid. Res.* 12: 4201–4209.

Rousseau, P., Normand, C., Loot, C., Turlan, C., Alazard, R., Duval-Valentin, G., and Chandler, M. (2002). Transposition of TS 911. In: *Mobile DNA II.* Craig, N.L., Craigie, R., Gellert, M., and Lambowitz, A.M. (eds.) ASM Press, Washington, DC. pp. 367–383.

Royo, J., Nass, N., Matton, D.P., Okamotot, S., Clarke, A.E., and Newbigin (1996) A retrotransposon-like sequence linked to the S-locus of *Nicotiana alata* is expressed in styles in response to touch. *Mol. Gen. Genet.* 250: 180–188.

Ruan, K. and Emmons, S.W. (1984) Extrachromosomal copies of transposon Tc1 in the nematode *Caenorhabditis elegans. Proc. Natl. Acad. Sci. (USA)* 81: 4018–4022.

Rubin, E. and Levy, A.A. (1997) Abortive gap repair: underlying mechanism for *Ds* element formation. *Mol. Cell. Biol.* 17: 6294–6302.

Rubin, G.M. (1983) Dispersed repetitive dNAs in *Drosophila.* In: *Mobile Genetic Elements.* Shapiro, J.A. (ed.) Academic Press, NY. pp. 329–361.

Rubin, G.M., Kidwell, M.G., and Bingham, P.M. (1982) The molecular basis of P-M hybrid dysgenesis: The nature of induced mutations. *Cell* 29: 987–994.

Saedler, H. and Starlinger, P. (1967) O^O mutations in the galactose operon in *E. coli.* I. Genetic characterization. *Mol. Gen. Genet.* 100: 178–189.

Sandmeyer, S. (1998) Targeting transposition: At home in the genome. *Genome Res.* 8: 416–418.

Sandmeyer, S.B. and Menees, T.M. (1996) Morphogenesis at the retrotransposon-retrovirus interface: *Gypsy* and *Copia* families in yeast and *Drosophila. Curr. Topics Microbiol. Immun.* 214: 261–296.

Sandmeyer, S.B., Aye, M., and Menees, T. (2002) Ty3: A position specific, *gypsy*-like element in *Saccaromyces cerevisiae.* In: *Mobile DNA II.* Craig, N.L., Craigie, R., Gellert, M., and Lambowitz, A.M. (eds). ASM Press, Washington, DC. pp. 663–683.

Sanger, F., Nicklen, S.M., and Coulson, A.R. (1977) Sequencing with chain-terminating inhibitors. *Proc. Natl. Acad. Sci. (USA)* 44: 5463–5467.

SanMiguel, P. and Bennetzen, J.L. (1998) Evidence that a recent increase in maize genome size was caused by the massive amplification of intergene retrotransposons. *Ann. Bot.* 82: 37–44.

SanMiguel, P., Gaut, B.S., Tikhonov, A., Nakajima, Y., and Bennetzen, J.L. (1998) The paleontology of intergene retrotransposons of maize. *Nature Genet.* 20: 43–45.

SanMiguel, P., Tikhonov, A., Jin, Y.-K., Motchoulskaia, N., Zakharov, D., Melake-Berhan, A., Springer, P.S., Edwards, K.J., Lee, M., Avramova, Z., and Bennetzen, J.L. (1996) Nested retrotransposons in the intergenic regions of the maize genome. *Science* 274: 765–768.

Satoh, N. and Jeffery, W.R. (1995) Chasing tails in ascidians: developmental insights into the origin and evolution of chordates. *Trends Genet.* 11: 354–359.

Scadden, A.D.J. and Smith, C.W.J. (2001) RNAi is antagonized by A>I hyper-editing. *EMBO Reports* 2: 1107–1111.

Schiefelbein, J.W., Furtek, D.B., Dooner, H.K., and Nelson, O.E. (1988) Two mutations in a maize bronze-1 allele caused by transposable elements of the *Ac-Ds* family alter the quantity and quality of the gene product. *Genetics* 120: 767–777.

Schläppi, M., Raina, R., and Fedoroff, N. (1996) A highly sensitive plant hybrid protein assay system based on the *Spm* promoter and TnpA protein for detection and analysis of transcription activation domains. *Plant Mol. Biol.* 32: 717–725.

Schmid, C.W. (1996) Alu: Structure, origin, evolution, significance, and function of one-tenth of human DNA. *Nucl. Acid. Res.* 53: 283–319.

Schmidt, R. and Willmitzer, L. (1989) The maize autonomous element *Activator (Ac)* shows a minimal germinal excision frequency of 0.2%–0.5% in transgenic *Arabidopsis thaliana* plants. *Mol. Gen. Genet.* 220: 17–24.

Schmidt, T. (1999) Lines, Sines and repetitive DNA: non-LTR retrotransposons in plant genomes. *Plant Mol. Biol.* 40: 903–910.

Schmidt, T., Kubis, S., and Heslop-Harrison, J.S. (1995) Analysis and chromosomal localization of retrotransposons in sugar beet (*Beta vulgaris* L.): LINEs and Ty1-*copia*-like elements as major components of the genome. *Chromosome Res.* 3: 335–345.

Schmitz, J., Ohme, M., and Zischler, H. (2001) SINE insertions in Cladistic analyses and the phylogenetic affiliations of *Tarsius bancamus* to other primates. *Genetics* 157: 777–784.

Schnable, P.S., Peterson, P.A., and Saedler, H. (1989) The bz-rcy allele of the Cy transposable element system of *Zea mays* contains a *Mu*-like element insertion. *Mol. Gen. Genet.* 217: 459–463.

Schnable, P.S. and Peterson, P.A. (1988) The *Mutator*-related *cy* transposable element of *Zea mays* L. behaves as a near-Mendelian factor. *Genetics* 120: 587–596.

Scholz, S., Lörz, H., and Lütticke, S. (2001) Transposition of the maize transposable element *Ac* in barley (*Hordeum vulgare* L.). *Mol. Genet. Genomics* 274: 653–661.

Schwartz, D. and Dennis, E. (1986) Transposase activity of the *Ac* controlling element in maize is regulated by its degree of methylation. *Mol. Gen. Genet.* 205: 476–482.

Schwartz, D.S., Hutvagner, G., Haley, B., and Zamore, P.D. (2002) Evidence that siRNAs function as guides, not primers, in the *Drosophila* and human RNAi pathways. *Mol. Cell.* 10: 537–548.

Schwarz-Sommer, Z., Gierl, A., Cuypers, H., Peterson, P.A., and Saedler, H. (1985) Plant transposable elements generate the DNA sequence diversity needed in evolution. *EMBO J.* 4: 591–597.

Schwarz-Sommer, Z., Gierl, A., Klosgen, R.B., Wieland, U., Peterson, P.A., and Saedler, H. (1984) The *Spm (En)* transposable element controls the excision of a 2-kb DNA insert at the wxm8 allele of *Zea mays. EMBO J.* 3: 1021–1028.

Schwarz-Sommer, Z., Huijser, P., Nacken, W., Saedler, H., and Sommer, H. (1990) Genetic control of flower development by homeotic genes in *Antirrhinum majus. Science* 250: 931–936.

Schwarz-Sommer, Z., Leclercq, L., Gobel, E., and Saedler, H. (1987) Cin4, an insert altering the structure of the A1 gene in *Zea mays*, exhibits properties of nonviral retrotransposons. *EMBO J.* 6: 3873–3880.

Scofield, S.R., English, J.J., and Jones, J.D.G. (1993) High level expression of the *Activator* transposase gene inhibits the excision of *Dissociation* in tobacco cotyledons. *Cell* 75: 507–517.

Scofield, S.R., Harrison, K., Nurrish, S.J., and Jones, J.D.G. (1992) Promoter fusion to the *Activator* transposase gene cause distinct patterns of *Dissociation* excision in tobacco cotyledons. *Plant Cell* 4: 573–582.

Searles, L.L., Jokerst, R.S., Bingham, P.M., Voelker, R.A., and Greanleaf, A.L. (1982) Molecular cloning of sequences from *Drosophila* RNA polymerase II locus by P element transposon tagging. *Cell* 31: 585–592.

Segal, Y., Peissel, B., Renieri, A., de Marchi, M., Ballabio, A., Pei, Y., and Zhou, J. (1999) LINE-1 elements at the sites of molecular rearrangements in alport syndrome-diffuse leiomyomatosis. *Am. J. Hum. Genet.* 64: 62–69.

Sekine, Y., Aihara, K., and Ohtsubo, E. (1999) Linearization and transposition of circular molecules of insertion sequence IS3. *J. Mol. Biol.* 294: 21–34.

Shapiro, J.A. (1969) Mutations caused by the insertion of genetic material into the galactose operon of *Escherichia coli. J. Mol. Biol.* 40: 93–105.

Shapiro, J.A. (1979) Molecular model for the transposition and replication of bacteriophage *Mu* and other transposable elements. *Proc. Natl. Acad. Sci. (USA)* 76: 1933–1937.

Sharp, P.A. and Zamore, P.D. (2000) RNA interference. *Science* 297: 2431–2433.

Sharp, P.A., Cohen, S.N., and Davidson, N. (1973) Electron microscope heteroduplex studies of sequence relations among plasmids of *Escherichia coli*. II. Structure of drug resistance (R) factors and F. factors. *J. Mol. Biol.* 75: 235–255.

Shepherd, N.S., Schwarz-Sommer, Z., Blumberg Vel Spalve, J., Gupta, M., Wienand, U., and Saedler, H. (1984) Similarity of the *CinI* repetitive family of *Zea mays* to eukaryotic transposable elements. *Nature* 307: 185–187.

Shepherd, N.S., Schwarz-Sommer, Z., Wienand, U., Sommer, H., Deumling, B., Peterson, P.A., and Saedler, H. (1982) Cloning of a genomic fragment carrying the insertion element *CinI* of *Zea mays*. *Mol. Gen. Genet.* 188:266-271.

Sherratt, D. (1989) Tn3 and related transposable elements: site-specific recombination and transposition. In: *Mobile DNA*. Berg, D.E. and Howe, M.M. (eds.) ASM Press, Washington, DC. pp. 163–184.

Shimamoto, K., Miyazaki, C., Hashimoto, H., Izwa, T., Itoh, K., Terada, R., Inagaki, Y., and Iida, S. (1993) Trans-activation and stable integration of the maize transposable element *Ds* cotransfected with the *Ac* transposase gene in transgenic rice plants. *Mol. Gen. Genet.* 239: 354–360.

Shore, D. (2001) Telomeric chromatin: replicating and wrapping up chromosome ends. *Curr. Opin. Genet. Devel.* 11: 189–198.

Shure, M., Wessler, S., and Fedoroff, N. (1983) Molecular identification and isolation of the waxy locus in maize. *Cell* 35: 225–233.

Simmen, M.W. and Bird, A. (2000) Sequence analysis of transposable elements in the sea squirt *Ciona intestinalis*. *Mol. Biol. Evol.* 17: 1685–1694.

Simmen, M.W., Leitgeb, S., Charlton, J., Jones, S.J.M., Harris, B.R., Clark, V.H., and Bird, A. (1999) Nonmethylated transposable elements and methylated genes in a chordate genome. *Science* 283: 1164–1167.

Simmons, M.J., Haley, K.J., and Tompson, S.J. (2002) Maternal transmission of P element transposase activity in *Drosophila melanogaster* depends on the last P intron. *Proc. Natl. Acad. Sci. (USA)* 14: 9306–9309.

Singer, M. and Berg, P. (1991) *Genes and Genomes – A Changing Perspective*. University Science Books, Mill Valley, CA, 929 p.

Singer, M.F. (1982a) Highly repeated sequences in mammalian genomes. *Intl. Rev. Cytol.* 76: 67–112.

Singer, M.F. (1982b) SINEs and LINEs: highly repeated short and long interspersed sequences in mammalian genomes. *Cell* 28: 433–434.

Singer, M.F. and Skowronski, J. (1985) Making sense out of LINES: long interspersed repeat sequences in mammalian genomes. *Trends Biochem. Sci.* 10: 119–122.

Singer, T., Yordan, C., and Martienssen, R.A. (2001) Robertson's mutator transposons in *A. thaliana* are regulated by the chromatin-remodeling gene decrease in DNA methylation (DDM1). *Genes Devel.* 15: 591–602.

Skowronski, J. and Singer, M.F. (1984) Expression of a cytoplasmic LINE-1 transcript is regulated in a human teratocarcinoma cell line. *Proc. Natl. Acad. Sci. (USA)* 82: 6050–6054.

Smit, A.F.A. (1996) The origin of interspersed repeats in the human genome. *Curr. Opin. Genet. Devel.* 6: 743–748.

Smit, A.F.A. and Riggs, A.D. (1996) Tiggers and other DNA transposon fossils in the human genome. *Proc. Natl. Acad. Sci. (USA)* 93: 1443–1448.

Smith, C.W. and Valcarcel, J. (2002) Alternative pre-mRNA splicing: the logic of combinational control. *Trends Biochem. Sci.* 25: 381–388.

Smith, P.A. and Corces, V.G. (1995) The suppressor of hairy-wing protein regulates the tissue-specific expression of the *Drosophila gypsy* retrotransposon. *Genetics* 139: 215–228.

Sommer, H., Beltran, J.-P., Huijser, P., Pape, H., Lonnig, W.-E., Saedler, H., and Schwarz-Sommer, Z. (1990) Deficiens, a homeotic gene involved in the control of flower morphogenesis in *Antirrhinum majus*: the protein shows homology to transcription factors. *EMBO J.* 9: 605–613.

Sommer, H., Carpenter, R., Harrison, B.J., and Saedler, H. (1985) The transposable element *Tam3* of *Antirrhinum majus* generates a novel type of sequence alteration upon excision. *Mol. Gen. Genet.* 199: 225–231.

Soriano, P., Meunier-Rotival, M., and Bernardi, G. (1983) The distribution of interspersed repeats is nonuniform and conserved in the mouse and human genomes. *Proc. Natl. Acad. Sci. (USA)* 80: 1816–1820.

Souer, E., Quattrocchio, F., de Vetten, N., Mol, J., and Koes, R. (1995) A general method to isolate genes tagged by a high copy number transposable element. *Plant J.* 7: 677–685.

Southern, E.M. (1975) Detection of specific sequences among DNA fragments separated by gel electrophoresis. *J. Mol. Biol.* 98: 503–517.

Speulman, E., Metz, P.L., van Arkel, G., te Lintel Hekkert, B., Stiekema, W.J., and Pereira, A. (1999) A two-component enhancer-inhibitor transposon mutagenesis system for functional analysis of the *Arabidopsis* genome. *Plant Cell* 11: 1853–1866.

Springer, M.S., Tusneem, N.A., Davidson, E.H., and Britten, R.J. (1995) Phylogeny, rates of evolution and patterns of codon usage among sea urchin retroviral-like elements, with implications for the recognition of horizontal transfer. *Mol. Biol. Evol.* 12: 219–230.

Springer, P.S., Edwards, K.J., and Bennetzen, J.L. (1994) DNA class organization on maize *Adh1* yeast artificial chromosomes. *Proc. Natl. Acad. Sci. (USA)* 91: 863–867.

Springer, P.S., McCombie, W.R., Sundaresan, V., and Martienssen, R.A. (1995) Gene trap tagging of Prolifera, an essential MCM2-3-5-like gene in *Arabidopsis*. *Science* 268: 877–880.

Starlinger, P. and Saedler, H. (1972) Insertion mutations in microorganisms. *Biochimie* 54: 177–185.

Starlinger, P. and Saedler, H. (1976) IS elements in microorganisms. *Curr. Topics. Microbiol. Immun.* 75: 111–152.

Steinemann, M. and Steinemann, S. (1997) The enigma of Y chromosome degeneration: *TRAM*, a novel retrotransposon is preferentially located on the *Neo-Y* chromosome of *Drosophila miranda*. *Genetics* 145: 261–266.

Strahl, B.D. and Allis, C.D. (2000) The language of covalent histone modifications. *Nature* 403: 41–45.

Strommer, J.N., Hake, S., Bennetzen, J., Taylor, W.C., and Freeling, M. (1982) Regulatory mutants of the *Adh1* gene caused by DNA insertion. *Nature* 300: 542–544.

Struhl, K. (1999) Fundamentally different logic of gene regulation in eukaryotes and prokaryotes. *Cell* 98: 1–4.

Sturtevant, A.H. and Beadle, G. (1939) *An Introduction to Genetics*. Sanders, Philadelphia, PA.

Stuurman, J., Nijkamp, H.J.J., and Van Haaren, M.M.M. (1998) Molecular insertion-site selectivity of *Ds* in tomato. *Plant J.* 14: 215–223.

Sullivan, K.F. (2001) A solid foundation: functional specialization of centromeric chromatin. *Curr. Opin. Genet. Devel.* 11: 182–188.

Sulston, J.E. and Brenner, S. (1974) The DNA of *Caenorhabditis elegans*. *Genetics* 77: 95–104.

Sundaresan, V., Springer, P., Volpe, T., Haward, S., Jones, J.D.G., Dean, C., Ma, H., and Martienssen, R. (1995) Patterns of gene action in plant development revealed by enhancer trap and gene trap transposable elements. *Genes Devel.* 9: 1797–1810.

Sutton, W.D., Gerlach, W.L., Schwatz, D., and Peacock, W.J. (1984) Molecular analysis of *Ds* controlling element mutations at the *Adh1* locus of maize. *Science* 223: 1265.

Svejstrup, J.Q. (2002) Chromatin elongation factors. *Curr. Opin. Genet. Devel.* 12: 156–161.

Swinburne, J., Balcells, L., Scofield, S.R., Jones, J.D.G., and Coupland, G. (1992) Elevated levels of *Activator* transposase mRNA are associated with high frequencies of *Dissociation* excision in *Arabidopsis*. *Plant Cell* 4: 583–595.

Syomin, B.V., Kandror, K.V. Semakin, A.B., Tsuprun, V.L., and Stepanov, A.S. (1993) Presence of the *gypsy* (MDG4) retrotransposon in extracellular virus-like particles. *FEBS Lett.* 323: 285–288.

Tabara, H., Grishok, A., and Mello, C.C. (1998) RNAi in *C. elegans*: soaking in the genome sequence. *Science* 282: 430–431.

Tabara, H., Sarkissian, M., Kellyt, W.G., Fleenor, J., Grishok, A., Timmons, L., Fire, A., and Mello, C.C. (1999) The *rde-1* gene, RNA interference, and transposon silencing in *C. elegans*. *Cell* 99: 123–132.

Tabara, H., Yigit, E., Siomi, H., and Mello, C.C. (2002) The dsRNA binding protein RDE-4 interacts with RDE-1, DCR-1, and a DexH-Box helicase to direct rNAi in *C. elegans*. *Cell* 109: 861–871.

Takeda, S., Sugimoto, K., Otsuki, H., and Hirochika, H. (1999) A 13-bp *cis*-regulatory element in the LTR promoter of the tobacco retrotransposon Tto1 is involved in responsiveness to tissue culture, wounding, methyl jasmonate and fungal elicitors. *Plant J.* 18: 383–393.

Takken, F.L.W., Schipper, D., Nijkamp, H.J.J., and Hille, J. (1998) Identification and *Ds*-tagged isolation of a new gene at the *Cf-4* locus of tomato involved in disease resistance to *Cladosporium fulvum* race 5. *Plant J.* 14: 401–411.

Takumi, S. (1996) Hygromycin-resistant calli generated by activation and excision of maize *Ac/Ds* transposable elements in diploid and hexaploid wheat cultured cell lines. *Genome* 39: 1169–1175.

Tamaru, H. and Selker, E.U. (2001) A histone H3 methyltransferase controls DNA methylation in *Neurospora crassa*. *Nature* 414: 277–283.

Tanda, S., Mullor, J.L., and Corces, V.G. (1994) The *Drosophila* tom retrotransposon encodes an envelope protein. *Mol. Cell. Biol.* 14: 5392–5401.

Taylor, L.P. and Walbot, V. (1985) A deletion adjacent to the maize transposable element Mu-1 accompanies loss of Adh1 expression. *EMBO J.* 4: 869–876.

Temin, H.M. and Mizutani, S. (1970) RNA-dependent DNA polymerase in various of Rous Sarcoma virus. *Nature* 226: 1211–1213.

Thomas, C.L., Jones, L., Baulcombe, D.C., and Maule, A.J. (2001) Size constraints for targeting post-transcriptional gene silencing and for RNA-directed methylation in *Nicotiana benthamiana* using a potato virus X vector. *Plant J.* 25: 417–425.

Thykjaer, T., Stiller, J., Handberg, K., Jones, J., and Stougaard, J. (1995) The maize transposable element *Ac* is mobile in the legume *Lotus japonicus*. *Plant Mol. Biol.* 27: 981–993.

Timmons, L. and Fire, A. (1998) Specific interference by ingested dsRNA. *Nature* 395: 854.

Tissier, A.F., Marillonnet, S., Klimyuk, V., Patel, K., Torres, M.A., Murphy, G., and Jones, J.D.G. (1999) Multiple independent defective suppressor-mutator transposon insertions in *Arabidopsis*: A tool for functional genomics. *Plant Cell* 11: 1841–1852.

Ton-Hoang, B., Betermier, M., Polard, P., and Chandler, M. (1997) Assembly of a strong promoter following IS911 circularization and the role of circles in transposition. *EMBO J.* 16: 3357–3371.

Ton-Hoang, B., Polard, P., and Chandler, M. (1998) Efficient transposition of IS911 circles *in vitro*. *EMBO J.* 17: 1169–1181.

Ton-Hoang, B., Polard, P., Haren, L., Turlan, C., and Chandler, M. (1999) IS911 transposon circles give rise to linear forms that can undergo integration *in vitro. Mol. Microbiol.* 32: 617–627.

Trentmann, S.M., Saedler, H., and Gierl, A. (1993) The transposable element *En/Spm* encoded TNPA protein contains a DNA binding and a dimerization domain. *Mol. Gen. Genet.* 238: 201–208.

Tristem, M., Kabat, P., Herniou, E.H., Karpas, A., and Hill, F. (1995) Easel, a *Gypsy* LTR retrotransposon in the salmonidae. *Mol. Gen. Genet.* 249: 229–236.

Tsay, Y.-F., Frank, M.J., Page, T., Dean, C., and Crawford, N.M. (1993) Identification of a mobile endogenous transposon in *Arabidopsis thaliana. Science* 260: 342–344.

Tu, Z. (1997) Three novel families of miniature inverted-repeat transposable elements are associated with genes of the yellow fever mosquito, *Aedes aegypti. Proc. Natl. Acad. Sci. (USA)* 94: 7475–7480.

Tu, Z. (2000) Molecular and evolutionary analysis of two divergent subfamilies of a novel miniature inverted repeat transposable element in the yellow fever mosquito, *Aedes aegypti. Mol. Biol. Evol.* 17: 1313–1325.

Tu, Z., Isoe, J., and Guzova, J.A. (1998) Structural, genomic and phylogenetic analysis of *Lan*, a novel family of non-LTR retrotransposons in the yellow fever mosquito, *Aedes aegypti. Mol. Biol. Evol.* 15: 837–853.

Tudor, M., Lobocka, M., Goodell, M., Pettitt, J., and O'Hare, K. (1992) The *pogo* transposable element family of *Drosophila melanogaster. Mol. Gen. Genet.* 232: 126–134.

Turchich, M.P., Bokhari-Riza, A., Hamilton, D.A., He, C., Messier, W., Stewart, G.B., and Mascarenhas, J. (1996) *Prem-2*, a *copia* type retroelement in maize is expressed preferentially in early microscopes. *Sex. Plant Reprod.* 9: 65–74.

Turcich, M.P. and Mascarenhas, J.P. (1994) PREM-1, a putative maize retroelement has LTR (long terminal repeat) sequences that are preferentially transcribed in pollen. *Sex Plant Reprod.* 7: 2–11.

Turcotte, K., Srinivasan, S., and Bureau, T. (2001) Survey of transposable elements from rice genomic sequences. *Plant J.* 25: 169–179.

Tuschl, T. (2001) RNA interference and small interfering RNAS. *Chembiochem.* 2: 239–245.

Upadhyaya, K.C., Sommer, H., Krebbers, E., and Saedler, H. (1985) The paramutagenic line *niv*-44 has a 5kb insert, Tam2, in the chalcone synthase gene of *Antirrhinum majus. Mol. Gen. Genet.* 199: 201–207.

Vaistij, F.E., Jones, L., and Baulcombe, D.C. (2002) Spreading of RNA targeting and DNA methylation in RNA silencing requires transcription of the target gene and a putative RNA-dependent RNA polymerase. *Plant Cell* 14: 857–867.

Van den Broeck, D., Maes, T., Sauer, M., Zethof, J., De Keukeleire, P., D'Hauw, M., Van Montagu, M., and Gerats, T. (1998) Transposon display identifies individual transposable elements in high copy number lines. *Plant J.* 13: 121–129.

Van Luenen, H.G.A.M., Colloms, S.A., and Plasterk, R.H.A. (1994) The mechanism of transposition of Tc3 in *C. elegans. Cell* 79: 293–301.

Van Sluys, M.A., Tempe, J., and Fedoroff, N. (1987) Studies on the introduction and mobility of the maize *Activator* element in *Arabidopsis thaliana* and *Daucus carota. EMBO J.* 6: 3881–3889.

Van West, P. and Kamoun, S. (1999) Intranuclear gene silencing in *Phytophthora infestans. Mol. Cell* 3: 339–348.

Varagona, M.J., Purugganan, M., and Wessler, S.R. (1992) Alternative splicing induced by insertion of retrotransposons into the maize waxy gene. *Plant Cell* 4: 811–820.

Vodkin, L., Rhodes, P.R., and Goldberg, R.B. (1983). CA lectin gene insertion has the structural features of a transposable element. *Cell* 34: 1023–1031.

Vogelauer, M., Wu, J., Suka, N., and Grunstein, M. (2000) Global histone acetylation and deacetylation in yeast. *Nature* 408: 495–498.

Voinnet, O. (2001) RNA silencing as a plant immune system against viruses. *Trends Genet.* 17: 449–459.

Voinnet, O., Pinto, Y.M., and Baulcombe, D.C. (1999) Suppression of gene silencing: A general strategy used by diverse DNA and RNA viruses of plants. *Proc. Natl. Acad. Sci. (USA)* 96: 14147–14152.

Voinnet, O., Vain, P., Angell, S., and Baulcombe, D.C. (1998) Systemic spread of sequence-specific transgene RNA degradation in plants is initiated by localized introduction of ectopic promoterless DNA. *Cell* 95: 177–187.

Volff, J.-N., Körting, C., and Schartl, M. (2001b) Ty3/*gypsy* retrotransposon fossils in mammalin genomes: did they evolve into new cellular functions. *Mol. Biol. Evol.* 18: 266–270.

Volff, J.-N., Körting, C., Meyer, A., and Schartl, M. (2001c) Evolution and discontinuous distribution of *Rex3* retrotransposons in fish. *Mol. Biol. Evol.* 18: 427–431.

Volff, J.-N., Körting, C., Sweeney, K., and Schartl, M. (1999) The non-LTR retrotransposon *Rex3* from the fish *Xiphophorus* is widespread among telosts. *Mol. Biol. Evol.* 16: 1427–1438.

Volff, J.-N., Körting, G., Altschmied, J., Duschl, J., Sweeney, K., Wichert, K., Froschauer, A., and Schartl, M. (2001a) Jule from the fish *Xiphophorus* is the first complete vertebrate Ty3/*Gypsy* retrotransposon from the *Mag* Family. *Mol. Biol. Evol.* 18: 101–111.

Volff, J.-N., Körting, G., and Schartl, M. (2000) Multiple lineages of the non-LTR retrotransposon *Rex1* with varying success in invading fish genomes. *Mol. Biol. Evol.* 17: 1673–1684.

Volpe, T.A., Kidner, C., Hall, I.M., Teng, G., Grewal, S.I.S., and Martienssen, R.A. (2002) Regulation of heterochromatic silencing and histone H3 lysine-9 methylation by RNAi. *Science* 297: 1833–1837.

Vongs, A., Kakutani, T., Martienssen, R.A., and Richards, E.J. (1993) *Arabidopsis thaliana* DNA methylation mutants. *Science* 260: 1926–1928.

Vos, J.C., van Luenen, H.G.A.M., and Plasterk, R.H.A. (1993) Characterization of the *Caenorhabditis elegans* Tc1 transposase *in vivo* and *in vitro*. *Genes Devel.* 7: 1244–1253.

Voytas, D.F. and Ausubel, F.M. (1988) A *copia*-like transposable element family in *Arabidopsis thaliana*. *Nature* 336: 242–244.

Voytas, D.F. and Boeke, J.D. (2002) Ty1 and Ty5 of *Saccharomyces cerevisiae*. In: *Mobile DNA II*. Craig, N.L., Craigie, R., Gellert, M., and Lambowitz, A.M. (eds.) ASM Press, Washington, DC. pp. 631–662.

Voytas, D.F., Cummings, M.P., Konieczny, A., Ausubel, F.M., and Rodermel, S.R. (1992) *Copia*-like retrotransposons are ubiquitous among plants. *Proc. Natl. Acad. Sci. (USA)* 89: 7124–7128.

Voytas, D.F., Konieczny, A., Cummings, M.P., and Ausubel, F.M. (1990) The structure, distribution and evolution of the Ta1 retrotransposable element family of *Arabidopsis thaliana*. *Genetics* 126: 713–721.

Waincoast, J. (1987) Out of the garden of Eden. *Nature* 325: 13.

Walbot, V. (1986) Inheritance of mutator activity in *Zea mays* as assayed by somatic instability of the *bz2-mu1* allele. *Genetics* 114: 1293–1312.

Walbot, V. (1991) The mutator transposable element family of maize. *Genetic Eng.* 13: 1–37.

Walbot, V. (1992) Strategies for mutagenesis and gene cloning using transposon tagging and T-DNA insertional mutagenesis. *Ann. Rev. Plant Physiol. Plant Mol. Biol.* 43: 49–82.

Walbot, V. (2000) Saturation mutagenesis using maize transposons. *Curr. Opin. Plant Biol.* 3: 103–107.

Walbot, V. and Rudenko, G.N. (2002) *MuDR/Mu* transposable elements in maize. In: *Mobile DNA II*. Craig, N.L., Craigie, R., Gellert, M., and Lambowitz, A.M. (eds.) ASM Press, Washington, DC. pp. 533–564.

Wang, L. and Kunze, R. (1998) Transposase binding site methylation in the epigenetically inactivated *Ac* derivative *Ds-cy*. *Plant J.* 13: 577–582.

Wang, M.-B. and Waterhouse, P.M. (2001) Application of gene silencing in plants. *Curr. Opin. Plant Biol.* 5: 146–150.

Wang, S., Zhang, Q., Maughan, P.J., and Maroof, M.A.S. (1997) *Copia*-like retrotransposons in rice: sequence heterogeneity, species distribution and chromosomal locations. *Plant Mol. Biol.* 33: 1051–1058.

Wassenegger, M. (2000) RNA-directed DNA methylation. *Plant Mol. Biol.* 43: 203–220.

Watanaba, T. and Fukasawa, T. (1961) Episome-mediated transfer of drug resistance in Enterobacteriaceae. III. Transduction of resistance factor. *J. Bacteriol.* 82: 202–209.

Waterhouse, P.M., Graham, M.W., and Wang, M.-B. (1998) Virus resistance and gene silencing in plants can be induced by simultaneous expression of sense and antisense RNA. *Proc. Natl. Acad. Sci. (USA)* 95: 13959–13964.

Waterhouse, P.M., Wang, M.-B., and Lough, T. (2001) Gene silencing as an adaptive defence against viruses. *Nature* 411: 834–842.

Weck, E., Courage, U., Döring, H.-P., Fedoroff, N., and Starlinger, P. (1984) Analysis of *sh-m6233*, a mutation induced by the transposable element *Ds* in the sucrose synthase gene of *Zea mays*. *EMBO J.* 3: 1713–1716.

Weil, C.F. and Wessler, S.R. (1990) The effects of plant transposable element insertion on transcription initiation and RNA processing. *Ann. Rev. Plant Physiol. Plant Mol. Biol.* 41: 527–552.

Weinand, Y., Sommer, H., Schwarz, Z., Shepherd, H., Seadler, H., Kreuzaler, F., Ragg, H., Fautz, E., Hahlbrock, H., Harrison, B.J., and Peterson, P. (1982) A general method to identify plant structural genes among genomic DNA. DNA clones using transposable elements induced mutations. *Mol. Gen. Genet.* 187: 195–201.

Wesley, S.V., Helliwell, C.A., Smith, N.A., Wang, M., Rouse, D.T., Liu, Q., Gooding, P.S., Singh, S.P., Abbott, D., Stoutjesdijk, P.A., Robinson, S.P., Gleave, A.P., Green, A.G., and Waterhouse, P.M. (2001) Construct design for efficient, effective and high-throughput gene silencing in plants. *Plant J.* 27: 581–590.

Wessler, S.R. (1988) Phenotypic diversity mediated by the maize transposable elements *Ac* and *Spm*. *Science* 242: 399–404.

Wianny, F. and Zernicka-Goetz, M. (2000) Specific interference with gene function by double-stranded RNA in early mouse development. *Nature Cell Biol.* 2: 70–75.

Wilhelm, W. and Wilhelm, D.-X. (2001) Reverse transcription of retroviruses and LTR retrotransposons. *Cell. Mol. Life Sci.* 58: 1246–1262.

Will, B.M., Bayev, A.A., and Finnegan, D.J. (1981) Nucleotide sequence of terminal repeats of 412 transposable elements of *Drosophila melanogaster*. *J. Mol. Biol.* 153: 897–915.

Wilson, K., Long, D., Swinburne, J., and Coupland, G. (1996) A dissociation insertion causes a semidominant mutation that increases expression of *TINY*, an *Arabidopsis* gene related *APETALA1*. *Plant Cell* 8: 659–671.

Wilson, M.R., Marcuz, A., Van Ginkel, F., Miller, N.W., Clem, L.W., Middleton, D., and Warr, G.W. (1990) The immunoglobulin M heavy chain constant region gene of the channel catfish, *Ictalurus punctatus* an unusual mRNA splice pattern produces the membrane form of the molecule. *Nucl. Acid Res.* 18: 5227–5233.

Wisman, E., Cardon, G.H., Fransz, P., and Saedler, H. (1998) The behavior of the autonomous maize transposable element *En/Spm* in *Arabidopsis thaliana* allows efficient mutagenesis. *Plant Mol. Biol.* 37: 989–999.

Woods-Samuels, P., Wong, C., Mathias, S.L., Scott, A.F., Kazazian, H.H., and Antonarakis, S.E. (1989) Characterization of a nondeleterious L1 insertion in an intron of the human factor VIII gene and further evidence of open reading frames in functional L1 elements. *Genomics* 4: 290–296.

Wright, D.A. and Voytas, D.F. (1998) Potential retroviruses in plants: Tat1 is related to a group of *Arabidopsis thaliana* Ty3/*gypsy* retrotransposons that encode envelope-like proteins. *Genetics* 149: 703–715.

Wright, D.A. and Voytas, D.F. (2001) *Athila* of *Arabidopsis* and *Calypso* of soybean define a lineage of endogenous plant retroviruses. *Genome Res.* 12: 122–131.

Wright, D.A., Ke, N., Smalle, J., Hauge, B.M., Goodman, H.M., and Voytas, D.F. (1996) Multiple non-LTR retrotransposons in the genome of *Arabidopsis thaliana*. *Genetics* 142: 569–578.

Wright, S.I., Le, Q.H., Schoen, D.J., and Bureau, T.E. (2001) Population dynamics of an *Ac*-like transposable element in self- and cross-pollinating *Arabidopsis*. *Genetics* 158: 1279–1288.

Wu-Scharf, D., Jeong, B.-R., Zhang, C., and Cerutti, H. (2000) Transgene and transposon silencing in *Chlamydomonas reinhardtii* by a DEAH-box RNA helicase. *Science* 290: 1159–1162.

Xiao, Y.-L. and Peterson, T. (2000) Intrachromosomal homologous recombination in *Arabidopsis* induced by a maize transposon. *Mol. Gen. Genet.* 263: 22–29.

Xiao, Y.-L. and Peterson, T. (2002) *Ac* transposition is impaired by a small terminal deletion. *Mol. Genet. Gen.* 266: 720–731.

Xiao, Y.-L., Li, X., and Peterson, T. (2000) *Ac* inserted site affects the frequency of transposon-induced homologous recombination at the maize *p1* locus. *Genetics* 156: 2007–2017.

Xiong, Y. and Eickbush, T.H. (1988) The site-specific ribosomal DNA insertion element R1Bm belongs to a class of non-long-terminal repeat retrotransposons. *Mol. Cell Biol.* 8: 114–123.

Xiong, Y. and Eickbush, T.H. (1990) Origin and evolution of retroelements based upon their reverse transcriptase sequences. *EMBO J.* 10: 3353–3362.

Xiong, Y., Burke, W.D., and Eickbush, T.H. (1993) *Pao*, a highly divergent retrotransposable element from *Bombyx mori* containing long terminal repeats with tandem copies of the putative R region. *Nucl. Acid Res.* 21: 2117–2123.

Yamashita, S., Takano-Shimizu, T., Kitamura, K., Mikami, T., and Kishima, Y. (1999) Resistance to gap repair of the transposon *Tam3* in *Antirrhinum majus*: A role of the end regions. *Genetics* 153: 1899–1908.

Yamazaki, M., Tsugawa, H., Miyeo, A., Yano, M., Wu, J., Yamamoto, S., Matsumoto, T., Sasaki, T., and Hirochika, H. (2001) The rice retrotransposon Tos17 prefers low-copy-number sequences as integration targets. *Mol. Genet. Genomics* 265: 336–344.

Yanez, M., Verdugo, I., Rodriguez, M., Prat, S., and Ruiz-Lara, S. (1998) Highly heterogeneous families of Ty1/*copia* retrotransposons in the *Lycopersicon chilense* genome. *Gene* 222: 223–228.

Yang, G., Dong, J., Chandrasekaran, M.B., and Hall, T.C. (2001) *Kiddo* a new transposable element family closely associated with rice genes. *Mol. Genet. Genomics* 266: 417–434.

Yang, J., Malik, H.S., and Eickbush, T.H. (1999) Identification of the endonuclease domain encoded by R2 and other site-specific, non-long terminal repeat retrotransposable elements. *Proc. Natl. Acad. Sci. (USA)* 96: 7847–7852.

Yant, S.R., Ehrhardt, A., Mikkelsen, J.G., Meuse, L., Pham, T., and Kay, M.A. (2002) Transposition from a gutless adenotransposon vector stabilizes transgene expression *in vivo*. *Nature Biotechnol.* 20: 999–1005.

Yoder, J.I. (1990) Rapid proliferation of the maize transposable element activator in transgenic tomato. *Plant Cell* 2: 723–730.

Yoder, J.I., Payls, J., Alpert, K., and Lassner, M. (1988) *Ac* transposition in transgenic tomato plants. *Mol. Gen. Genet.* 213: 291–296.

Yoshioka, Y., Matsumoto, S., Kojima, S., Ohshima, K., Okada, N., and Machida, Y. (1993) Molecular characterization of a short interspersed repetitive element from tobacco that exhibits sequence homology to specific tRNAs. *Proc. Natl. Acad. Sci. (USA)* 90: 6562–6566.

Yoshiyama, M., Tu, Z., Kainoh, Y., Honda, H., Shono, T., and Kimura, K. (2001) Possible horizontal transfer of a transposable element from host to parasitoid. *Mol. Biol. Evol.* 19: 1952–1958.

Youngman, S., van Luenen, G.A., and Plasterk, R.H.A. (1996) Rte-1, a retrotransposon-like element in *C. elegans. FEBS Lett.* 380: 1–7.

Yu, J., Hu, S., Wang, J., et al. (2002) A draft sequence of the rice genome (*Oryza sativa* L. ssp. *indica*). *Science* 296: 79–92.

Yu, Z., Wright, S.I., and Bureau, T.E. (2000) Mutator-like elements in *Arabidopsis thaliana*: Structure, diversity and evolution. *Genetics* 156: 2019–2031.

Zakian, V.A. (1995) Telomeres: beginning to understand the end. *Science* 270: 1601–1607.

Zamore, P.D. (2002) Ancient pathways programmed by small RNAs. *Science* 296: 1265–1269.

Zamore, P.D., Tuschl, T., Sharp, P.A., and Bartel, D.P. (2000) RNAi: Double-stranded RNA directs the ATP-dependent cleavage of mRNA at 21 to 23 nucleotide intervals. *Cell* 101: 25–33.

Zelentsova, H., Poluectova, H., Mnjoian, L., Lyozin, G., Veleikodvorskaja, V., Zhivotovsky, L., Kidwell, M.G., and Evgen, M.B. (1999) Distribution and evolution of mobile elements in the *virilis* species group of *Drosophila. Chromosoma* 108: 443–456.

Zhang, J. and Peterson, T. (1999) Genome rearrangements by nonlinear transposons in maize. *Genetics* 153: 1403–1410.

Zhang, Q., Arbuckle, J., and Wessler, S.R. (2000) Recent, extensive, and preferential insertion of members of the miniature inverted-repeat transposable element family *Heartbreaker* into genic regions of maize. *Proc. Natl. Acad. Sci. (USA)* 97: 1160–1165.

Zhang, Y. and Reinberg, D. (2001) Transcription regulation by histone methylation: interplay between different covalent modifications of the core histone tails. *Genes Devel.* 15: 2343–2360.

Zhou, J.H. and Atherly, A.G. (1990) *In situ* detection of transposition of the maize controlling element (*Ac*) in transgenic soybean tissues. *Plant Cell Rep.* 8: 542–545.

Zietkiewicz, E., Richer, C., Sinnett, D., and Labuda, D. (1998) Monophyletic origin of Alu elements in primates. *J. Mol. Evol.* 47: 172–182.

Index

7SL RNA, 137
35S CaMV promoter, 180, 181, 203, 220
α-glucan phosphorylase, 127
β-galactosidase, 279
β globin, 133, 151, 241
β-glucuronidase (GUS), 126, 180, 181, 183, 194, 205, 219, 220, 282,
β-lactamase, 48
Ac, 19, 21, 22, 115, 163–184, 195–200, 206–208, 213–216, 218–224, 237, 244, 256, 272, 274, 277, 278, 280–282
Ac/Ds, 22, 115, 163–167, 169–171, 173–179, 181–184, 195, 197–200, 206, 207, 213–216, 219, 221, 223, 237, 256, 277, 278, 280–282
acetylation, 267–270, 272–274
acetylation of chromatin, 270
Activator (*AC*), 19, 21, 165, 273
adenovirus infection, 138
adenovirus vector, 280
adenyltransferase, 62
Adh1, 116, 119, 120, 171–173, 175, 199, 200
Adineta vaga, 80
Adoxophyes honmai, 240
ADP-glucose-glucosyl transferase, 116
Aedes aegypti, 61, 140, 145, 233, 240
African green monkey (AGM), 132–134
Afut, 93
AGAMOUS, 264
Agrobacterium, 41, 54, 115, 126, 179, 182, 220, 223, 232, 275, 281, 282
Agrobacterium-mediated genetic transformation, 54, 126, 182, 220, 232, 275, 282
Agrobacterium rhizogenes, 179
Agrobacterium tumefaciens, 115, 179, 223
Alaska native, 137
Alcaligenes, 41
aleurone, 18, 19, 165–167, 178, 184, 186, 187, 190, 198, 199, 204–207
Alexander (The Great), 275
algae, 112, 129, 155, 263
Alien, 223, 235
allele, 13, 16, 116, 125, 166, 167, 169, 170, 172, 173, 178, 179, 182, 184, 186, 187, 190, 92–194, 198, 199, 205, 211, 213–215, 217, 218, 221, 226, 234, 236, 247, 248
allelomorph, 13
allyl alcohol, 172
Alport syndrome (AS), 149
alternative splicing, 255
Alu, 132–134, 137–139, 141, 149, 151
Amathia convoluta, 80
Amerindian137
amphibia, 103, 108, 145, 240, 241
amphitetraploid, 125
ampicillin, 48, 69, 252
Amy-Bm, 140
amylopectin, 19, 166

amylose, 19, 116, 117, 166
Analogy of the Cave, 259
analytical ultracentrifuge, 25
Andropogoneae, 122
angiosperm plants, 155, 232, 234, 265
angiosperms, 54, 79, 112, 114, 163, 195
Anopheles gambiae, 104, 140, 144, 145
Anopheles stephensi, 233
anthocyanin, 19, 184, 206, 211, 212, 236
antibiotic compound, 47, 54, 62, 277
antibiotic resistance, 54, 62
Antirrhinum majus, 209, 212, 256
APC gene, 148
APETALA, 264
Aphis, 80
Arabidopsis, 76, 106, 112, 113, 120, 126, 127, 140, 141, 156, 160, 174, 176, 179, 181, 194, 218, 221, 222, 232, 235, 264, 269, 271, 277, 280–282
Arabidopsis lyrata, 222
Arabidopsis thaliana, 112–115, 120, 126, 128, 140, 141, 156, 179, 181, 218–224, 232, 277, 281,
ARGONAUTE, 263
Arthropod, 80, 103, 111, 143, 162, 242
Ascaris, 103, 104, 160
Ascaris lumbricoides, 104, 105
Ascogaster reticulatus, 240
ascomycete fungi, 154
Aspergillus, 93, 116, 219, 236–238
Aspergillus fumigatus, 93
Athila, 1, 113, 114
attTn7 site, 33, 62–67, 71
Aurelia aurita, 80
Australian aborigine, 137
auxins, 126
Aveneae, 122
Avena sativa, 121
Avena spp., 122

B5, 117, 118
Bacillus, 36, 41, 45, 49
Bacillus thuringiensis, 45, 49
bacterial genophore, 25, 41, 45, 47, 53, 60–62, 66, 67, 69
bacteriophage Mu, 32, 50, 75, 251
bacteriophage λ, 36, 45, 53, 247
Bactrocera tryoni, 233
baculovirus, 180
baleen whales, 139
*Bam*H1, 132, 135, 174
*Bam*H1 elements, 135
Bari, 227
barley, 116, 118, 120–122, 156, 181, 183
barley stripe mosaic virus (BSMV), 116
Basho, 222
Bateson, 11–13
Battrachocottus baikalensis, 146
Bauer, 13

Bdelloidea, 77, 80
Beadle, 15, 17, 154, 246
beans (*Phaseolus vulgaris*), 128
Becker/Duchenne muscular dystrophy, 148
BEL, 130
Berkeley, George, 173, 246, 257, 259
Beta, 155, 273
Beta vulgaris, 155
binomial equation, 9
Biomphalaria glabrata, 140
Bis1, 121
BNR, 155
Bombyx, 103, 160
Bombyx mori, 103, 106, 109, 140, 144
Bos taurus, 80, 243
Botrytis cinerea, 93
Boty, 93
Boveri, 13
Brachionus calyciflorus, 80
Brachionus plicatilis, 80
Bradyrhizobium japonicum, 44
Brachydanio rerio (Danio rerio), 80, 130, 233, 241, 242
Brahe, 2
Branchiostoma floridae, 80
brassinosteroids, 221
breakage-fusion-bridge (BFB), 17–21, 166
Bridges, 14
broad bean, 128, 129
bromodomain (BrD), 269
Brünn, 2, 8, 10
Bs1, 116
Bs-Dm, 140
budding yeast, 81

Caenorhabditis elegans, 105, 224–225, 231–233, 262, 264, 265, 277, 279
Caenorhabditis briggsae (TCb1), 130, 225
camel, 139
CaMV, 130, 180, 181, 194, 203, 206, 207, 220, 223
CAMV 35S promoter, 180, 181, 203, 206, 220, 223
Candida albicans, 91, 92
capping protein, 160
Capsicum annuum, 235
carnivore, 135
carp, 146, 243, 580
carrot, 179
Castaway, 1, 235
catfish, 241, 242
cattle (*Bos taurus*), 142, 243
Caucasian, 137
caulimovirus, 76, 109
CbPat1, 130
central cell, 18, 21
centromere, 18, 20, 21, 121, 122, 155, 271, 273
centromeric chromatin, 273
centromeric region, 122, 241, 271
Cephalochordate, 107, 108
Cer1, 105, 110, 130
Cer12, 105

cerebra, 122
Cer element, 105
Certitis capitata, 233
Cetacea, 139
CfT-1, 93
CgT1, 140
chalcone synthase, 211–213, 215, 263
chevrotains, 139
Chinese hamster (*Mesocricetus auratus*), 243
Chione cancellata, 80
Chironomus tentans, 140
Chlamydomonas reinhardtii, 112, 265
Chlamydomonas thummi, 140
Chlonorchis sinensis, 106
chloramphenicol, 69
chlorate, 116, 125, 219, 236, 237
Chlorella, 155, 160, 162
Chlorella vulgaris, 155
Chlorideae, 122
chlorite, 219, 236
chlorobenzoate catabolic gene, 49
Chordata, 80, 103
chordates, 107
chromatin, 1, 47, 79, 83, 87, 88, 100, 101, 121, 138, 159, 162–164, 175, 192, 221, 222, 230, 250, 252–254, 261, 264, 266–275
chromatin acetylation, 268
chromatin modification, 164, 268, 271, 274, 275
chromatin modulation, 222, 252, 254
chromatin remodeling, xii, 192, 261, 264, 266, 269, 270, 273–275
chromatin silencing, 261
chromatin structure, 47, 88, 159, 192, 267, 268, 271–274
chromodomain (ChrD), 269
chromosomal abberations, 2, 11, 247, 253
chromosomal rearrangement, 96, 167, 179, 247
chromosome breakage (dissociation), 8, 19, 21, 22, 163, 165, 166, 169, 173, 179, 196
chromosome theory, 13, 16
chromosome walking, 120
chrystalographic analyses, 56
Cigr, 107, 108
ciliated protozoa, 161, 238, 239
CiLINE2, 146, 147
Cin, 154, 211
Cin1, 154, 211
Cin4, 76, 140, 155
Cin4-Zm, 140
Cinful, 119, 120
Cinful-2, 120
Ciona (sea squirt), 107
Ciona intestinalis, 107
circular transposon, 40
circumcision, 5
citron, 6
Citrus, 6
Cladosporium fulvum, 93, 281, 282
CLAVATA3, 264
Clupea pallasi, 109

cob, 178, 196
cointegrase, 42, 43
cointegration, 36, 40, 50, 183
Cold Spring Harbor, 18, 21, 22, 165, 166, 169, 184, 185, 197
Colletotrichum gloeosporioides, 140
Colletotrichum lagenarium, 93
combination theory, 8
Compositae, 141
composite transposon, 49, 53, 61, 69
compound bacterial transposon, 44
compound transposon, 32, 33, 45
Condylactus, 80
conjugating plasmid, 62, 63, 66
conjugative genophore transfer, 41
conjugative transposon, 26, 45
conservative transposition, 34
controlling element, 1, 14, 21–23, 26, 75, 163–166, 184, 185, 197, 259
Copernican Revolution, 2
Copernicus, 2
Copia, 76, 81, 89, 91, 92, 96–99, 103, 104, 109, 112–114, 116, 118, 121–125, 127, 128, 143
Correns, 11–13
cotton, 45
CpG dinucleotides, 173
Cpr1, 110
Cr1-Gg, 140
Cr1-Ps, 140
Cre1, 140
Cre2, 140
Cretaceous era, 141
Crithidia fasciculate, 140
cross breeding, 6
Cruciferae, 141, 235
Cryphonectria parasitica (Crypt1), 237
CsRn1, 106
Cucumis sativus, 141
Cucurbitaceae, 141
Culex pipens, 140
cut, 77, 79, 99, 153, 168, 193, 194, 203, 206, 222, 229, 252, 254, 276
cut and paste, 40–42, 50, 53, 55, 62, 70, 119, 168, 176, 200, 204, 206–208, 215, 238, 243, 245, 252
cyanidin, 211
Cyclops, 129
Cyclops-2, 129
Cy/rcy, 197
cytoplasmic male sterility (CMS), 262

Dam methylation, 54, 60
Darwin, 10, 209
DDE, 29, 31, 32, 26–38, 41, 42, 50, 56, 63, 64, 73, 129, 230
DDE concensus, 31
DDE domain, 29, 32, 37
DDE motif, 29, 32, 36, 42, 56, 63, 64, 73
DD(35)E domain, 242
D.D(35)E motif, 29, 228, 229, 239, 242

DDM1, 1, 221, 264
ddm1, 115, 221
deacetylation of histones, 268
deer mouse, 150
del2, 155
demethylation, 177, 195, 196, 221, 222, 270, 274
dendrogram, 44
density gradient centrifugation, 132
deuterostome, 108
de Vries, 10–12, 209
Diced Defense, 263
dicentric chromatid, 21, 166
Dicer, 263
Dictyostelium discoideum, 129
diffuse leiomyomatosis (DL), 149
dihydroflavonol 4-reductase, 211
directly repeated (DR) short sequences, 27
DIRS1, 129, 130
dissociation (*Ds*), 19, 21, 101, 165–167, 169–175, 177–183, 198, 200, 206, 215, 216, 223, 277, 280–282
Ditto, 235
DMTase, 269
DNA methylation, 114, 115, 135, 183, 195, 221, 260, 264, 265, 269–272, 275
DNA replication, 28, 66, 70, 118, 160, 168, 176, 266, 274
DNA supercoiling, 28
Doc-Dm, 140
dog, 148, 280
Dolomite mountains, 1
dominant trait, 9, 10
Doppler, 1
double-end break complex (DEBC)70, 71
double fertilization, 11, 18
double-strand break, 51, 63, 88, 111, 161
Down's syndrome, 276
downstream control region (DCR), 195, 196
DR, 27, 28, 33, 36, 37, 40
DrDirs 1, 130
DRE, 76
Drosophila, 72, 76, 80, 93–97, 99–103, 109, 111, 114, 116, 134, 137, 143, 144, 159, 161162, 226, 227, 231, 232, 240, 241, 244, 246, 247, 250–252, 254–257, 262, 269, 270, 278
Drosophila hydei, 233, 240
Drosophila melanogaster, 14, 61, 79, 80, 81, 93, 94, 96–99, 101, 102, 111, 140, 143, 147, 225, 227, 233, 240, 246, 247, 255, 258, 265, 266, 277, 279
Drosophila mauritiania, 233
Drosophila miranda, 140
Drosophila pseudoobscura, 102
Drosophila simulans, 226
Drosophila subobscura, 101
Drosophila teissieri, 140
Drosophila virilis, 80, 97, 111, 233
Dr Watson, 120
Ds, 19, 21, 22, 101, 165–167, 169–173, 175, 177, 178–184, 200, 216, 223, 277, 280–282

dSpm, 185–196, 200, 207, 216
dSpm elements, 186–190, 200, 207
DsRNA, 203, 262–266
dTph1, 235, 281
dTph3, 281
dTph4, 281
Dugesia tigrina, 80

ears, 18, 165, 168, 185
Easel, 110
Echinacea, 107
Echinodermata, 80, 103
echinoids, 106
Echinometra mathaei, 80
egg cell, 9, 10, 13, 18, 21, 254
eggplant, 156
electroporation, 60
Emerson, 15, 16, 18, 169, 178, 197, 209
Emigrant, 235
EN, 145, 152, 183
En-crown, 188
endosperm, 18–22, 116, 117, 165–167, 169, 176, 186, 196, 198, 199, 204, 205
En (enhancer) element, 164, 183, 184, 187
enhancer trapping, 257, 279, 280, 282
En/Spm, 163, 184, 280
Enterobacteria, 35, 37
env gene, 101
ENV (envelope) proteins, 101, 107, 110, 113
epidermal growth factors, 279
Eptatretus stouti, 241
Equus, 44
Equus przewalskii, 44
Erwinia chrysantemi, 126
Escherichia, 33, 35, 37, 41, 204
Escherichia coli, xii, 25, 37, 38, 42, 53, 54, 61, 62, 66–68, 72, 84, 96, 230, 233, 251, 262
euchromatin, 100, 121, 271
Euperipatoides rowelli, 80
eukaryotes, 2, 22, 23, 30, 60, 61, 75, 76, 79, 83, 87, 153, 154, 161, 163, 194, 230, 261, 266, 268, 269, 274
Euplotes, 238, 239
Euplotes crassus, 238
excision, 22, 46, 50, 61, 63, 95, 97, 168, 171, 175, 178, 180, 181, 183, 186–188, 190–192, 194, 198–200, 204–210, 213–217, 219, 220, 222, 223, 225, 229, 231, 237, 239, 245, 250, 252, 256, 278
exons, 117, 118, 130, 138, 147, 149, 151, 172, 174, 189, 190, 192, 200, 202, 203, 248, 249, 255, 278, 279
Explorer, 1, 235
Ezekiel, 1, 220

Factor VIII, 147, 148
Father Adam, 44
F-Dm, 140
F-element, 143
Fenzl, 8, 11

ferns, 112
filamentous fungi, 92, 269
fishes, 79, 103, 108–111, 146, 147, 228, 240–243
fishing, 76, 79, 108–111, 123, 130, 222, 223, 234–236, 240, 241
fission yeast, 91, 261, 271, 272
Flamenco (flam), 100, 101
flowering plants (angiosperms), 79
Fluorescent *In Situ* Hybridization (FISH), 122
Foldback (FB), 92, 107
footprint, 175, 193
Foret, 92
forked, 98, 99
forked locus, 98
Formica polyctenum, 80
forward genetics, 124
forward mutations, 197–199, 204, 278
Fot1, 237, 238
Fourf, 119, 120
frame shift, 27, 34–38, 83, 84, 89, 91, 156, 190
frog, 130, 240, 241, 244
fruit fly, 14, 93, 233, 246, 257, 265, 270, 277
Fugu, 108, 109, 146
FuguBEL, 130
Fugu rubripes, 108, 146
fungal organisms, 79, 87, 91–93, 141, 154, 160, 163, 236–238
fungi, 87, 92, 142, 154, 160, 236, 238, 261, 262, 275
Fusarium, 92, 93, 114, 237, 238
Fusarium moniliforme, 238
Fusarium oxysporum, 92, 93, 114, 237, 238

G, 36, 80, 117, 118, 143, 159, 173, 176
G1, 122
G2, 122
G3, 122
G4, 122
G418, 53
GAG protein, 82, 84
Gaijin, 235
gal, GAL, 25, 26, 82, 84
galactose, 82
Galileo Galilei, 6
Gallus gallus, 140
Galton, 4
gap repair, 59, 70, 208, 216, 217, 252, 253
G-element, 143
gene conversion, 85
gene tagging, xii, 163, 164, 179, 181, 183, 199, 207, 209, 217, 219, 220, 223, 231, 232, 238, 243, 261, 275–282
gene therapy, 131, 280
genetic aberration, 3
genetic revolution, 2
genetic tagging, 124, 163, 164, 179, 181, 183, 199, 207, 209, 217, 219, 220, 223, 231, 232, 238, 243, 261, 275–282

genetic transformation, 54, 115, 124, 126, 162, 175, 176, 179, 181–183, 186, 206, 218, 223, 232, 243, 261, 276, 277, 282
genophore, 25, 33, 41, 47, 53, 60–63, 66, 67, 69,
genophore site, 33, 67
genotype, 13, 154
germinal cell, 18
G_F, 150
Giardia lamblia, 80
Glottidea pyramidata, 80
Glycera, 80
graft connections, 261
Gramineae, 118, 122, 141, 156, 220, 235
Grande, 119, 120
grasshopper, 92, 93
green-fluorescent-protein (GFP), 60, 87, 232
group II introns, 76, 130, 131
Gulliver, 105, 106
gymnosperms, 112
Gypsy, 1, 76, 77, 80, 81, 89, 91–93, 95–101, 103–114, 117, 118, 122, 127–130, 143, 144, 273
gypsy-like RTase, 80

Habrotrocha constricta, 80
Haematobia irritans, 228, 231, 233, 241
haemophilia, 147, 148
hagfish, 241
Hairy-wing *(su(Hw))*, 99
Halichondria bowerbankii, 80
hAT superfamily, 237, 238, 256
HC-Pro, 264
Heartbreaker (Hbr), 236
Hebrew bible, 5
Heine, 12, 102
Helena, 97, 102
Helianthus annuus, 141
helix-turn-helix (HTH), 37, 38, 230
hemophilia, 5
hepadnaviruses, 76, 107
Heraclitus, xi
hereditary nephritis, 149
Herring, 109, 110
HeT-A, 162
heterocaryon, 154
heterocaryon hyphae, 154
heterochromatin, 100, 101, 162, 221, 250, 253, 271–273
heteroduplex analyses, 170, 171, 212
heteroduplex hybridization, 25, 133, 188
heterokaryons, 263
heterozygote, 13, 125, 211
Himar, 241
Himar1, 227–233
*Hind*III element-families of man, 134
Hippopotamidae, 139
hippopotamuses, 139,
histone, 1, 175, 208, 260, 261, 266–275
histone-acetyltransferase (HAT), 268
histone code, 260, 269, 272, 274

histone deacetylase (HDAC), 272
histone deacetylation, 272
histone methylation, 270, 274
histone modification, 175, 268–272, 274, 275
histone protein, 266–268, 270, 273
histones, 79, 87, 164, 192, 196, 204, 230, 268–270, 272
hMuDR, 203, 209
Hobo, 239, 240, 255, 256
homeotic mutations, 218
homing, 131, 144
homologous recombination, 42, 63, 67, 149, 151, 178, 179, 181, 252, 257
Homo sapiens (see man), 137, 140, 243
homozygote, 13
Hooke, 6
Hordeum vulgare, 121
horizontal transmission, 108, 232, 233
Hsr1, 108
HTH motif, 37
Huck, 119, 120
Hum1, 244
Hum7, 244
human diseases, 138, 145
human gene therapy, 280
human genome, 1, 108, 109, 130, 132–134, 137, 138, 149–151, 244, 245
human immunodeficiency virus (HIV), 29, 84, 85, 92, 228
human maladies, 148
husks, 19, 175
hybrid dysgenesis, 93–97, 102, 143, 225, 226, 240, 246, 247, 250, 255, 279
hybrid corn, 16
Hydra littoralis, 80
Hydromantes, 108
hygromycin, 182, 223, 281
Hyp1-Cte, 140
Hyp2-Cth, 140
hypermethylation, 177
Hyphal fungi, 92, 154, 236, 238

Ictalurus punctatus, 241
I element, 95–97, 186, 262
I factor, 143, 147
IHF, 35, 43, 60, 68, 70, 72
immunity, 31, 42, 48, 50, 66, 69, 216
immunodepressive treatments, 92
Impala, 227, 237, 238
imprinting, 253
Indica rice, 183, 186
Ingi, 140, 142
Ingi-Tb, 140
initiator methionine tRNA, 82, 85
in planta transformation, 115, 218
*ins*A, 33–35
*ins*AB', 34, 35
*ins*AB'/*ins*A, 34
*ins*B', 33–35

insects, 45, 61, 95, 103, 111, 112, 132, 144, 160–163, 180, 233, 239, 240, 243, 244, 246
insertional mutagenesis, 123, 124, 148, 151, 166, 170, 199, 218, 277, 278, 282
in situ hybridization, 95–97, 121, 122, 128, 155, 279
integrase (IN), 29, 32, 37, 42, 43, 45, 49, 50, 56, 67, 75, 76, 79, 81, 82, 84, 86–89, 91, 92, 96, 103, 104, 106, 107, 109–111, 113, 118, 129, 130, 163, 164, 203, 251
integration, 33, 36, 40, 43, 45, 50, 60, 61, 64, 66, 68, 69, 76–79, 83, 85–87, 89, 91, 92, 94, 101, 106, 120, 122, 124, 144, 145, 151, 155, 179, 182, 183, 186, 199, 207, 214, 215, 232, 252, 256, 257, 280
integration host factor (IHF), 35, 43, 60, 68, 70, 72
interspecific hybridization, 5, 6
intermolecular transposition, 50, 69
internal inverted repeats, 248, 249
intrachromosomal recombination, 181
intrachromosomal transpositions, 167, 168
intramolecular transposition, 69
intron, 82, 86, 95, 98, 99, 108, 117, 124, 127, 130, 131, 138, 148, 150, 172, 174, 175, 178, 179, 189–191, 200, 202, 203, 213, 214, 219, 220, 226, 236–238, 255, 263
inverse PCR (IPCR), 276
inverse repeat (IR), 26, 27, 33, 35, 37, 40, 42, 43, 49, 50, 54, 60, 62, 174, 219
IRL, 27, 33–35, 37, 39–41, 43
Irma, 187
IRR, 27, 33, 35, 37, 39–41, 43
I transposable elements, 95, 132
IXz, 241

Jacob, xi, 8, 48, 166, 184, 260
Janssen, 6
Japonica rice, 183
Ji, 119
Ji-1, 120
Ji-2, 120
Ji-3, 120
Ji-4, 120
Ji-6, 120
jittery, 208
Jockey, 140, 144, 145, 162
Jockey-Dm, 140
John Innes, 13, 121, 155, 182, 210, 211, 218, 219, 263
Juan-A, 145
Juan-Aa, 140
Juan-Cp, 140
Jule, 110

Kabuki, 106
Kake, 119
Kake-1, 120
Kake-2, 120
Kamikaze, 103, 104
kanamycin, 45, 53, 54, 281

Kepler, 2
Kiddo, 124
King Gordius, 275
Klebsiella, 37
Kölreuter, 6
*Kpn*I, 132, 134, 135
*Kpn*I elements, 134, 148
KP proteins, 249, 252

L1, 43, 76, 95, 134, 147–152
L1$_{\beta\text{-thals}}$, 151
L1 elements, 144, 147–152
L1Md, 135
L1$_{RP}$, 151
Laban, xi, 8
lac, 25, 26
Lactococcus, 41
lagomorphs, 135
leader, 173, 174, 176, 180, 190, 194, 203, 205, 227
Lemi1, 223
lemors, 139
Lasius niger, 80
Leishmania major, 233, 239, 243
Lepidodermella, 80
leucine zipper (LZ), 37, 38
L. hirsutum, 282
Lian-Aa1, 140, 145
Lilium speciosum, 155
Lily, 1, 106, 108, 155
Limited Head Capacity, 25
LINE elements, 134, 135, 142, 143, 145–147, 151, 154, 156, 157
LINE1-Bg, 140
LINE1Mm, 140
LINE-Hs, 140
LINE-like element, 76, 134, 135, 144, 145, 147, 152, 154–157
Lineus, 80
linkage, 14–18, 51, 86, 130
linkage group, 17
linkage map, 14, 18
Linnaeus, 5
Lissomyema mellita, 80,
L1Tc, 142, 143
long-distance transmission of silencing, 262
long interspersed nucleotide element (LINE), 76, 77, 80, 132, 134, 135, 138, 140, 142–147, 149–152, 154–157
long terminal repeat (LTR), 45, 75–79, 82, 83, 87, 88, 91–93, 95, 96, 98, 102–132, 134, 137, 138, 141–147, 153, 154, 156
Lotus, 181, 183
LRE2, 149, 150
Lucilia cuprina, 233
Lycopersicon chilense, 127
Lycopersicon esculentum, 125, 281
Lysenko, 97, 98
lysine methyltransferase (HMT), 272
LZ motif, 37

macrogametophyte, 205
macronucleus, 161, 239
Macrotrachela quadricornifera, 80
mag, 130
Magellan, 117, 118
MAGGY, 93, 109, 110, 130
Magnaporthe grisea, 92, 93
maize, 1, 2, 11, 14–19, 22, 23, 26, 45, 101, 106,
    112–120, 127, 134, 137, 154   156, 163–165,
    167–171, 173, 177–187, 189, 191, 194–198,
    200, 203–207, 209, 211, 214, 215, 219–221,
    235, 236, 244, 256, 266, 276, 277, 280, 281
major groove, 65
mammals, 3, 92, 95, 103, 131–139, 142, 143, 147,
    149–151, 153, 154, 159, 194, 232, 233,
    239–241, 243–246, 254, 265, 266, 269–271,
    275, 279
man (*Homo sapiens*), 44, 47, 61, 76, 106, 134, 135,
    142, 143, 147, 148, 229, 240, 243–246, 266,
    270, 276, 277, 280
mandarin, 6
Mariner, 1, 80, 222, 224, 226–234, 237, 239,
    243–246, 280
Mariner-like elements (MLE), 222, 228, 229, 231,
    234, 238, 240, 243, 245
matrix attachment region (MAR), 119
McClintock, xii, 1, 2, 7, 14–23, 26, 101, 159,
    163–167, 169, 170, 172, 184–188, 194, 197,
    259
M cytotype, 247, 253
Mdg1, 96
Mdg3, 96
Mdg4, 96
Mediator, 187
Mendel, Gregor, xii, 2, 5, 6–15, 178, 246, 259
Mendel's laws of inheritance, 12–14, 246, 259
Mesorhizobium, 36
MET1, 264, 265
Metaviridae, 81
metazoa, 103–105, 130, 146, 239, 240, 244, 246,
    265
methylation, 60, 114, 115, 138, 164, 173, 176, 177,
    183, 188, 191, 194–196, 199, 205, 209, 221,
    222, 261, 264, 265, 268–275
methyl jasmonate, 126
methyltransferases, 195, 269, 271, 272
methionine primer tRNA, 84
micronucleus, 239
micro RNAs, 266
microsynteny, 277
Micro-Tom, 282
Miescher, 7, 110, 147
Milnesium, 80
Milt, 119, 120
MINEs, 163
Miniature Inverted-repeat Transposable Element
    (MITE), 222, 223, 227, 234–236, 240
minichromosomes, 161
minor groove, 52, 65

Minos, 163, 232, 233, 239, 240
mitochondrial genome, 114, 131, 156
MLE, 222, 228–234, 240, 243, 244
mobile DNA, 3, 15, 29, 30, 75, 263
mobile introns, 130
Moloney Murine Leukemia virus, 92
Moniliformis moniliformis, 80
monkeys, 132, 135, 137, 147
monoembryonic, 6
Monostyla, 80
Morgan, 13, 14, 93, 244, 246, 281
Mos, 228
Mos1, 228, 229, 231–233, 237, 239, 243
mosquitoes, 104, 144, 145, 233
Mother Eve, 44
mouse (*Mus musculus*), 76, 103, 135, 150, 233,
243,   245, 277, 280
Mp element, 167, 168
Mrs, 1
M strains, 94, 250
mtanga-Y, 104
Mu1, 199–201, 205–208
MuDR, 1, 196, 197, 201–209, 217, 219, 221, 260,
    278
mudrA, 201, 203–207
mudrB, 201, 203–207, 209
MuDR/Mu elements, 196, 209, 217
MuDR/Mu family, 260
MULE, 219, 221, 222
Muller, 14, 161
mung bean, 129
MURA, 204–206, 208, 209
MURB, 205, 208, 209
Mus musculus (mouse), 80, 140, 243
Mutator (Mu), 1, 100, 163, 164, 167, 196–201,
    203–206, 208, 209, 214, 217, 221, 280
Mutator effect, 186
Mycobacterium, 36, 41, 44, 233
Mycobacterium smegmatis, 233
Mycobacterium tuberculosis, 44
Mycoplasma, 36

nematode, 72, 103–105, 129, 130, 143, 224, 226,
    233, 242, 261, 262, 265, 270, 277, 278
neomycin, 53, 243
Neptune, 111, 112
nested retroelement structure, 120
neurofibromatosis, 138
Neurospora, 17, 140, 154, 160, 236, 262, 269, 270,
    272
Neurospora crassa, 17, 140, 154, 236, 262, 269
Newton, 9
nia, 125, 237
nia1, 125
nia2, 125
Nicotiana, 6
Nicotiana alata, 127
Nicotiana plumbaginifolia, 180, 181
Nicotiana tabacum, 141, 189
Nigerian people, 137

nitrate-reductase (NR), 125, 236–238
niv⁺ (Nivea), 210–215
non-homologous end-joining (NHEJ), 252, 253
non-LTR retrotransposon, 75–78, 88, 95, 106, 114,
    129–132, 137, 138, 141–147, 153, 154, 156,
    161, 162
nos promoter, 180
nuclear localization signal (NLS), 87, 177, 242
nuclein, 7, 110
nucleo capsid (NC), 105, 128
nucleophile residue, 29
nucleophilic attack, 32, 39, 51, 55, 56, 58, 71
nucleosome, 79, 87, 267, 268, 270, 271, 273

oats, 120, 122
Odysseus, 102, 129
Oenothera lamarckiana, 11
Onchorhynchus keta, 80
open-reading-frame (*orf*), 27, 68, 72, 76, 89, 91,
    100, 101, 103, 106, 107, 114, 118, 125,
    129–131, 143–145, 150, 152, 154, 161, 162,
    176, 231, 237, 238, 242, 244, 245, 248, 249,
    251
Opie-1, 119, 120
Opie-2, 119, 120
Opie-3, 119, 120
Opie-4, 119, 120
Oreochromis niloticus, 147
Oryzeae, 122
Oryza spp., 123
Oryzias latipes, 242
Oryza sativa, 123, 234
Osmar1, 234
ovo, 100, 101
Oxytricha, 160, 238, 239
Oxytricha fallax, 238

pal⁺ (Pallida), 211, 213–215
Panagrellus redivivus, 129
pangenesis, 11
Paniceae, 122
Pantoea, 37
Pao, 103–105, 130
Paris, 97, 102, 236
PAT, 130
pathogenic bacteria, 47
pathogenic yeast, 91, 92
pCal, 92
PCR, 80, 110, 123, 126, 128, 186, 200, 207, 221,
    223, 234, 241, 276, 277, 281
PCR amplification, 110, 118, 243, 244, 276
PCR methods, 186, 256
P cytotype, 247, 250, 253
pea, 7, 8, 10, 129, 268
peanut, 129
peccaries, 139
P element, 72, 93–95, 155, 162, 167, 168, 201, 232,
    240, 246–257, 279
Penelope, 96, 97, 102, 109, 111, 112
penicillin, 47

peptidyl tRNA, 84
PERIANTHIA, 264
pericarp, 165, 168, 169, 178, 196
Peromyscus, 150
petunia, 125, 181, 183, 235, 262, 276, 281
Petunia hybrida, 125, 235
P factor, 247
pHANNIBAL, 263
phase shift, 109
pHELLSGATE, 263
phenotype, 13, 99, 125, 149, 165, 166, 168, 172,
    185–187, 200, 207, 213, 214, 247, 277
Philodina rapida, 80
Philodina roseola, 80
Phoenician, 263, 264
Phoronis architecta, 80
phosphinotricin resistances, 182
phosphodiester bond, 32, 45, 46, 229, 251
phosphorylation (HK), 269
phosphoryltransferase, 29, 37
photoperiodism, 262
Phycomyces blakesleeanus, 129
Phytophthora, 93, 263
Phytophthora infestans, 263
PIF, 236
pigs, 139
Pisum sativum, 129
plasmid rescue, 206, 256, 277
Platemys spixii, 140
Plato, 259
Platyhelminthes, 80, 106
Pogo, 223, 227, 228, 239, 240, 244, 245
Pogo-like elements (*PLE*), 234
pol, 76, 89, 91, 118, 128, 129, 161
PolII holoenzyme, 274
polar mutation, 25, 26, 33
pollen grains, 116–118, 173
POL protein, 82, 84, 89
poly (A) retrotransposons, 76
poly (A) tail, 142, 146, 152, 155, 161, 176
polyembryogenic, 6
polygalacturonase, 119
polyposis coli, 148
polytenic chromosomes, 95, 96
pommelo, 6
Pooideae, 121, 122
Poseidon, 111, 112
post-transcriptional gene silencing (PTGS),
    261–266
potato, 127, 156, 179–181, 194, 218, 265
PPT, 82, 106
preintegration complex (PIC), 86–88
PREM-1, 118, 119
presetting, 188
Priapulus caudatus, 80
primates, 95, 133–135, 137–139, 141, 147, 149,
    154, 245
primer binding site (PBS), 82, 84, 85, 92, 104, 129
Populus, 183
programmed-rearrangement of DNA, 3

prokaryotes, 2, 29, 33, 34, 76, 194, 274, 277
prokaryotic organisms, 274
promoter, 27, 28, 34, 35, 37–40, 43, 49, 52, 54, 68,
    72, 78, 82, 87, 89, 91, 93, 98, 99, 126, 137, 138,
    150–152, 161, 174, 176, 177, 180, 181, 190,
    194, 195, 200, 201, 203, 205–208, 214, 215,
    219–221, 223, 224, 226, 227, 231, 232, 238,
    251–253, 257, 261, 263, 269, 274, 276, 278,
    279, 282
promoter trapping, 282
protease (PR), 82, 84, 89, 91, 106, 107, 109, 110
protozoa, 238–240, 242, 243, 265
Prt1, 130
Pseudomonas, 36, 41
Pseudomonas aeruginosa, 41
Pseudomonas putida, 49
Pseudoviridae, 81
P strain, 94, 247, 248, 253
P transposase, 95, 249, 251, 256

Q-Ag, 140
quail, 280
quelling, 261, 262, 264

R1, 43, 102, 140, 144
R1-Bm, 140
R1-Dm, 140
R2, 43, 67, 78, 97, 102, 140, 144, 153, 244
R2-Bm, 140
R2-DM, 140
rabbit, 135, 141, 280
RAD proteins, 85
Rana esculenta, 241
Randolph, 17
rat (*Rattus norvegicus*), 149, 150, 243
recessive trait, 9, 10
reciprocal recombination, 85
Reina, 119, 120
renaturation under thermal gradients, 132
replicative pathway, 48
replicative transposition, 30, 51, 69, 116, 200, 204,
    208
reptiles, 103
res, 48, 49, 51–53
rescueMu, 206, 207
resolvase, 29, 45, 49–53
resolvase synaptic complex, 52
Restless, 238
restriction endonuclease, 67, 94, 132–134, 182,
    188, 195, 212
restriction-endonuclease map, 133
restriction-endonuclease mapping, 188
resurrection approach, 221
retroelements, 76, 79, 85, 89, 91–93, 95, 99, 102,
    105, 109, 110, 113, 114, 118, 120–133, 137,
    138, 142–147, 152, 153, 162, 167
retrohoming, 131
retroposons, 76, 112, 139

retrotransposons, 45, 47, 75–83, 88, 91–93, 95–98,
    102–134, 138, 141–147, 153–156, 161–163,
    220, 222, 224, 240, 261, 273
retroviral integrase, 29, 32, 27, 45, 56, 63, 203, 251
retrovirus, 32, 45, 75, 78, 79, 82, 83, 85, 92,
    98–101, 105, 109, 110, 113, 114, 128, 135, 148,
    154
reverse genetics, 275, 277, 280, 282
reverse splicing, 131
reverse transcriptase (RT), 75–77, 79, 82, 84, 85,
    89, 91, 103–111, 113, 118, 122–124, 127,
    129–131, 138, 140, 142–146, 148, 152, 155,
    156, 161, 162
reverse transcription, 77–79, 82, 84, 85, 89, 96, 97,
    99, 104, 118, 119, 123, 126, 133, 135, 145, 146,
    148, 152, 153, 162
Rex1, 146, 147
Rex2, 146
Rex3, 146, 147
Rex retroelements, 146
Rhine, 7
Rhoades, 15–17, 23, 128, 178
ribonucleo-protein particle (RNP), 131, 152
ribosome binding site (RBS), 42, 43
rice, 122–124, 126, 127, 156, 174, 181–183, 186,
    203, 220, 234, 235, 277
RISC, 264, 266
Rle, 120
RNA-binding (RB) domain, 82
RNA-dependent DNA polymerase (RdDP), 75
RNA-dependent RNA polymerase (RdRP), 262,
    266
RNAi knock-out, 265
RNA interference (RNAi), 164, 203, 209, 215, 231,
    241, 250, 254, 261, 262, 264–266, 275, 280
RNA intermediate, 2, 75, 81, 88, 135, 163
RNA Pol II, 87
RNA polyadenylation, 97
RNA polymerase II (RPII), 279
RNA polymerase III (RNA Pol III), 87, 133
RNase III, 266
RNase H (RH), 81, 82, 84, 89, 103, 106, 107
RNA silencing, xii, 164, 260, 261, 263, 264, 266
rodents, 133–135, 137, 149, 150
rotifers, 77
Rrts, 124
RT1-Ag, 140
RTBV, 130, 140
RT-Ce, 140
RTE, 143
Rte-1, 143
RT-RH, 84, 89
Ruminantia, 139
rye, 118, 121, 122
ryegrass, 156

S1, 141
Saccharomyces cerevisiae, 61, 81, 88, 90, 93, 204
Saccharomyces paradoxus, 81
Sacchoglossus kowalevskii, 80

Sagitta, 80
salicylic acid, 126
Salmo, 110, 241
Salmon, 109, 241, 242
Salmonella, 41
Salmonella typhimurium, 68
Salmo salar, 110, 241
SALT1, 242
Salvelinus, 110
Sancho, 143
Sancho 1, 143
Sancho 2, 143
Sart1-Bm, 140
Schistosoma japonicum, 105
Schizosaccharomyces pombe, 90, 91, 93, 261
scute, 99, 143
SDSA, 217, 253
sea urchin, 106, 107, 110, 130
Secale cereale, 121, 122
seminiferous epithelium, 150
serine resolvase, 49
SF females, 96
Sharp, 16, 67, 266
sheep (*Ovis aries*), 243, 280
Sherlock Holmes, 120, 198
Shigella, 33, 36, 41, 67
Shigella dysenteriae, 36
Shigella flexeri, 67
short interspersed nucleotide element (SINE), 76,
    107, 132–139, 141, 156, 222, 244
silence, 1, 83, 209, 262–264, 266, 269–271, 275,
    280
silencing, 164, 208, 209, 215, 231, 241, 260–266,
    269–272, 274, 275
silkworm, 103, 110
Sinantherina socialis, 80
single-end break complex (SEBC), 70, 71
single-stranded RNA, 262
SIRE-1, 128
sister chromatids, 18, 21, 168, 217, 253
site-specific recombination system, 29, 42, 45,
    48–50, 182, 245
skeletons, 1, 75, 102, 103, 105, 114, 203, 220–222,
    234, 243
Skipper, 130
Skippy, 93
SLACS, 140
Sleeping Beauty(SB), 227, 228, 243, 245, 280
sleeping sickness, 142
small-interfering RNA (siRNA), 264–266
snapdragon, 189, 209–218, 280
Solanaceae, 124, 125, 141
Solanum tuberosum, 125, 141, 189
solo LTR, 82, 113
somaclonal variation, 114, 115, 123
somatic excision (SE), 207
sorghum, 118, 120, 122, 235
Southern blot (hybridization), 25, 94, 129,
    132–134, 170, 182, 199, 212, 213, 223, 225
soybean (*Glycine max)*, 128, 129, 183, 234

Soymar1, 234
SpD1, 130
SpD2, 130
SpD3, 130
SpD4, 130
spectinomycin, 61, 62
spinach, 156
Spinoza, 6
spliceosomes, 254, 255
splicing, 82, 94, 95, 101, 117, 130, 131, 175, 189,
    191, 200, 241, 249, 254, 255
splicing complex, 254
Spm, 22, 115, 163, 164, 183–200, 206, 207, 213,
    214, 216, 219, 221, 224, 280
Spm-c, 187, 188
Spm-dependent allele, 190, 193
Spm/dSpm, 186, 196, 207
Spm/En elements, 183, 184
Spm^i, 187
Spm-s, 187, 188
Spm-suppressible allele, 190, 193
Spm-w, 187–189
Spongilla, 80
Sprague, 15
ssRNA, 265
staggered dsDNA cut, 27, 77, 79, 86, 145, 153, 193,
    229
Staphylococcus, 36
states of *Ac*, 169
stem-loop structure, 130, 189
stepping stone, 5
Stonor, 117, 118
Stowaway, 1, 234, 235
strand capture, 40
strand transfer, 28–30, 38–40, 42, 46, 50, 51, 55,
    56, 59, 64, 70–72, 229, 252
streptomycin, 47, 54, 61, 62, 180, 181, 223
streptomycin-resistance, 54, 180, 181, 223
streptothricin, 61, 62
Strongylocentrotus purpuratus, 80, 130
Sturtevant, 14, 15, 246
subterminal repetitive regions, 189, 193, 195
sucrose synthase, 19, 166, 170
Suiformes, 139
supercoiled DNA, 53, 59
Suppressor-mutator (Spm), 22, 115, 163, 164,
    183–200, 207, 213, 214, 216, 221, 224, 280
SURL, 107, 110
Sushi, 109, 110, 130
Sushi-ichi, 109
Sushi-ni, 109
Sutton, 13, 171
SV40 virus, 133
synapsis, 51, 52, 56
synaptic complex, 28, 38, 40, 52, 53, 55, 56, 59, 60,
    177
synaptosome, 53
Synechocystis, 33, 36
systemic acquired silencing, 262

T1-Ag, 140
Ta1, 113, 128, 130
Ta2, 113
Ta3, 113
Tad, 154, 236, 237
Tad1, 140, 154
Tad1-Nc, 140
Tag1, 219, 220
Tal1-Ag, 140
Tal1-1, 156
Tal1-1-At, 140
Tal-2, 113
Tal-3, 113
Talmud, 5, 11
Tam, 189, 212, 214–218
Tam2, 213–215
Tam3, 73, 213–218
target capture, 50, 56, 71
target DNA, 27–31, 33, 38–40, 42, 46, 48, 50, 51, 56, 58, 59, 63–65, 70, 71, 77, 86–88, 176, 200, 229, 249
target immunity, 42, 50
target-primed reverse transcription (TPRT), 77, 78, 145, 152–154
target site duplication (TSD), 87, 152
TART, 161, 162
Tart-Dm, 140
TAS, 103–105, 250, 253
TAS repeats, 250, 253
tassel, 165
Tat1, 113
TATA box, 98, 106, 213, 214, 251, 274
TBE, 238, 239
Tca1, 92
Tca2, 130
Tc1, 72, 80, 224–234, 237–246, 262, 278–280
Tc1-like elements (*TLE*), 229, 230, 232, 234
Tc1/Mariner, 224, 226–230, 233, 234, 237–239, 243, 244, 280
Tc1/Mariner superfamily, 224, 226, 228–230, 237, 239, 278, 280
Tc1 transposase (TcA), 226, 228
Tc3, 226–233
T. cruzi, 142, 143
T-DNA, 115, 182, 276, 281, 282
T-DNA-mediated gene tagging, 276
Tdr1, 242, 243
Tdr2, 242
Tec1, 238, 239
Tec2, 238, 239
Tekay, 120
Telemac, 102
telomerase, 159–162
telomere-associated sequence (*TAS*), 103, 250, 253
telomeres, 3, 121, 122, 155, 159–162, 245, 271
Teosinte guerrero, 116
terminal inverted-repeat (TIR), 189, 193, 194, 200, 201–203, 205, 206, 208, 209, 221, 223, 227, 231, 235–237, 244

terminator, 203, 232
tetracycline, 68
tetracycline-resistance, 45, 68, 69, 251, 252
Tetrahymena, 160, 161
Tetraodon nigroviridis, 130
T_F, 150
Tf1, 91, 92
Tf2, 91
TFIID, 274
Tfo1, 237
Tgm, 128
Themiste alutacea, 80
Thermoplasma, 36
Tigger, 227, 244, 245
Tigger1, 244, 245
Tigger2, 244, 245
TLC1 to TLC4, 127
TLE, 229, 230, 232, 234
Tn*7*, 26, 32, 33, 46, 47, 61–67, 71, 204, 253
Tn*10*, 26, 45–47, 59, 67–73
TnDirs1, 130
*tnp*A, 48–50, 196
TNPA, 189, 193–196
TNPB, 190
TNPC, 190
TNPD, 189, 193, 194
*tnp*R, 48, 49, 51, 52
tnsA, 61–63, 204
Tnt1, 115, 125, 126
tobacco, 6, 115, 123–127, 141, 156, 176, 178–182, 192, 194, 218, 219, 262
Tol2, 243
Tom, 130, 282
tomato (*Lycopersicon esculentum*), 45, 126, 127, 130, 176, 181, 182, 218, 277, 281, 282
toothed whales, 139
Tos1 to Tos10, 123
Tos17, 123, 124
Tourist, 1, 235, 236
TPRT, 77, 78, 145, 152–154
Tpv2, 128
TRAM, 102
transesterification, 51
transesterification reaction, 29, 46, 51, 64
transgenic *A. thaliana*, 115, 181, 220, 223, 224
transgenic barley, 183
transgenic maize, 203, 205–207
transgenic plants, 45, 115, 126, 176, 179–183, 192, 194, 204, 207, 216, 220, 223
transgenic potato, 180
transgenic tobacco, 179, 180, 194
transgenic tomato, 181, 182
translational frame shift, 27, 34, 35, 37, 38
transposase (Tpase), 26–29, 33, 34, 36–38, 40, 42, 174, 176, 177, 180, 181, 183
transposition, 1, 21–23, 26–31, 34, 35, 37–42, 45, 47–50, 53–56, 59–63, 65, 66, 68–73, 75, 78, 81–89, 91, 94–98, 100, 102, 105, 113, 115, 116, 118, 120, 121, 123, 124, 126–128, 130, 131,

133, 134, 137, 138, 141, 142, 144, 147–157,
161, 164, 169, 171, 174–183, 185, 187–189,
191, 192, 194–196, 199, 200, 203, 204,
207–209, 214, 215, 219–221, 223, 225, 226,
229–233, 239, 242, 243, 245, 248–252, 254,
256, 260–262, 269, 270, 277, 280
transposition complex, 176, 194, 196
transposome, 70, 71, 177
transposon tagging, 243, 280, 282
transposon-yeast-1 (Ty1), 81–89, 91, 92, 98, 103,
104, 112–114, 116, 118, 122–125, 127–129,
148
Tras1-Bm, 140
Trichoderma viride, 126
Trim-Dmi, 140
trimethoprim, 61, 62
Tripsacum, 118
Tripsacum andersonii, 203
trisomic chromosomes, 276
trisomics, 17, 276
Triticeae, 122
Triticum, 118, 127, 183
Triticum aestivium, 183
Triticum monococcum, 183
tRNA, 78, 82–86, 89, 91, 92, 107, 123, 141
tRNA genes, 89, 91
tRNA^{Lys}, 99, 141
tRNA^{Met}, 104, 113, 123, 127, 129
Trypanosoma, 132, 142, 160
Trypanosoma brucei, 140, 142, 265
TS elements, 141
Tse-Tse, 142
Tto1, 115, 124, 126, 127
Tto2, 126
turnip, 156
Tx1, 76, 129, 130
Tx1D, 145
Tx1L, 145
Tx1-X1, 140
TXr, 241
Ty1/*copia*, 76, 91, 92, 103, 104, 112, 114,
122–124, 127, 128, 130
Ty1/*copia* elements, 112–114, 118, 122, 125
Ty2, 81, 84, 88, 89, 93
Ty3, 81, 84, 88, 89, 91, 93, 95, 98, 105, 107, 130
Ty3/*gypsy*, 76, 91, 103, 104, 106, 111
Ty3/*gypsy*-like elements, 91, 93, 107, 108–110,
117, 127–129, 273
Ty3/*gypsy*-like retrotransposons, 92, 107, 108, 113,
129, 273
Ty3/*gypsy*-type, 109, 110, 112, 114, 118, 122
Ty4, 130
Ty5, 81, 82, 87–89, 130
Tylopoda, 139
type I repressor, 249
type II repressor, 249
tyrosine-recombinase, 49
Tzf, 240, 242

ubiquitination, 268–270
Ulysses, 96, 97, 102, 110
upstream control region (UCR), 195, 196
Uri, 111
urochordates, 103, 107

van Leeuwenhoek, 6
Vavilov, 96, 97
vertebrates, 79, 85, 98, 99, 105, 107–109, 111, 137,
141–143, 153, 154, 160, 163, 239–243, 265,
270, 280
Vicia, 128
Vicia faba, 128, 183
Vicia melanops, 128
Vicia sativa, 128
Victim, 119, 120
virus-like particles (VLP), 82–87, 89, 91, 100, 148
viviparous state, 124
Volvox carteri, 112
von Ettingshausen, 8
von Gärtner, 7
von Nägeli, 13
von Tschermak, 11

Wanderer, 235, 283
Waxy (Wx), 19–22, 116–118, 167, 169–171, 173,
175, 188
Whales, 138, 139
Weldon, 12
wheat, 114, 115, 118, 120–122, 127, 156, 183, 220,
277
worms, 103–105, 110, 162, 224, 225, 240, 262,
265, 278, 279
Wx gene, 21, 117, 118
Wx protein, 118

X chromosome, 14, 101, 147, 149, 252, 271, 275,
276, 279
X chromosome inactivation, 275
Xena, 109
Xenopus, 130, 240, 280
Xenopus laevis, 80, 140, 145, 240
Xiphophorus, 109, 110, 146
Xiphophorus maculatus, 110, 147
Xmrk, 110, 146
X-ray irradiation, 96, 273
XtD1, 130

Yamato, 103, 104
Yeast, 3, 45, 61, 76, 79, 81–89, 91, 92, 98, 114,
116, 130, 131, 137, 148, 154, 159, 160, 162,
194, 204, 257, 261, 269–272
Yeast debranching enzyme (Dbr1p), 86
yeast mitochondria, 130, 131
yellow, 9, 10, 14, 56, 58, 99, 145, 196, 198, 233,
240
Yersinia, 41

Zea, 118, 120, 127, 203, 278
Zea mays, 118, 140, 141, 165

zebrafish (*Danio rerio*), 130, 146, 233, 240–243, 277
Zeon, 118
Zeon-I, 118
Zepp, 76, 155
zinc finger, 99, 142, 220
zinc finger motif, 105
Zorro, 76